Chitin and Chitosan

Chitin and Chitosan
Physical and Chemical Properties

edited by

Shakeel Ahmed | Soumeia Zeghoud
Hadia Hemmami | Ilham Ben Amor

JENNY STANFORD
PUBLISHING

Published by

Jenny Stanford Publishing Pte. Ltd.
101 Thomson Road
#06-01, United Square
Singapore 307591

Email: editorial@jennystanford.com
Web: www.jennystanford.com

British Library Cataloguing-in-Publication Data
A catalogue record for this book is available from the British Library.

Chitin and Chitosan: Physical and Chemical Properties

ISBN 978-981-5129-39-7 (Hardcover)
ISBN 978-1-003-58977-8 (eBook)

Contents

Preface xv

1. Polysaccharides: Chitin and Chitosan **1**

Ilham Ben Amor, Soumeia Zeghoud, Hadia Hemmami, and Shakeel Ahmed

1.1 Introduction 2
1.2 Polysaccharides 2
1.3 Storage Polysaccharides 3
1.3.1 Starch 4
1.3.2 Glycogen 6
1.4 Structural Polysaccharides 6
1.4.1 Cellulose 6
1.4.2 Chitin 8
1.4.3 Mucosubstances 10
1.5 Chitin and Its Derivatives 10
1.5.1 Availability of Chitin 10
1.5.2 Structure of Chitin 11
1.5.3 Properties of Chitin 13
1.5.4 Chitin Industrial Relevance 13
1.5.5 Derivatives of Chitin 15
1.6 Chitosan and Its Derivatives 16
1.6.1 Availability of Chitosan 16
1.6.2 Structure of Chitosan 17
1.6.3 Properties of Chitosan 18
1.6.4 Chitosan Derivatives 21
1.7 Conclusion 24

2. Chitosan and Chitin Sources and Their Isolation 33

Elena Tafi

2.1 Introduction 34
2.2 Sources of Chitin and Chitosan 36
2.2.1 Crustaceans 36
2.2.2 Fungi 39
2.2.3 Insects 42
2.3 Chitin Purification 43
2.3.1 Chemical Method 44
2.3.1.1 Demineralization 45
2.3.1.2 Protein removal 46
2.3.1.3 Decolorization 47
2.3.2 Biotechnological Method 48
2.4 Chitin Deacetylation 51
2.4.1 Chemical Method 52
2.4.2 Enzymatic Deacetylation 53
2.5 Conclusion and Future Prospects 53

3. Preparation of Chitin and Chitosan 63

Ali Alnazza Alhamad, Ilham Ben Amor, Soumeia Zeghoud, Noura Gheraissa, Salah Eddine Laouini, Mohammed Fouad Ferhat, Nani Sadok, and Shakeel Ahmed

3.1 Introduction 64
3.2 Extraction Methods 66
3.2.1 Chemical Techniques 66
3.2.1.1 Demineralization 66
3.2.1.2 Deproteination 68
3.2.1.3 Deep eutectic acids and ionized liquids 69
3.2.2 Biological Processes 73
3.2.2.1 Enzymatic processes 74
3.2.2.2 Fermentation processes 74
3.2.3 Comparing Extraction Techniques 75
3.3 Deacetylation 77

3.4 Nanochitin Synthesis 79
3.5 Applications 82
3.6 Conclusion and Future Prospects 83

4. Physical Properties of Chitin and Chitosan 95

Ahmed Barhoum, Ilham Ben Amor, and Shakeel Ahmed

4.1 Introduction 96
4.2 Molecular Weight 97
4.3 Viscosity 99
4.4 Degree of Deacetylation and Chitosan Dissociation Constant 100
4.5 Techniques for Estimating Deacetylation of Degree 100
4.5.1 Fourier Transforms Infrared Spectroscopy 101
4.5.1.1 Drawbacks 102
4.5.1.2 Advantages 102
4.5.2 ^{1}H-NMR Spectroscopy 102
4.5.3 UV-Vis Spectroscopy 104
4.5.4 Potentiometric Titration 105
4.5.5 Raman Spectroscopy 106
4.5.5.1 Elemental analysis 108
4.5.5.2 Method of capillary zone electrophoresis 108
4.5.5.3 Base-acid titration 109
4.6 Degradation of the Polymer or Stability 109
4.7 Thermal Properties 111
4.8 Swelling Capacity and Hygroscopicity Chitosan and Chitin 112
4.9 Conclusions and Future Prospects 113

5. Chemical Properties of Chitin and Chitosan 119

Olfa Ouled Ltaief, Wiem Hamza, Mabrouk Eloussaief, and Mourad Benzina

5.1 Introduction 120

5.2 Solubility 121

5.2.1 Chitin's Solubility 121

5.2.1.1 NaOH/urea aqueous solution 122

5.2.1.2 $CaCl_2$/solvent 123

5.2.1.3 N,N-dimethylacetamide/lithium chloride solvent 124

5.2.1.4 Ionic liquid 125

5.2.2 Chitosan's Solubility in a Solution 127

5.3 Mucoadhesion 128

5.4 Forming Properties of Films 128

5.5 Biocompatibility and Biodegradability 129

5.6 Reactivity 130

5.7 Bioactivity 131

5.8 Adsorption Capacity 133

5.9 Conclusion 136

6. Biological Characteristics of Chitin and Chitosan **143**

Mohamed Belkacem, Hadjer Bouosmat, Asmaa Benettayeb, Mustapha Hadj Brahim, and Boumediene Haddou

6.1 Introduction 144

6.2 Chitosan and Chitin Structure 145

6.3 Biological Properties 145

6.3.1 Biodegradability 146

6.3.2 Biocompatibility 147

6.3.3 Hemostatic 148

6.3.4 Analgesic Effect 149

6.3.5 Antitumor Activity 150

6.3.6 Mucoadhesion 150

6.3.7 Permeation Enhancing Effect 151

6.3.8 Anticholesterolemic 152

6.3.9 Antimicrobial Activity 154

6.3.9.1 Chitosan concentration effects on antibacterial effects 156

6.3.9.2 Influence of environment pH on chitosan's antibacterial effect 156

6.3.9.3 Contribution of bacteria type to the antibacterial influence of chitosan 158
6.3.10 Antifungal Properties 161
6.3.11 Antioxidative Activity 162
6.4 Relationship Between Biological Activities and Chemical Structure 164
6.5 Conclusion and Prospects 166

7. Physicochemical Properties of Chitosan Products 175

Thiago Luis Aguayo de Castro, Giovanna Eduarda Machado, Juliana Ketlin Moro da Silva, Jéssica Bassani de Oliveira, Rosangela Maria Ferreira da Costa e Silva, and Claudia Andrea Lima Cardoso

7.1 Introduction 176
7.2 Chitosan Powder and Liquid 178
7.3 Chitosan Gel 183
7.4 Films 185
7.5 Fibers 186
7.6 Membranes 186
7.7 Chelates 191
7.8 Flame Retardant 197
7.9 Physical and Chemical Characterization Methods 200
7.10 Conclusion 202

8. Chitin and Chitosan Characterization: Spectroscopic Methods 221

Hadia Hemmami, Soumeia Zeghoud, and Ilham Ben Amor Shakeel Ahmed

8.1 Introduction 222
8.2 Spectroscopic Methods for Structural Analysis 224
8.2.1 X-Ray Spectrometry 224
8.2.1.1 Typical X-ray measurement circumstances 224
8.2.1.2 Chitosan and chitin X-ray spectrum 225
8.2.2 Infrared Spectroscopy 227

8.2.2.1 Conditions typical for FTIR spectroscopy of chitosan and chitin 228
8.2.2.2 Use of infrared spectroscopy to characterize chitin and chitosan 228
8.2.3 UV-Visual Spectroscopy 230
8.2.3.1 Typical UV-Vis measurement circumstances 231
8.2.3.2 Chitosan and chitin UV-Vis spectroscopy 231
8.2.4 Mass Spectrometry 232
8.2.4.1 Standard settings for chitin/chitosan mass spectrometric analysis 233
8.2.4.2 Measurement of chitosan and chitin polymerization levels by mass spectrometry 236
8.2.5 Nuclear Magnetic Resonance Spectroscopy 237
8.2.5.1 Explanation of characterization NMR methods 238
8.2.6 Thermogravimetric Analysis 240
8.2.7 Utilizing Energy-Dispersive Spectroscopy for X-Ray Microanalysis 242
8.2.8 Scanning Electron Microscope Analysis 242
8.2.9 Transmission Electron Micrograph 244
8.3 Connection between Characterization Techniques and Physical Characteristics 245
8.4 Conclusion 246

9. Application of Chitosan and Chitin in Food Packaging 255

Abdelmalek Zaater, Saoussan Temmar, Talha Bin Emran, Abdelkrim Rebiai, Asma Ben Amor, Mohammed Tayeb Oucif Khaled, and Ali Tliba

9.1 Introduction 256

9.2 Techniques for Enhancing the Characteristics of Films Made with Chitosan 259
9.2.1 Mechanical Properties 259
9.2.2 Barrier Properties 262
9.2.3 Antimicrobial Characteristics 266
9.2.4 Optical Properties 268
9.3 Chitosan Blends 269
9.3.1 Blends of Natural Biopolymers with Chitosan 270
9.3.2 Blends of Chitosan-Synthetic Polymers 270
9.4 Examples of Films Based on Chitosan 271
9.4.1 Examples of Applications for Active Films 271
9.4.2 Examples of Intelligent Films 272
9.5 Conclusions and Future Views 274

10. Chitin and Chitosan Applications in Medication Delivery Systems 283

Mohamed Sharaf and Chen-Guang Liu

10.1 Introduction 284
10.2 Oral Drug Delivery System 290
10.3 Colon-Specific Drug Delivery System 292
10.4 Ophthalmic Delivery System 294
10.5 Buccal Drug Delivery System 297
10.6 Transdermal Drug Delivery System 299
10.7 Periodontal Drug Delivery System 302
10.8 Nasal Drug Delivery System 303
10.9 Vaginal Drug Delivery System 307
10.10 Conclusion and Outlook 311

11. Application of Chitosan in Textiles 321

Deepak Chandran, Sureshkumar Rajalingam, Marthandan Viswanathan, Pran Mohankumar, Deepika Krishnan, Ankitha Indu Jisha, Abhirami Vinu Nair, and Aishwarya Prashanth

11.1 Introduction 322

11.2 Production of Fiber Using Chitosan 323
11.2.1 Spinning Technique 323
11.2.1.1 Wet spinning 323
11.2.1.2 Solvent system 324
11.2.2 Dry-Spinning Technique 325
11.2.3 Electrospun 326
11.3 Fibers' Mechanical Characteristics 328
11.4 Examples of Textile Applications of Chitosan 328
11.4.1 Yarn and Fibers 328
11.4.2 Nonwoven Textiles 330
11.4.3 Printing and Dyeing 330
11.4.4 Functional Finishing 331
11.4.4.1 Antimicrobial finishing 331
11.4.5 Agent for Shrink-Proof Finishing 335
11.4.6 Carriers for Textile Finishing 336
11.5 Processing of Textiles with Chitosan 337
11.5.1 Chitosan Is a Sizing and Desiring Agent Used in the Processing of Textiles 337
11.5.2 Chitosan: A Binder in the Printing Process 338
11.5.3 As a Finishing Agent for Sturdy Textile Presses 339
11.5.4 Chitosan: A Textile Finishing Agent with Antistatic Properties 340
11.5.5 Chitosan: A Textile Fiber That Is Flame- and Fire-Resistant 341
11.5.6 Chitosan Is a Hydrophobic Substance Used to Make Textiles That Repel Water 341
11.5.7 Chitosan: A UV-Blocking Textile Coating 342
11.6 Conclusion 342

12. Chitin and Chitosan in Wastewater Treatment 351

Atif Khurshid Wani, Hadia Hemmami, Atulika Sharma, and Sourbh Suren Garg

12.1 Introduction 352

12.2 Textile Industries—Synthetic Dyes and Effluent Wastewaters—Environmental Issues 355
12.3 Function of Chitosan and Nanocomposite Membranes Based on Chitosan in the Treatment of Wastewater 357
12.4 Chitosan as Bioadsorbents 358
12.5 Nano-Biocomposites Based on Chitosan 361
12.6 Removal of Various Contaminants Using Chitosan-Based Materials 364
12.6.1 Metal Ions Removal 364
12.6.2 Organic Pollutants 366
12.6.3 Desalination 367
12.6.4 Antibacterial Effect 369
12.6.5 Other Pollutants 370
12.7 Mechanism of Adsorption 371
12.7.1 Mechanism of Interaction and Methylene Blue Dye Removal 373
12.7.2 Fators Influencing Adsorption Effectiveness 375
12.8 Conclusion and Future Perspective 377

13. Other Applications of Chitosan **389**

Omaima Aidat and Asma Ben Amor

13.1 Introduction 390
13.2 Chitosan in Green Metal Nanoparticle Synthesis 391
13.2.1 As a Stabilizer 391
13.2.2 As a Green Reducing Agent 393
13.2.3 As a Size-Controlling Agent 394
13.2.4 As a Shape-Directing Agent 395
13.2.5 Chitosan as a Multifunctional Agent for Metal Nanoparticle Preparation 396
13.3 Chitosan's Function in Cosmetics 397
13.4 Food 400
13.5 Plastics and Filler Materials 401
13.6 Recent Use of Chitosan and Chitinin Plants 403

13.6.1 Chitin 403
13.6.2 Chitosan 406
13.7 Paper and Pulp Industry 410
13.8 Dentistry 411
13.8.1 Oral Prophylaxis 412
13.8.2 Dental Root Treatment 412
13.8.3 Periodontology 413
13.8.4 Implantology 413
13.9 Treatment for Burns and Wounds 413
13.10 Tissue Engineering 414
13.11 Use of Chitosan as Artificial Leather 415
13.12 Opthalmology 416
13.13 Conclusion 416

Index 429

Preface

Current standards for polysaccharides in many uses have generated increased interest in chitin and chitosan because of their biocompatible, biodegradable, non-toxic, and non-sensitizing properties. Chitin, which is an example of a structural polysaccharide, can be obtained from marine sources. Cellulose, being the most widespread natural raw material, is followed by cellulose in second place in terms of abundance. Chitosan is a derivative of chitin that has undergone partial deacetylation. Chitosan, a polysaccharide, has garnered considerable attention for its abundant availability and impressive array of properties, which include durability, biodegradability, inertness, and easy conversion into gels and films.

Chitin, chitosan, and their derivatives have found widespread use in several fields, including wastewater treatment, medicine, food industry, and other applications. Research on these materials is expanding rapidly. A great deal of knowledge is currently being produced and disseminated in the public domain on many aspects related to the characterization, synthesis, and use of chitosan and chitin.

This book contains 13 chapters. Chapter 1 deals with generalities about polysaccharides, chitin, and chitosan in general. Chapter 2 provides information on the natural sources of chitin and chitosan. Chapter 3 deals with the extraction of chitin and the processes of preparing chitosan with different chemicals (biological and other methods). Chapters 3, 4, and 5 provide a comprehensive source of information on the physical, and chemical properties of chitin and chitosan, while Chapter 6 discusses their biological characteristics. Chapter 7 focuses on the physicochemical properties of chitosan products. Chapter 8 summarizes the different methods for characterizing chitosan and chitin. The remaining five chapters are devoted to exploring the various applications of chitin and chitosan in diverse fields, such as food packaging, medication delivery, textiles, wastewater treatment, and other areas. This book covers works published from 2009 to 2023, mainly in the areas of textiles, wastewater treatment, biomedicine, and food applications. The subject matter examined within this book may hold significance for scholars and professionals engaged in academic and industry research pertaining to chitin and chitosan.

In conclusion, we are very pleased to thank all the authors who contributed to the realization of this book. We would also like to express our gratitude to the University of El Oued, Algeria, for their assistance and cooperation.

Dr Shakeel Ahmed
Dr Soumeia Zeghoud
Dr Hadia Hemmami
Dr Ilham Ben Amor

Winter 2024

Chapter 1

Polysaccharides: Chitin and Chitosan

Ilham Ben Amor,[a,b] Soumeia Zeghoud,[a,b] Hadia Hemmami,[a,b,c] and Shakeel Ahmed[d,e]

[a]*Department of Process Engineering and Petrochemical, Faculty of Technology, University of El Oued, El Oued 39000, Algeria*
[b]*Renewable Energy Development Unit in Arid Zones (UDERZA), University of El Oued, El Oued 39000, Algeria*
[c]*Laboratory of Applied Chemistry and Environment, Faculty of Exact Sciences, University of El Oued, 39000 El Oued, Algeria*
[d]*Postgraduate Department of Chemistry, Government Postgraduate College Rajouri, Jammu and Kashmir 185133, India*
[e]*University Centre for Research and Development, Chandigarh University, Mohali, Punjab, India*

ilhambenamor97@gmail.com

The scientific community has taken an interest in chitin, a polysaccharide that follows cellulose in abundance in nature and is widely available and inexpensive. It is produced mainly from aquatic crustaceans and the exoskeleton of insects. Although its deacetylated derivative, chitosan, was found to be soluble in protonated water media and is often utilized in different biomedical disciplines, it is frequently insoluble in typical solvents, which restricts its applicability. Chitosan has a fundamental characteristic that makes it simple to chemically convert it into

Chitin and Chitosan: Physical and Chemical Properties
Edited by Shakeel Ahmed, Soumeia Zeghoud, Hadia Hemmami, and Ilham Ben Amor

ISBN 978-981-5129-39-7 (Hardcover), 978-1-003-58977-8 (eBook)
www.jennystanford.com

various derivatives, thanks to the presence of the primary amino group on its backbone. The structural characteristics of chitin and its derivatives are discussed in this chapter.

1.1 Introduction

The most adaptable polysaccharide-based materials of the twenty-first century are chitin and chitosan. Basically, 2-acetamido-2-deoxy-D-glucopyranose and chitin are homopolymers. It is a biodegradable polymer with a non-toxic molecular weight (Mw). After cellulose, it is the second most prevalent polysaccharide in nature [1]. Chitin is a fiber with remarkable biological and chemical characteristics that may be used in various medical and industrial applications, including biocompatibility, biodegradability, nontoxicity, and adsorption. Chitin, present in large quantities in prawn shells, is the source of chitosan. Chitosan is the term for chitin that has been deacetylated (CH_3-CO) to a level of at least 50% and is soluble in weak acids. Deacetylation, a process that releases amine groups (NH), provides chitosan with cationic properties. β(1–4)-linked d-glucosamine and N-acetyl-d-glucosamine are randomly dispersed throughout the linear polysaccharide known as chitosan [2]. This is intriguing because most polysaccharides are usually negatively or neutrally charged in an acidic environment [3, 4].

1.2 Polysaccharides

In nature, polysaccharides are a common macromolecule, including biomaterials based on carbohydrates [5]. Monosaccharides $(CH_2O)_n$, a straightforward carbohydrate monomer, is typically used to create linear to highly branched-chain carbohydrate biopolymers. In a polysaccharide structure, monosaccharide units like glucose ($C_6H_{12}O_6$), fructose ($C_6H_{12}O_6$), and glyceraldehyde ($C_3H_6O_3$) are joined together by glycosidic linkage (covalent bond) [6]. Polysaccharides, such as starch $(C_6H_{10}O_5)_n$, glycogen ($C_{24}H_{42}O_{21}$), cellulose $(C_6H_{10}O_5)_n$, and chitin $(C_8H_{13}O_5)_n$, are long polymers of monosaccharides. The life cycles of both plants and animals in the natural world depend on polysaccharides [7]. Depending on how they are used, polysaccharides can be classified as structural, mucosubstances, or storage polysaccharides.

1.3 Storage Polysaccharides

Polysaccharides for storage are used as reserve food. Storage polysaccharides undergo hydrolysis when necessary, and the liberated sugars are then made accessible to the live cells for use in biosynthetic and energetic processes. Two primary storage polysaccharides include glycogen and starch (Fig. 1.1) [8].

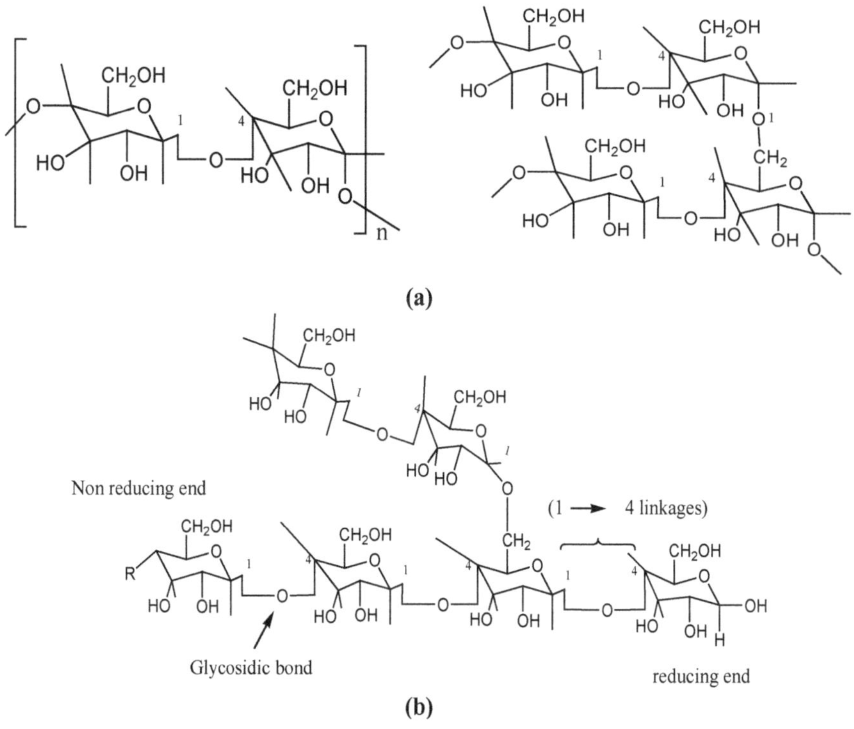

Figure 1.1 Polysaccharides for storage, (a) starch components (a portion of the molecule amylopectin and linear amylose), and (b) glycogen (*n*, the number of glucose groups is denoted as the degree of polymerization).

1.3.1 Starch

Starch is a polysaccharide composed of glucose monomers linked together in a linear fashion and is a major energy-related carbohydrate reserve in plants (Fig. 1.1). Simple or complex starch granules can exist alone as single simple grains or in groupings. Amylopectin, a highly branched polymer, and amylose, a linear polymer, make up starch [9]. More water is soluble in amylose than in amylopectin.

Figure 1.2 displays a diagram of amylopectin and amylose. Each round of the continuous straight amylose chain includes roughly 6 glucose units. In amylose, the repeating glucose units are connected by glycosidic bonds (1–4 linkages) and a few branches of α(1–6) links [10]. Amylopectin's structure is comparable to that of amylose. Every 10–60 glucose units, there are extra α(1–6) glycosidic branching sites on amylopectin. Additionally, as seen in Fig. 1.2, the straight-chain amylopectin exhibits many side chains that possess the potential to further branch.

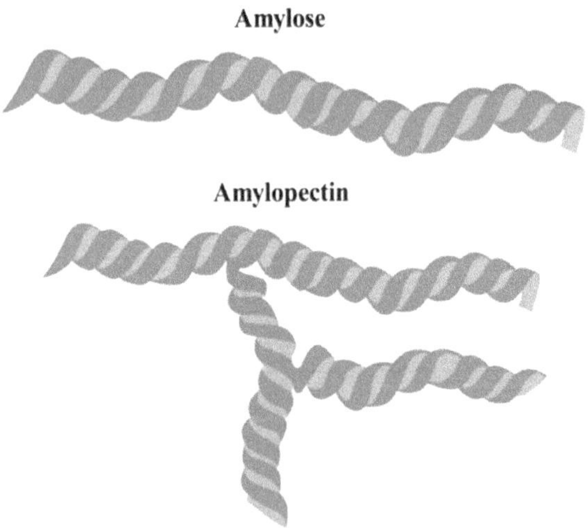

Figure 1.2 Schematic illustrating amylopectin and amylose.

In Fig. 1.3, a schematic diagram of starch granules' internal makeup is shown. Most of the amylopectin in starch granules, 65–90%, is found as amylose. The amorphous areas and crystalline that make up the amorphous and crystalline growth rings make up the interior structure of starch granules (Fig. 1.3). The amylopectin lamellae are the main structural element in the crystalline area. Crystalline lamellae can be formed by packing together neighboring amylopectin chains with similar branch structures. Figure 1.3b demonstrates that the branch point is located in amorphous lamellae. The starch granules matrix is naturally semi-crystalline, due to the alternation of crystalline and amorphous lamellae. Amylose does not participate in semi-crystalline granules and is mainly found in the amorphous zone. A proteinaceous point (the hilum) is surrounded by a crystalline kit and crystalline semi-shells or layers that are organized either concentrically or eccentrically in the starch granules. Some starches, including those derived from maize and some grains, are wholly composed of amylopectin. On the other side, pea starch sources may contain up to 98% amylose.

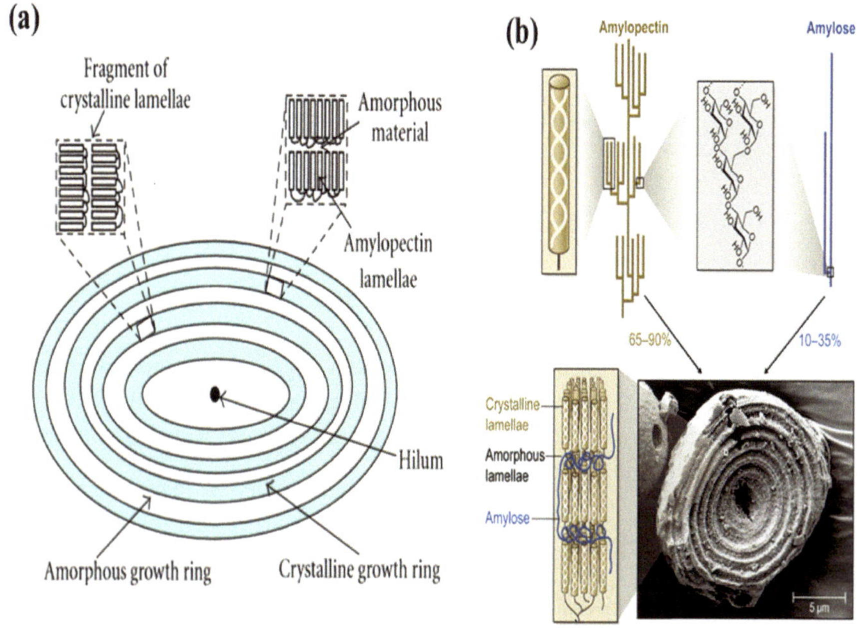

Figure 1.3 Schematic showing starch granules' structure and amylopectin's chemical structure (a) (reproduced with permission from Ning Lin et al. [10]); and the structure of starch granules' molecules (b) (reproduced with permission from Seung et al. [11]).

1.3.2 Glycogen

A glucose long-chain polymer is known as glycogen. For bacteria, mammals, and fungi, it is a major polysaccharide food source connected to energy. The storage polysaccharide glycogen's schematic structure is shown in Fig. 1.4. Animal starch is the common name, mostly stored in the muscles and liver.

Glycogen, a polysaccharide, has a longer, more branched chain than starch. However, like amylopectin, glycogen possesses α(1–6) branch points per 8–4 glucose remains in its chemical structure [12].

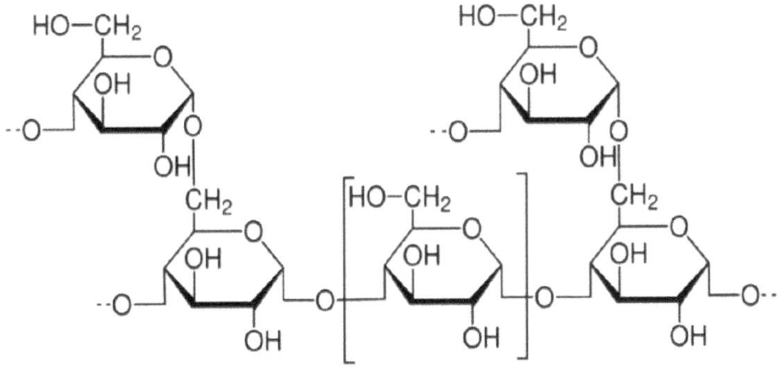

Figure 1.4 An illustration of the glycogen structure.

1.4 Structural Polysaccharides

Structural polysaccharides are essential for the development of walls of cells and skeleton walls in plants and animals. Instances of the two main categories of structural polysaccharides are chitin and cellulose.

1.4.1 Cellulose

A structural polysaccharide is cellulose. It serves as the main load-bearing element in plant cell walls. D-glucopyranosyl units, which consist of repeating monomers connected by a β(1–4) glycoside bond, give rise to cellulose [13, 14]. The plant cells' stiffness comes from cellulose. The chemical makeup of structural polysaccharides is displayed in Fig. 1.5.

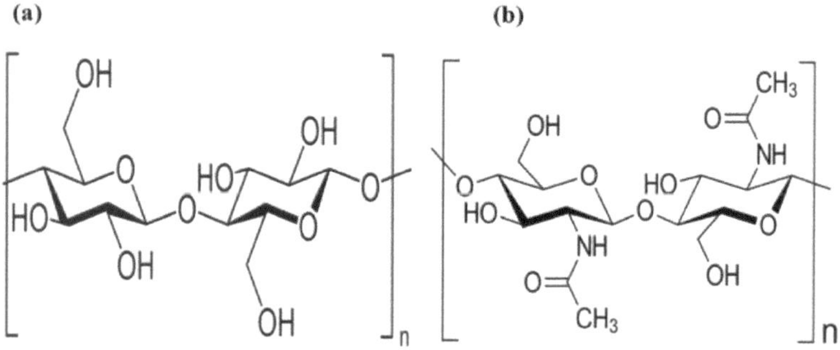

Figure 1.5 The basic chemical structure of structural sugars: cellulose (a) and chitin (b).

Strong linkages connect one cellulose molecule to the next. The chain's intermolecular strength is strengthened by establishing a hydrogen connection between the subsequent residue's oxygen atom and the third-position hydroxyl group (Fig. 1.5). Figure 1.6 displays a representative field-emission scanning electron micrograph depicting cellulose microfibrils. A microfibril is a bundle of roughly 2000 cellulose chains or molecules that may be seen under an electron microscope.

Figure 1.6 Cellulose microfibers that make up the surface structure (reproduced with permission from Butchosa et al. [15]).

1.4.2 Chitin

Like cellulose, chitin is a kind of structural polysaccharide. Except for the acetylamino group (–$NHCOCH_3$) in the location of the hydroxyl group at the carbon atom, chitin's chemical structure is comparable to cellulose [14]. Despite being the second most prevalent biopolymer, next to cellulose, chitin does not exist in organisms that produce cellulose. Heteropolysaccharides make up the complex carbohydrate known as chitin. It is extracted from cartilaginous fish tissues and crustacean bones. The chitin found in the fungal walls is called fungus cellulose. Because it is supple and rubbery, it offers both strength and flexibility. When chitin is impregnated with calcium carbonate and certain proteins, it hardens. The fundamental building block of chitin is N-acetyl glucosamine, a glucose derivative that contains nitrogen instead of glucose. (1–4)β-linkages hold the monomers together (Fig. 1.6). Hydrogen connections between parallel molecules keep its unbranched structure together. An illustration of a typical polysaccharide from natural sources may be seen in Table 1.1.

Table 1.1 Examples of sugars from natural resources [16–19]

Polysaccharide	Composition	Sugar function and content	Source
Chitin	N-acetylglucosamine residues create the long-chain, unbranched polysaccharide known as chitin	Animal cells from a crustacean structure	The skeleton of crustaceans and insects
Dextran	Certain bacteria produce a kind of glucose polymer known as dextran	Glucose/function is unknown	Bacteria, in the first place
Inulin	Glucose molecules that terminate chains and a repeating fructose moiety make up inulin	Fructose/natural storage poly-saccharides	Chicory, artichoke
Cellulose	A linear polymer is mostly formed by the D-glucopyranosyl units.	Plant cell walls' stability is influenced by glucose	Plant kingdom

another and are highly reactive to solvents. In γ-chitin, two of the three chitin chains are organized parallelly, though the third chain is orientated in a contradictory direction. Therefore, γ-chitins are viewed as a combination of β and α forms. Although the forms of chitin are linked to flexibility, motility, and toughness as well as various physiological functions other than support, structural characteristics associated with support, such as hardness, are determined by the γ and β forms [34].

1.5.3 Properties of Chitin

Chitin is a nitrogenous polysaccharide that is white, rigid, and semi-transparent. It has limited solubility in the majority of organic solvents and has a weak chemical reactivity [35]. It is offered in granule, sheet, and powder forms [36]. Pure chitin is readily accessible and reasonably priced.

Because of the strong intermolecular hydrogen bonding, chitin is not soluble in water [37]. However, some chitin-based derivatives, like carboxymethyl or chitosan, may be made from chitin and are soluble in water. One of these compounds' most significant benefits is that it allows for the creation of various shapes, including fibers, membranes, beads, hydrogels, and sponges. Chitin is naturally crystalline and has a far better tensile strength than most artificial materials. The tremendous tensile strength of chitin results from the substantial hydrogen bonding within its chains [38].

1.5.4 Chitin Industrial Relevance

Unique of the bioresources that are greatest readily available on Earth is chitin, which is mostly derived from marine sources as a cost-effective renewable biopolymer. Chitin is regarded as the primary biomass resource and produces 10^{11} tons per year in nature [39]. Its high pollution levels and perishability contribute to its buildup as biomass waste, which is dangerous for human health and the environment. It quickly causes eutrophication on land and water, poses an earnest threat to the environment, and is frequently assaulted by diseases and bacteria. Reducing pollution, energy consumption, and global climate change may all be addressed by preventing and reusing this enormous amount of biomass waste. The shrewd and logical use of by-products from seafood processing with a scientific approach to creating reusable, long-lasting components, and recyclables saves the integrity and biodiversity of society and the environment. According to research, occurring crustacean chitin contains minerals, colors, lipids, and proteins [40], and the processes for extracting it are as follows [29]: (i) removing calcium carbonate by demineralization; (ii) decoloration to remove pigments; and (iii) deproteinization to remove proteins. While the deproteinization process uses 1 M NaOH for 65–100 °C at a time of 1–72 h, the demineralization process uses 0.275–2 M HCl for a duration of 1–84 h at 0–100 °C. Similarly, decolorization is accomplished using solvents such as hydrogen peroxide, acetone, or ethanol [31, 41]. Due to the seasonal scarcity of waste from seafood and geographic constraints for fishing businesses, chitin can be facilely manufactured by fermenting fungal mycelium with no limits [42].

Due to its advantages over currently available sources, the manufacture of chitin from fungi has drawn more interest from researchers in recent years. Fungal chitin differs from crustacean chitin in several significant ways, most related to their chemical and physical characteristics; the former contains more consistently distributed properties and less inorganic material than crustacean wastes. Additionally, fungal chitin has greater efficiency in eliciting an immunological response in the plant, making it a better choice for use in agriculture. Using nanotechnology to manufacture nanofibrils of chitin and other organic polymers to create one-of-a-kind goods with superior characteristics raises the potential for additional industrial uses and usability of this natural biomass [43, 44]. As a result, a sufficient number of applications have been documented in various fields, including medication delivery mechanisms, bionanotechnology, microbiology, engineering with tissues, material science, agriculture, food technology, and wastewater treatment [45–47].

1.5.5 Derivatives of Chitin

Chitosan and chitin are less accessible as possible reactants than cellulose because they both display significant intermolecular interactional forces and, as a result, demonstrate resistance to carrying out chemical reactions in conventional organic solvents. The structural alteration of these polysaccharides determines how they are used in various application domains. Performing regio-selective substitution during chemical transformation can be challenging, due to the low reactivity, multi-functionality, and poor solubility of the compound. Other limitations include low reaction levels, partial degradation, the formation of structurally irregular products, and difficulty in modifying the substance under heterogeneous conditions. A clearly specified structure of derivatives is also necessary to create an intelligent polymer with certain functionality. Derivatives of chitin are used in several sectors based on structure-property connections. The two main ways to produce chitosan, one of the most significant chitin derivatives, are (i) a partial deacetylation process when NaOH conc. is present, and (ii) hydrolysis when the chitin deacetylase enzyme is present [48]. The second-most investigated chitin derivative is carboxymethyl chitin, a water-soluble anionic polymer [49, 50]. Other chitin derivatives have also been created and published in the literature, including ethyl chitin (diethylamino), fluorinated chitin [51], chitin carbamates, mercapto chitin, and phosphoryl chitin [52].

Similarly, several additional chitin derivatives have also been discovered in the literature. These derivatives were created by esterification and etherification processes, which include the C-3 and C-6-OH groups of chitin [53]. Contrary to this, chitin blends with organic or inorganic polymers, tangles with cellulose using agents including glutaraldehyde, epichlorohydrin, etc., and forms graft copolymers when ceric salt is present [54, 55].

Chitin, like other polysaccharides, has advantageous qualities such as nontoxicity, biocompatibility, biodegradability, and a few key biological functions. As a result, researchers have focused a lot of attention on their potential uses in the pharmaceutical, biotechnology, biomedical, and food industries. Additionally, they have been noted during lysozyme activity testing. The anticancer, bactericidal, and fungicidal properties of oligo chitin, an oligomer of chitin, are well recognized [56, 57]. These qualities have led to their recognition in the packaging and food preservation industries, where their use eliminates the need for preservatives made of chemicals.

1.6 Chitosan and Its Derivatives

1.6.1 Availability of Chitosan

About 40 years after chitin was discovered, chitosan, a significant chitin variation, was found. The discovery of chitosan may be attributed to Charles Rouget's research from 1859, which involved heating chitin in an alkaline solution to create the substance. However, the term "chitosan" was first used in 1894 by Felix Hoppe-Seyler, and the molecular structure was not identified until 1950 [58]. Even after this earlier discovery, industrial processing and beginning in the 1970s, chitin and chitosan were commercialized [59]. Since the late 1970s, chitosan has aroused the interest of scientists and industry in China, Korea, the United States, Norway, Canada, France, Australia, Germany, Poland, and the United Kingdom, due to its biodegradability, biocompatibility, and unique macromolecular structure, among other basic chemical elements [60]. Japan had notable growth in 2015, controlling around 35% of the worldwide market. Since the 1980s, Japan has unquestionably achieved major advancements in these biopolymers' technology, marketing, and applications [7, 61]. This sector consumes about 700–800 tons of chitosan annually [62].

The exoskeletons of insects, mollusks, crustaceans, and arthropods exhibit a high concentration of chitosan, which is a cationic polymer derived from the deacetylation of chitin [63]. Shaala et al. [64] have proposed a chemical extraction platform for the separation of chitin, which comprises demineralization, decoloration, and deproteinization phases. Similarly, Ehrlich et al. [65] have also discussed this topic.

Chitosan can only be made by deacetylating chitin. These polymers can also be treated using a biological extraction technology based on microorganisms. According to the application, chitosan and its derivatives can create sponges, beads, films, membranes, or gels [66]. Reusing waste from the maritime sector is not advised, and a significant proportion of waste biomass is released unprocessed into the atmosphere [67]. Approximately 106 tons of waste are produced by the fish industry each year. Most of it is composted or turned into low-cost items like animal feed and fertilizers. Chitosan is processed in quantities of roughly 2000 tons annually, with the majority of the material coming from discarded prawn and crab shells [68].

1.6.2 Structure of Chitosan

The physical and chemical composition of chitosan has also been identified. As illustrated in Fig. 1.8 [69], it is a linear polysaccharide comprised of D-glucosamine units that have been acetylated and deacetylated connected by β-(1, 4) glycosidic connections. Chitin is deacetylated to yield acetate ions and an NH_2 group, created when the acetamide group is hydrolyzed. The ratio of glucosamine to N-acetyl glucosamine often measures the degree of deacetylation in chitosan. In contrast to chitosan, which contains a higher proportion of glucosamine than N-acetyl glucosamine, chitin has a higher amount of N-acetyl glucosamine. Depending on the experimental settings, the degree of deacetylation affects both chemical and biological characteristics, including biodegradability, antioxidant, adsorption enhancer, biocompatibility, and bioavailability. Solubility, conductivity, surface area, tensile strength, flexibility, viscosity, and porosity are chemical qualities [70, 71].

Chitosan from prawn shells had more excellent antibacterial action against *Escherichia coli*, a gram-negative bacterium, compared to commercial biopolymers [72]. Additionally, β(1–4)-linked D-glucosamine and N-acetyl-D-glucosamine were removed from chitin with minimum deacetylation to form the linear polymer known as chitosan. The high nitrogen content of chitin and chitosan, which ranges from 2% to 8%, makes them more desirable for several industrial applications [73].

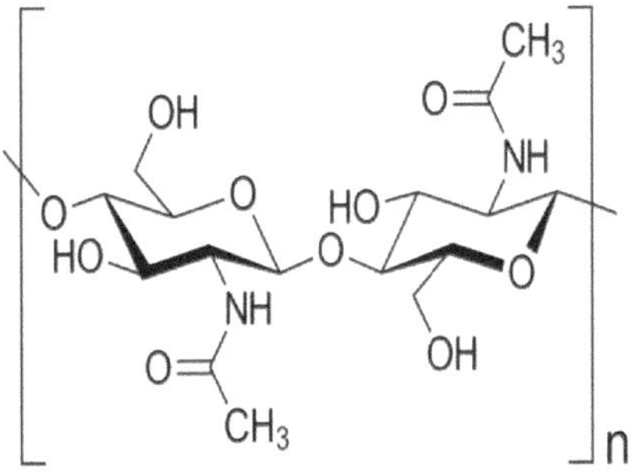

Figure 1.8 General structure of chitosan.

1.6.3 Properties of Chitosan

Due to its unique qualities, including biodegradation, biocompatibility, and nontoxicity (among many others), chitosan has gained more attention in the past ten years. One of the most famous and often-used properties is its antibacterial activity, which helps it be useful in the field of biomedicine by preventing the growth of a wide range of bacteria, yeasts, and fungi. Additionally, it can bind harmful metal ions, making it helpful for applications such as air and water filtration. The manifestation of these features is a result of the protonation of NH_2 groups on the chitosan backbone [74].

Chitosan is the second most prevalent natural polymer, behind cellulose [75], which has a structure strikingly similar to cellulose. The quantity of protonated amino groups in the polymer's chain determines the substance's solubility, biodegradability, reactivity, and capacity for the adsorption of substrates. Water, organic solvents, and aqueous bases cannot dissolve chitosan; while, nitric, acetic, perchloric, hydrochloric, and phosphoric acids can cause it to dissolve [76].

In an alkaline or neutral environment, chitosan's amino group is not protonated, making it insoluble in water; nevertheless, in an acidic pH, it produces a soluble protonated polysaccharide with both organic and inorganic acids, such as ascorbate, malate, tartrate, pyruvate, malonate, acetate, glycolate, lactate, and citrate

[77], and chitosan forms water-soluble salts. Whenever the solution's pH is less than 6.5, inherent chitosan becomes soluble in organic acids. The neutralization of chitosan with acids like formic acid, hydrochloric acid, lactic acid, or acetic acid may likely result in the formation of the water-soluble salts of chitosan.

Researchers may create distinct classes of chitosan that differ mainly in the degree of deacetylation (DD), crystallinity, and molecular weight thanks to various additional variables that may alter the physicochemical characteristics of chitosan (Fig. 1.9). The physical properties of the finished chitosan product might change throughout processing from the raw material depending on factors like the kind and concentration of chemicals used, the time and temperature utilized, and others. Additionally, a substance's solubility, viscosity, elasticity, and tearing strength affect its molecular weight.

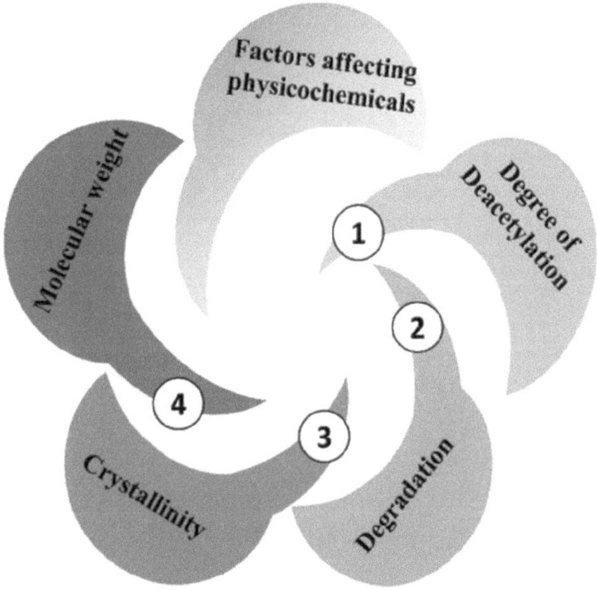

Figure 1.9 Influencing factors in physicochemical.

In acidic settings, the pseudoplastic substance chitosan works wonders at increasing viscosity. The molecular weight, pH, level of deacetylation, concentration, temperature, ionic potency, and concentration of the chitosan solution all impact its viscosity. An increase in chitosan concentration often causes the solution's viscosity to rise. And as the temperature increases, it falls. The specific type of acid determines how the pH affects viscosity [78].

DD and molecular weight of chitosan determine the possessions needed for a certain application. However, an extrinsic characteristic, increasing the temperature or acidity of the alkaline solution, can increase the DD of the chitin molecular chain. The biological characteristics of chitosan, such as its capacity to heal wounds and its susceptibility to lysozyme biodegradation, are also influenced by its viscosity.

Chitosan may form gels at acidic pH levels since it is naturally hydrophilic. The use of these gels as a slow-release medication delivery mechanism is possible. Chitosan can be cross-linked with covalent bonds using glutaraldehyde to reduce its solubility. Chitosan's capacity to swell is diminished when the concentration of the cross-linking agent rises [77].

The following are some of chitosan's many chemical and biological characteristics (Fig. 1.10):

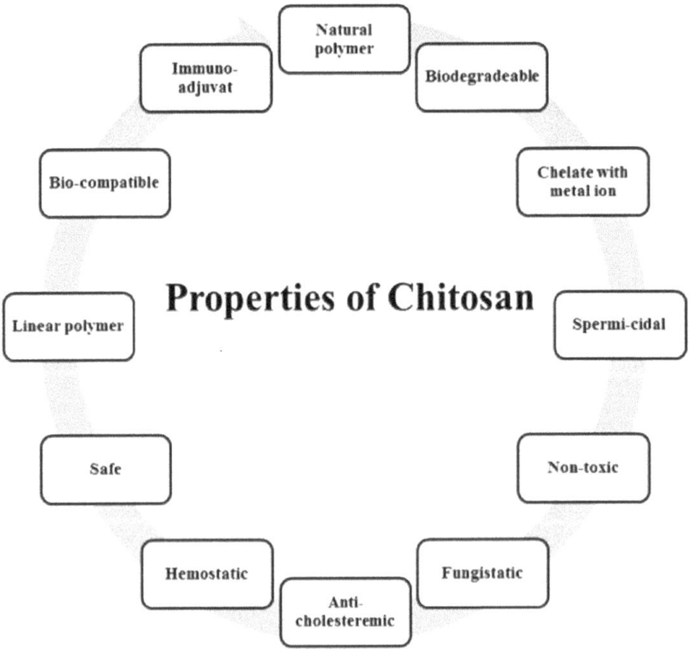

Figure 1.10 Showing different properties of chitosan.

- Linear, natural polyamine with reactive hydroxyl and amino groups 38.
- Relates to transitional metal ions in chelates
- Biocompatible with natural bodily components and biodegradable
- Safe to use and non-toxic
- Clings to both human and microbial cells
- Agent, which is hemostatic, fungistatic, and spermicidal
- Anti-inflammatory and anti-tumor agents
- Promote faster bone regrowth
- Drug delivery and immunoadjuvant

1.6.4 Chitosan Derivatives

Chitosan's usefulness is restricted by its insolubility in common organic solvents and water. Therefore, efforts have been undertaken to chemically modify chitosan's reactive functional groups, $-NH_2$ and $-OH$, to overcome its restricted water solubility [79]. By maintaining its antimicrobial, antibacterial, anticancer, and antiviral activities, adapted chitosan derivatives enhance the physicochemical characteristics of chitosan. It is frequently claimed that derivatives have increased bioactivity, biodegradability, and biocompatibility [80].

The most often reported derivatization is acylation, and N-acyl derivatives of chitosan are said to increase its solubility while maintaining its degradability and bioactivity. The natural solubility of N-acyl derivatives varies with acylation level and rises with acyl substitution number. The hydrophobicity of chitosan is increased by O-acylation, and as a result, according to reports, these compounds are soluble in fats and other solvents that are nonpolar [80]. Alkylation is a different chemical modification of chitosan that is thought to solubility improvement. Nevertheless, as alkyl groups are naturally hydrophobic, adding too many of them has increased the polymer's hydrophobicity [80]. Chitosan that has been carboxylated is said to have better water solubility and an extensive range of uses in the medicinal, agricultural, and biochemical fields. It is noted that chitosan's water solubility and antibacterial properties are improved by quaternization.

N,N,N-trimethyl ethyl ammonium chitosan, N-4-pyridyl methyl chitosan, N-4,N,N-dimethyl amino benzyl chloride, etc., are some of the leading quaternized derivatives. According to reports, some chitosan N-aryl derivatives have increased aqueous solubility and antibacterial and antifungal activity. It has been observed that graft co-polymerization with hydrophilic polymers, including polyethylene glycol (PEG), increases chitosan's solubility without compromising its biocompatibility. The increased solubility in this situation will be contingent on the molecular weight of the PEG utilized for grafting. The solubility of chitosan can be enhanced by N-acetylation, which involves the copolymerization of chitosan with acetic anhydride [81]. Creating a carboxymethyl derivative of chitosan is another way to increase its solubility. According to reports, the resulting carboxymethyl chitosan has exceptional physical-chemical characteristics and is biocompatible, making it perfect for the cosmetic and food industries. The antibacterial function of chitosan is said to be improved by grafting it with appropriate groups, such as vinyl imidazole, acetyl, benzoyl, or chloroacetyl group. Chitosan derivatives thiourea and thiosemicarbazone are claimed to have better antifungal efficacy in contradiction of *R. solani* and *S. solani* [82].

Chitosan's muco-adhesiveness, tissue-specific targetability, and pH sensitivity are said to be improved by derivatization [82]. When compared to pure chitosan, thiolated chitosan derivatives are said to have 1.7 times greater muco-adhesiveness. Similarly, it has been claimed that quaternary ammonium compounds enhance the intra-nasal administration of loaded medications that are beneficial for treating respiratory tract infections [83]. Grafting chitosan onto pH-sensitive polymers can make it more pH sensitive. Methyl chitosan derivatives are commonly acknowledged to exhibit swelling behavior at a pH of 5. Conversely, carboxymethyl chitosan derivatives have the ability to swell and then release medications specifically at the pH levels found in the intestines. This characteristic renders them well-suited for therapeutic applications in the treatment of illnesses such as Crohn's disease.

Chitosan with quaternary ammonium fatty acid modification is claimed to target the liver when a pharmaceutical delivery to the liver is desired to address various clinical conditions [84]. It has been demonstrated that glycyrrhizinic acid-conjugated chitosan is effective at targeting the liver through targeted administration of doxorubicin in a mouse model of hepatocellular cancer [85]. It is asserted that folic acid-grafted PEG chitosan copolymer exhibits lung targeting behavior in a mouse lung cancer model and enhances taxol's pulmonary delivery. Table 1.2 displays many chitosan derivatives with recognized advantages.

Table 1.2 Chitosan derivatives with realized benefits

Derivatization	Chemicals used in the derivation	Achieved results	Ref.
Quaternization	N-(3-chloro-2-hydroxypropyl) Trimethyl ammonium chloride	More effective antibacterial and water solubility	[79]
Acylation	Hydrochloric acid and N-hydroxysulfosuccinimide (NHS) and acetic acid, specifically, 1-ethyl-3-(3-dimethyl aminopropyl)-1-carbodiimide hydrochloride (EDC) are two coupling systems	Improved water solubility due to n-acylation O-acylation enhanced the hydrophobic quality	[80]
Quaternary ammonium and fatty acid	N-(3-(di-methyl amino) propyl)-N′-ethyl carbodiimide hydrochloride N-hydroxysuccinimide (NHS), Glycidyltrimethylammonium chloride, oleic acid, lauric acid	Targeting the liver	[81]
PLGA and folic acid	Folic acid, methanol, dichloromethane poly [D,L-lactide-co-glycolide] (PLGA), dimethyl sulfoxide, acetonitrile	Lung targeting	[86]
Thiolation	Thioglycolic acid Cysteine CS-4-thio-butyl-amine	Improved much-adhesiveness	[84]

1.7 Conclusion

Structural polysaccharides, storage polysaccharides, and mucosubstances are different terms for polysaccharides. Polysaccharides are widely recognized for helping support the life cycle of animals and plants. For instance, structural polysaccharides are crucial in creating the skeletal system of mammals and the structural framework of plant cell walls.

Polysaccharides are beneficial natural compounds for food and therapeutic purposes and may be found in marine and terrestrial plants, animals, and plants. Chitin and chitosan are the two most noteworthy structural polysaccharides derived from marine sources. There are several uses for chitosan, chitin, and their derivatives. The market price and the kinds of uses for chitosan goods are controlled by the quality of such items.

Despite having several drawbacks, chitin and its derivatives are intriguing materials for upcoming uses. New methods of extracting materials from shells can be created using rising technologies like nanotechnology. Although it readily forms complexes with minerals, it may be used to filter out excess and dangerous minerals from drinking water to lessen water contamination. Given the paucity of previous studies, it would make for fascinating research. It also functions as an antibacterial agent; this may be used to create a water disinfection device. In addition, its qualities, such as renewability, biodegradability, and nontoxicity, demonstrate its promise as a material for food preservation packaging.

References

1. Amor IB, Hemmami H, Laouini SE, Abdelaziz AG, Barhoum A (2023) Influence of chitosan source and degree of deacetylation on antibacterial activity and adsorption of AZO dye from water. *Biomass Conversion and Biorefinery*. 1–11. DOI: https://doi.org/10.1007/s13399-023-03741-9.
2. Ben Amor I, Hemmami H, Laouini SE, Zeghoud S, Benzina M, Achour S, Naseef A, Alsalme A, Barhoum A (2023) Use of insect-derived chitosan for the removal of methylene blue dye from wastewater: Process optimization using a central composite design. *Materials* 16: 5049.

3. Elieh-Ali-Komi D, Hamblin MR (2016) Chitin and chitosan: Production and application of versatile biomedical nanomaterials. *International Journal of Advanced Research* 4: 411.
4. Di Nardo T, Hadad C, Van Nhien AN, Moores A (2019) Synthesis of high molecular weight chitosan from chitin by mechanochemistry and aging. *Green Chemistry* 21: 3276–3285.
5. Mohammed ASA, Naveed M, Jost N (2021) Polysaccharides: Classification, chemical properties, and future perspective applications in fields of pharmacology and biological medicine (a review of current applications and upcoming potentialities). *Journal of Polymers and the Environment* 29: 2359–2371.
6. Darge HF, Andrgie AT, Tsai H-C, Lai J-Y (2019) Polysaccharide and polypeptide based injectable thermo-sensitive hydrogels for local biomedical applications. *International Journal of Biological Macromolecules* 133: 545–563.
7. Yang L, Zhang L-M (2009) Chemical structural and chain conformational characterization of some bioactive polysaccharides isolated from natural sources. *Carbohydrate Polymers* 76: 349–361.
8. Buckeridge MS (2010) Seed cell wall storage polysaccharides: Models to understand cell wall biosynthesis and degradation. *Plant Physiology* 154: 1017–1023.
9. Willfahrt A, Steiner E, Hötzel J, Crispin X (2019) Printable acid modified corn starch as non-toxic, disposable hydrogel-polymer electrolyte in supercapacitors. *Applied Physics A* 125: 1–10.
10. Lin N, Huang J, Chang PR, Anderson DP, Yu J (2011) Preparation, modification, and application of starch nanocrystals in nanomaterials: A review. *Journal of Nanomaterials* 2011: 1–13.
11. Seung D (2020) Amylose in starch: Towards an understanding of biosynthesis, structure and function. *New Phytologist* 228: 1490–1504.
12. Singh AV (2011) Biopolymers in drug delivery: A review. *Pharmacologyonline* 1: 666–674.
13. Joseleau J-P, Pérez S (2016) *The Plant Cell Walls: Complex Poly-saccharide Nano-Composites.* Glycopedia.
14. Beeson WT, Vu VV, Span EA, Phillips CM, Marletta MA (2015) Cellulose degradation by polysaccharide monooxygenases. *Annual Review of Biochemistry* 84: 923–946.
15. Butchosa N, Leijon F, Bulone V, Zhou Q (2019) Stronger cellulose microfibril network structure through the expression of

cellulosebinding modules in plant primary cell walls. *Cellulose* 26: 3083–3094.

16. Rosenau T, Potthast A, Hell J (2018) *Cellulose Science and Technology: Chemistry, Analysis, and Applications.* John Wiley & Sons.
17. Bajpai VK, Rather IA, Lim J, Park Y-H (2014) Diversity of bioactive polysaccharide originated from marine sources: A review.
18. Piontek J, Lunau M, Händel N, Borchard C, Wurst M, Engel A (2010) Acidification increases microbial polysaccharide degradation in the ocean. *Biogeosciences* 7: 1615–1624.
19. Hafdani FN, Sadeghinia N (2011) A review on application of chitosan as a natural antimicrobial. *International Journal of Pharmacological and Pharmaceutical Sciences* 5: 46–50.
20. Kiernan J (2010) Chapter 9, Carbohydrate histochemistry. Education Guide: Special Stains and H&E: pp. 75–92.
21. Sahoo S (2023) Chitin/chitosan based superabsorbent polymers, in *Bio-Based Superabsorbents: Recent Trends, Types, Applications and Recycling*. Springer, pp. 77–91.
22. Thakur N, Nath AK, Chauhan A, Parmar SC, Pandey H, Thakur K (2019) Chitinases from microbial sources, their role as biocontrol agents and other potential applications. *Journal of Entomology and Zoology Studies* 7: 837–843.
23. Rathore AS, Gupta RD (2015) Chitinases from bacteria to human: Properties, applications, and future perspectives. *Enzyme Research* 2015: 791907.
24. Amin A (2013) Characterization of chitinolytic bacterial strains isolated from local habitat. *Mycopath* 9: 51–55.
25. Gherbawy Y, Elhariry H, Altalhi A, El-Deeb B, Khiralla G (2012) Molecular screening of Streptomyces isolates for antifungal activity and family 19 chitinase enzymes. *Journal of Microbiology* 50: 459–468.
26. Usman A, Zia KM, Zuber M, Tabasum S, Rehman S, Zia F (2016) Chitin and chitosan-based polyurethanes: A review of recent advances and prospective biomedical applications. *International Journal of Biological Macromolecules* 86: 630–645.
27. Narayanan D, Jayakumar R, Chennazhi K (2014) Versatile carboxymethyl chitin and chitosan nanomaterials: A review. *Wiley Interdisciplinary Reviews: Nanomedicine and Nanobiotechnology* 6: 574–598.

28. Chen J-K, Shen C-R, Liu C-L (2010) N-acetylglucosamine: Production and applications. *Marine Drugs* 8: 2493–2516.

29. Younes I, Rinaudo M (2015) Chitin and chitosan preparation from marine sources: Structure, properties and applications. *Marine Drugs* 13: 1133–1174.

30. Hoffmann K, Daum G, Köster M, Kulicke W-M, Meyer-Rammes H, Bisping B, Meinhardt F (2010) Genetic improvement of Bacillus licheniformis strains for efficient deproteinization of shrimp shells and production of high-molecular-mass chitin and chitosan. *Applied and Environmental Microbiology* 76: 8211–8221.

31. Khan FI, Rahman S, Queen A, Ahamad S, Ali S, Kim J, Hassan MI (2017) Implications of molecular diversity of chitin and its derivatives. *Applied Microbiology and Biotechnology* 101: 3513–3536.

32. Deguchi S, Tsujii K, Horikoshi K (2015) In situ microscopic observation of chitin and fungal cells with chitinous cell walls in hydrothermal conditions. *Scientific Reports* 5: 11907.

33. Radji M, Agustama RA, Elya B, Tjampakasari CR (2013) Antimicrobial activity of green tea extract against isolates of methicillin–resistant Staphylococcus aureus and multi–drug resistant Pseudomonas aeruginosa. *Asian Pacific Journal of Tropical Biomedicine* 3: 663–667.

34. Ahmad SI, Ahmad R, Khan MS, Kant R, Shahid S, Gautam L, Hasan GM, Hassan MI (2020) Chitin and its derivatives: Structural properties and biomedical applications. *International Journal of Biological Macromolecules* 164: 526–539.

35. Arbia W, Arbia L, Adour L, Amrane A (2013) Chitin extraction from crustacean shells using biological methods–a review. *Food Technology and Biotechnology* 51: 12–25.

36. Ehrlich H, Ilan M, Maldonado M, Muricy G, Bavestrello G, Kljajic Z, Carballo J, Schiaparelli S, Ereskovsky A, Schupp P (2010) Threedimensional chitin-based scaffolds from Verongida sponges (Demospongiae: Porifera). Part I. Isolation and identification of chitin. *International Journal of Biological Macromolecules* 47: 132–140.

37. Ifuku S, Nogi M, Abe K, Yoshioka M, Morimoto M, Saimoto H, Yano H (2011) Simple preparation method of chitin nanofibers with a uniform width of 10–20 nm from prawn shell under neutral conditions. *Carbohydrate Polymers* 84: 762–764.

38. Muzzarelli RA, Boudrant J, Meyer D, Manno N, DeMarchis M, Paoletti MG (2012) Current views on fungal chitin/chitosan, human chitinases, food preservation, glucans, pectins and inulin: A tribute

to Henri Braconnot, precursor of the carbohydrate polymers science, on the chitin bicentennial. *Carbohydrate Polymers* 87: 995–1012.

39. Mincea M, Negrulescu A, Ostafe V (2012) Preparation, modification, and applications of chitin nanowhiskers: A review. *Reviews on Advanced Materials Science* 30: 225–242.
40. Kaya M, Baran T, Karaarslan M (2015) A new method for fast chitin extraction from shells of crab, crayfish and shrimp. *Natural Product Research* 29: 1477–1480.
41. Xu Y, Bajaj M, Schneider R, Grage SL, Ulrich AS, Winter J, Gallert C (2013) Transformation of the matrix structure of shrimp shells during bacterial deproteination and demineralization. *Microbial Cell Factories* 12: 1–12.
42. Wu T, Zivanovic S, Draughon FA, Conway WS, Sams CE (2005) Physicochemical properties and bioactivity of fungal chitin and chitosan. *Journal of Agricultural and Food Chemistry* 53: 3888–3894.
43. Liu D, Zhu Y, Li Z, Tian D, Chen L, Chen P (2013) Chitin nanofibrils for rapid and efficient removal of metal ions from water system. *Carbohydrate Polymers* 98: 483–489.
44. Muzzarelli RA (2011) Biomedical exploitation of chitin and chitosan via mechano-chemical disassembly, electrospinning, dissolution in imidazolium ionic liquids, and supercritical drying. *Marine Drugs* 9: 1510–1533.
45. Manikandan A, Sathiyabama M (2016) Preparation of chitosan nanoparticles and its effect on detached rice leaves infected with Pyricularia grisea. *International Journal of Biological Macromolecules* 84: 58–61.
46. Kashyap PL, Xiang X, Heiden P (2015) Chitosan nanoparticle-based delivery systems for sustainable agriculture. *International Journal of Biological Macromolecules* 77: 36–51.
47. Nishad PA, Bhaskarapillai A, Velmurugan S (2014) Nano-titaniacrosslinked chitosan composite as a superior sorbent for antimony (III) and (V). *Carbohydrate Polymers* 108: 169–175.
48. Synowiecki J, Al-Khateeb NA (2003) Production, properties, and some new applications of chitin and its derivatives.
49. Jayakumar R, Prabaharan M, Nair S, Tokura S, Tamura H, Selvamurugan N (2010) Novel carboxymethyl derivatives of chitin and chitosan materials and their biomedical applications. *Progress in Materials Science* 55: 675–709.

50. Azuma K, Nishihara M, Shimizu H, Itoh Y, Takashima O, Osaki T, Itoh N, Imagawa T, Murahata Y, Tsuka T (2015) Biological adhesive based on carboxymethyl chitin derivatives and chitin nanofibers. *Biomaterials* 42: 20–29.

51. Pestov A, Bratskaya S (2016) Chitosan and its derivatives as highly efficient polymer ligands. *Molecules* 21: 330.

52. Yamamoto C, Hayashi T, Okamoto Y (2003) High-performance liquid chromatographic enantioseparation using chitin carbamate derivatives as chiral stationary phases. *Journal of Chromatography A* 1021: 83–91.

53. Seoud OAE, Nawaz H, Arêas EP (2013) Chemistry and applications of polysaccharide solutions in strong electrolytes/dipolar aprotic solvents: An overview. *Molecules* 18: 1270–1313.

54. Gaikwad U, Pande S (2013) A review of biopolymer chitosan blends in polymer system. *International Research Journal of Science and Engineering* 1: 13–16.

55. Khouri J, Penlidis A, Moresoli C (2019) Viscoelastic properties of crosslinked chitosan films. *Processes* 7: 157.

56. Lodhi G, Kim Y-S, Hwang J-W, Kim S-K, Jeon Y-J, Je J-Y, Ahn C-B, Moon S-H, Jeon B-T, Park P-J (2014) Chitooligosaccharide and its derivatives: Preparation and biological applications. *BioMed Research International* 2014: 654913.

57. Schmitz C, González Auza L, Koberidze D, Rasche S, Fischer R, Bortesi L (2019) Conversion of chitin to defined chitosan oligomers: Current status and future prospects. *Marine Drugs* 17: 452.

58. Ruiz GAM, Corrales HFZ (2017) Chitosan, chitosan derivatives and their biomedical applications. *Biological Activities and Application of Marine Polysaccharides* 87.

59. Hahn T, Roth A, Ji R, Schmitt E, Zibek S (2020) Chitosan production with larval exoskeletons derived from the insect protein production. *Journal of Biotechnology* 310: 62–67.

60. Ferraro V, Cruz IB, Jorge RF, Malcata FX, Pintado ME, Castro PM (2010) Valorisation of natural extracts from marine source focused on marine by-products: A review. *Food Research International* 43: 2221–2233.

61. Badawy ME, Rabea EI (2017) Chitosan and its modifications as biologically active compounds in different applications. *Advances in Physicochemical Properties of Biopolymers, Part 2*: 1–108.

62. Morin-Crini N, Lichtfouse E, Torri G, Crini G (2019) Applications of chitosan in food, pharmaceuticals, medicine, cosmetics, agriculture, textiles, pulp and paper, biotechnology, and environmental chemistry. *Environmental Chemistry Letters* 17: 1667–1692.

63. Klinger C, Żółtowska-Aksamitowska S, Wysokowski M, Tsurkan MV, Galli R, Petrenko I, Machałowski T, Ereskovsky A, Martinović R, Muzychka L (2019) Express method for isolation of ready-to-use 3D chitin scaffolds from Aplysina archeri (Aplysineidae: Verongiida) Demosponge. *Marine Drugs* 17: 131.

64. Shaala LA, Asfour HZ, Youssef DT, Żółtowska-Aksamitowska S, Wysokowski M, Tsurkan M, Galli R, Meissner H, Petrenko I, Tabachnick K (2019) New source of 3D chitin scaffolds: The Red Sea demosponge Pseudoceratina arabica (Pseudoceratinidae, Verongiida). *Marine Drugs* 17: 92.

65. Ehrlich H, Shaala LA, Youssef DT, Żółtowska-Aksamitowska S, Tsurkan M, Galli R, Meissner H, Wysokowski M, Petrenko I, Tabachnick KR (2018) Discovery of chitin in skeletons of nonverongiid Red Sea demosponges. *PloS One* 13: e0195803.

66. Wysokowski M, Bazhenov VV, Tsurkan MV, Galli R, Stelling AL, Stöcker H, Kaiser S, Niederschlag E, Gärtner G, Behm T (2013) Isolation and identification of chitin in three-dimensional skeleton of Aplysina fistularis marine sponge. *International Journal of Biological Macromolecules* 62: 94–100.

67. Wang S-L (2019) Production of potent antidiabetic compounds from shrimp head powder via Paenibacillus conversion. *Process Biochemistry* 76: 18–24.

68. Santos VP, Marques NS, Maia PC, Lima MABd, Franco LdO, Campos Takaki GMD (2020) Seafood waste as attractive source of chitin and chitosan production and their applications. *International Journal of Molecular Sciences* 21: 4290.

69. Casadidio C, Peregrina DV, Gigliobianco MR, Deng S, Censi R, Di Martino P (2019) Chitin and chitosans: Characteristics, eco-friendly processes, and applications in cosmetic science. *Marine Drugs* 17: 369.

70. Fan W, Yan W, Xu Z, Ni H (2012) Formation mechanism of monodisperse, low molecular weight chitosan nanoparticles by ionic gelation technique. *Colloids and Surfaces B: Biointerfaces* 90: 21–27.

71. Ben Amor I, Hemmami H, Laouini SE, Mahboub MS, Barhoum A (2022) Sol-gel synthesis of ZnO nanoparticles using different

chitosan sources: Effects on antibacterial activity and photocatalytic degradation of AZO Dye. *Catalysts* 12: 1611.

72. Nowak V, Persijn D, Rittenschober D, Charrondiere UR (2016) Review of food composition data for edible insects. *Food Chemistry* 193: 39–46.
73. Chae K-S, Shin C-S, Shin W-S (2018) Characteristics of cricket (Gryllus bimaculatus) chitosan and chitosan-based nanoparticles. *Food Science and Biotechnology* 27: 631–639.
74. Sun K, Li Z (2011) Preparations, properties and applications of chitosan-based nanofibers fabricated by electrospinning. *Express Polymer Letters* 5: 342–361.
75. Amor IB, Hemmami H, Laouini SE, Temam HB, Zaoui H, Barhoum A (2023) Biosynthesis MgO and ZnO nanoparticles using chitosan extracted from Pimelia Payraudi Latreille for antibacterial applications. *World Journal of Microbiology and Biotechnology* 39: 19.
76. de Alvarenga ES (2011) Characterization and properties of chitosan. *Biotechnology of Biopolymers* 91: 48–53.
77. Bansal V, Sharma PK, Sharma N, Pal OP, Malviya R (2011) Applications of chitosan and chitosan derivatives in drug delivery. *Advances in Biological Research* 5: 28–37.
78. Dash M, Chiellini F, Ottenbrite RM, Chiellini E (2011) Chitosan—A versatile semi-synthetic polymer in biomedical applications. *Progress in Polymer Science* 36: 981–1014.
79. Mohamed NA, Sabaa MW, El-Ghandour AH, Abdel-Aziz MM, Abdel-Gawad OF (2013) Quaternized N-substituted carboxymethyl chitosan derivatives as antimicrobial agents. *International Journal of Biological Macromolecules* 60: 156–164.
80. Wang W, Meng Q, Li Q, Liu J, Zhou M, Jin Z, Zhao K (2020) Chitosan derivatives and their application in biomedicine. *International Journal of Molecular Sciences* 21: 487.
81. Li H, Zhang Z, Bao X, Xu G, Yao P (2018) Fatty acid and quaternary ammonium modified chitosan nanoparticles for insulin delivery. *Colloids and Surfaces B: Biointerfaces* 170: 136–143.
82. Milosavljević NB, Milašinović NZ, Popović IG, Filipović JM, Kalagasidis Krušić MT (2011) Preparation and characterization of pH-sensitive hydrogels based on chitosan, itaconic acid and methacrylic acid. *Polymer International* 60: 443–452.
83. Kumar Singh Yadav H, Shivakumar H (2012) In vitro and in vivo evaluation of pH-sensitive hydrogels of carboxymethyl chitosan for

intestinal delivery of theophylline. *International Scholarly Research Notices* 2012: 763127.

84. Shastri D (2017) Thiolated chitosan: A boon to ocular delivery of therapeutics. *MOJ Bioequivalence & Bioavailability* 3: 34–37.
85. Tian Q, Zhang C-N, Wang X-H, Wang W, Huang W, Cha R-T, Wang C-H, Yuan Z, Liu M, Wan H-Y (2010) Glycyrrhetinic acid-modified chitosan/poly (ethylene glycol) nanoparticles for liver-targeted delivery. *Biomaterials* 31: 4748–4756.
86. Zhao P, Wang H, Yu M, Liao Z, Wang X, Zhang F, Ji W, Wu B, Han J, Zhang H (2012) Paclitaxel loaded folic acid targeted nanoparticles of mixed lipid-shell and polymer-core: in vitro and in vivo evaluation. *European Journal of Pharmaceutics and Biopharmaceutics* 81: 248–256.

Chapter 2

Chitosan and Chitin Sources and Their Isolation

Elena Tafi

CREA Research Centre for Agriculture and Environment, via di Corticella 133, 40128 Bologna, Italy

elena.tafi@crea.gov.it

Chitin is a natural polymer, a component of the exoskeleton of arthropods and the cell walls of fungi and yeasts. Due to their inherent valuable properties, chitin and its primary derivative, chitosan, have discovered extensive applications across diverse fields, including agriculture, food production, medicine, bioengineering, cosmetics, and wastewater treatment. Currently, chitin is industrially produced primarily from fishing industry waste by chemical processes that involve treatments for removing minerals, proteins, and pigments using acids and bases. Alternative sources and more environmentally friendly extraction processes are still under development. This chapter discusses the primary sources of chitosan, chitin, and the applied extraction methods.

Chitin and Chitosan: Physical and Chemical Properties
Edited by Shakeel Ahmed, Soumeia Zeghoud, Hadia Hemmami, and Ilham Ben Amor

ISBN 978-981-5129-39-7 (Hardcover), 978-1-003-58977-8 (eBook)
www.jennystanford.com

2.1 Introduction

Chitin is the most plentiful natural biopolymer, second only to cellulose [1]. It is a β-(1-4)-*N*-acetyl-D-glucosamine polysaccharide (Fig. 2.1a) found in the exoskeleton of arthropods as the main carbohydrate component, in mollusks shells and radulae, cephalopod beaks, poriferans, bryozoans, as well as in the cell walls of fungi and yeasts [2–4]. Chitin has a structure similar to cellulose, differing in the presence of acetamido groups at the C2 hydroxyl position of the glucose unit [1]. It was first isolated in 1799 from mollusk shells, but its large-scale production and commercialization started only in the 1970s [5].

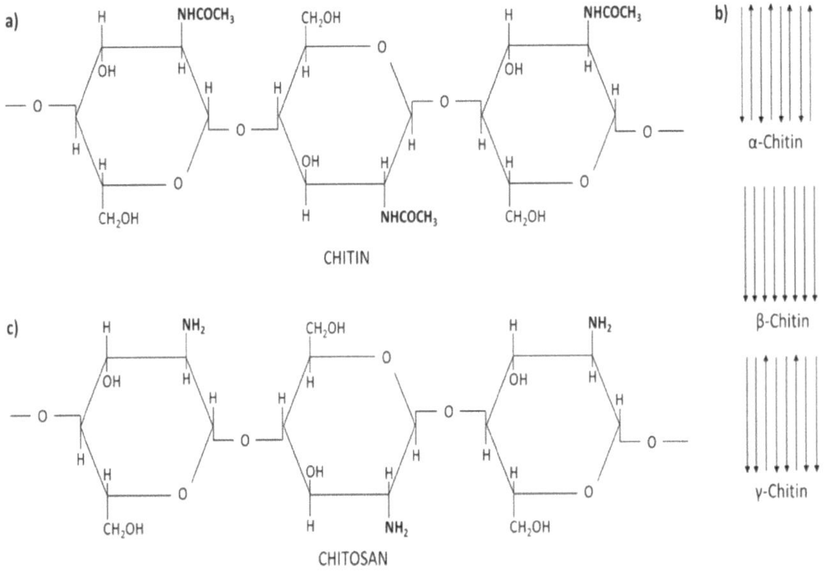

Figure 2.1 Structure of a chitin chain (a), its three natural allomorphs (b), and structure of its main deacetylated derivative chitosan (c).

Chitin exists in three allomorphs based on the different orientations of its microfibrils: α, β, and γ (Fig. 2.1b). The most abundant form in nature, with anti-parallel chain orientation, responsible for the hardness and rigidity of the polymer, is the α chitin [6]. β-chitin is less common, mainly found in squid pens, diatom spines, and pogonophora. It has parallel-oriented chains and weak intra- and inter-molecular hydrogen bonding. γ-chitin is found in yeasts, fungi, *Loligo* stomach, and *Ptinus* beetle cocoon fibers. It is a mixture of the α and β forms, with parallel and anti-parallel lipo-chitooligosaccharide units [7].

Despite its useful properties, the direct application of chitin is limited by its insolubility. To overcome this constraint and widen its application range, chitin can be converted into chitosan, its primary deacetylated derivative. Chitosan was obtained for the first time in 1859 by eating chitin in alkali, but its chemical structure was determined in 1950 [8]. Chitin can be converted into chitosan by alkaline hydrolysis, resulting in a linear cationic polysaccharide with a reduced content of N-acetyl groups (acetylation degree < 50%), lower molecular weight (from more than 1000 kDa of chitin to 100–1000 kDa of chitosan) [9] and less crystallinity. The increased content of free primary amine groups makes chitosan soluble in mildly acid-aqueous solutions and even in water.

Due to valuable features like non-toxicity, biocompatibility, biodegradability, bioadhesiveness, high porosity, low immunogenicity, and antimicrobial activity, chitin and chitosan have attracted increasing interest over the last decade for various uses. Chitosan is especially finding applications in several fields, including agriculture, the food industry, wastewater treatment, biomedicine, bioengineering, and cosmetics [10]. Its proven antimicrobial activity against essential pathogens, including *Staphylococcus aureus* [11], *Escherichia coli*, *Aspergillus niger*, and *Pseudomonas aeruginosa* [12], is one of the most used properties. Beyond biomedical applications, this feature has been successfully exploited for food preservation using chitosan as a coating material to replace traditional synthetic packaging [13]. Furthermore, the chitosan chelation and adsorption capacities of heavy metals and other waste molecules make it suitable for wastewater treatment [14]. Chitosan has also been applied as a hemostatic dressing for wound healing, as a finishing agent for fabrics [15], and in cosmetics for skin hydration and protection against UV.

The applicability of chitin and chitosan strictly depends on their physicochemical properties, primarily the acetylation degree and molecular weight, which, in turn, vary depending on the source and the purification conditions. It is, therefore, crucial to optimize the extraction by adjusting the process parameters to the polymer source and the characteristics required for the desired final application.

2.2 Sources of Chitin and Chitosan

Chitin and chitosan are industrially produced primarily from waste from the fishing industry, mainly crustacean exoskeletons [16]. The debate over the sustainability of fishing, along with the seasonality of this activity, which may not guarantee continuous availability of raw materials throughout the year, has, however, highlighted the need for new sources of chitin. According to the Allied Market Research estimation, the global chitosan market size was 1.7 billion dollars in 2019, and it is expected to grow to 4.7 billion dollars by 2027 [17]. To meet this large growing market demand, the primary alternative source to fishery waste that is being explored is mushrooms. The fungal cell wall contains up to 15% chitin with a structure similar to that of marine sources. Beyond chitin, chitosan can also be directly isolated from the fungal cell wall of certain species (e.g., *A. niger*, *Mucor rouxii*, *Lentinus edodes*, and *Rhizopus oryzae*), avoiding the chitin deacetylation process [18].

In parallel with the global growth of facilities for their large-scale breeding for feed production and organic waste management, interest has emerged in insects as an additional source of chitin and chitosan. Insects can be bred on a continuous cycle, without seasonality, and in large masses, due to their high fertility and reproductive rate [19]. Insect farms generate large quantities of waste biomass rich in chitin that might be exploited for the extraction of this polymer.

Despite the potential of these sources, the purification of chitin and chitosan from fungi and insects has not yet been established on an industrial scale.

2.2.1 Crustaceans

The annual global production of crustaceans for human consumption is about 8 million tons, estimated in 2016 [20], and it will increase by 36–74% in the next 30 years [21]. Waste exoskeletons amount to 40% of this production. This material can become an environmental pollutant if not properly disposed of. Chitin constitutes, on average, 15–40% of crustacean exoskeleton. Thus, exploiting this waste for producing valuable polymers can maximize the economic value of fishery products while minimizing environmental risk.

Shrimps, crabs, and lobsters are the primary marine organisms used for chitin purification, although the most significant source was suggested to be zooplankton, with more than 350 million tons of krill available globally but not commercially viable [22].

Chitin in the α-form can be extracted from crustacean shells with yields ranging from 5% to 32% [23]. However, the chitin content is variable, depending not only on the species but also on the body part and developmental stage of the organism. For instance, a higher chitin yield from crab legs compared to the carapace was reported by Thirunavukkarasu et al. [24].

The crustacean exoskeleton is made of a cuticle lying over the epidermis (Fig. 2.2). The cuticle, in turn, comprises three layers: endocuticle, exocuticle, and epicuticle, from the inside to the outside. The epicuticle is the outermost and thinnest layer, a barrier to the external environment. It is coated with waxes and lipids that waterproof it, and it is normally not mineralized. The exocuticle is calcified, with mineral deposition in pore canals and between and within chitin-protein fibrils. The innermost endocuticle is the thickest mineralized layer of the exoskeleton, consisting of numerous layered lamellae of chitin-protein complex fibers. The chitin-protein fibers are arranged in planes with different orientations. In one plane, they are parallel to the surface of the exoskeleton; others are perpendicular to the cuticle and have a honeycomb-like structure, while others form a helicoidal architecture. This structure provides mechanical strength to the cuticle [25]. The major components of the cuticle are calcium carbonate, protein, and chitin; the latter is contained in the endo- and exocuticle. Other minor components include lipids, proteoglycans, and other inorganic materials. These constituents are present in different proportions depending on the species and body part of the individual. Both chitin and proteins are synthesized in the epithelial cell and released into the cuticle. The architecture of chitin fibers constitutes the scaffold for the deposition of calcium carbonate.

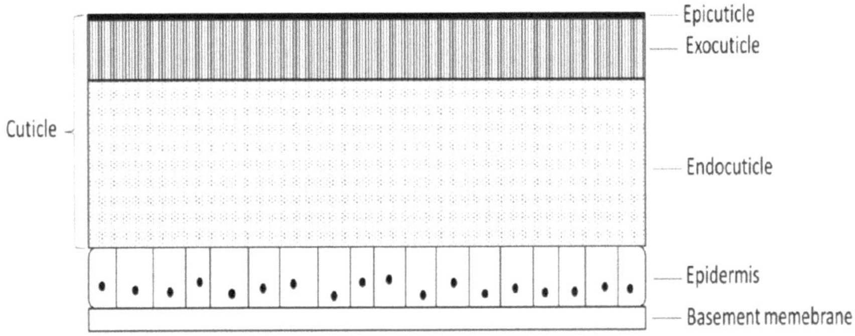

Figure 2.2 Structure of arthropods' exoskeleton.

Beyond the crustacean exoskeleton, squid pens are an essential source of β-chitin, the more soluble, reactive, and deacetylate form than α-chitin. β-chitin is also more susceptible to enzymatic degradation and appears safer than α-chitin for human medical applications. Allergenic proteins, potentially dangerous to shellfish allergy sufferers, may indeed reside in the α-chitin derived from crustaceans. Allergy to mollusks such as squid is less frequent, so beta chitin extracted from them can be considered safer [26]. β-chitin in the squid pen can amount to 35% of the dry weight. In it, chitin crystallites are enveloped in a protein layer forming nanofibrils. These are the building blocks of thicker fibers that eventually form the pen [27]. In 2002, squid production was above 2 million tons. This can represent a further huge supply of waste material for β-chitin purification, whose applications range from regenerative medicine to nanotechnology.

Some authors have begun to evaluate freshwater crustaceans as a source of chitin and chitosan. For the freshwater crab *Potamon algeriense*, a species found in North Africa, chitin was produced and characterized with a standardized chemical method. This is the only freshwater species for which this has been done. Many other freshwater organisms, including water fleas, crayfish, clams, tadpole shrimps, branchiopods, copepods, ostracods, amphipods, and isopods, could be explored for chitin production [5].

2.2.2 Fungi

Compared to crustaceans, fungi are a source of chitin and chitosan that is not seasonal and not influenced by geographical factors. Chitin extraction from this source requires less use of chemicals, as there is usually no need for mineral removal and decoloration. In fungal chitin and chitosan, there are generally no heavy metal contaminants or allergenic proteins that can be found in marine sources [28]. Furthermore, fungal chitin and chitosan have more consistent characteristics and properties, due to their production by controlled fermentation, and they seem to be more suitable for agricultural applications in enhancing plant immune response [29]. Despite these advantages, the availability and quantity of fungal biomass for the production of chitin and chitosan are much lower than marine sources. Furthermore, the extraction process has not yet been optimized for fungi, and lower yields are obtained [28].

The major components of the fungal cell wall are chitin and chitosan, glucans, glycoproteins, and mannans (Fig. 2.3). Although not ubiquitous in the fungi kingdom, chitin is widely distributed in many fungal classes including Basidiomycetes, Ascomycetes, Zygomycetes, and Chytridiomycetes, comprising 1–15% of the cell wall mass [30]. The components of the fungal cell wall are all covalently cross-linked to each other. Chitin forms interchain hydrogen bonds with glucans, creating a glucan/chitin matrix that can assemble into microfibrils that form a scaffold around the cell [30]. Different from crustaceans, chitosan can also be directly extracted from the cell walls of some species of fungi. For this purpose, the most investigated species include *Mucor rouxii*, *Absidia* spp., and *Rhizophus oryzae*, all belonging to Zygomycetes, *Aspergillus niger* (Ascomycetes) and *Lentinus edodes* (Basidiomycetes).

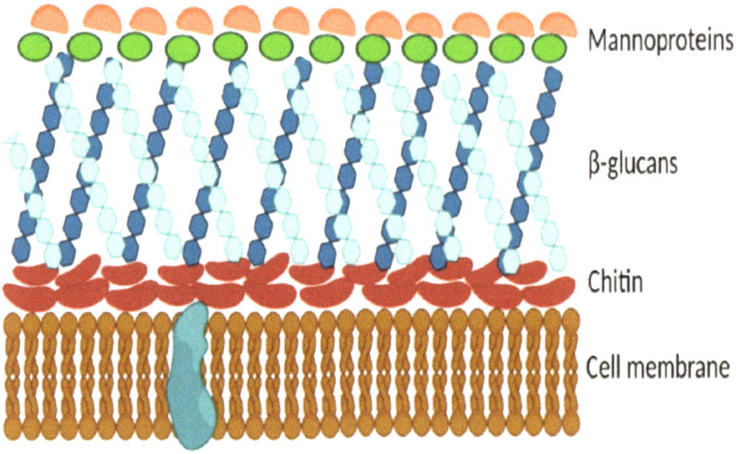

Figure 2.3 Structure of fungal cell wall (imagine created with Biorender.com).

Three categories of potential fungal sources for the commercial production of chitin and chitosan have been described: waste biomass from biotech industries, fungus fermentation, and exploitation of existing mycothec products [31]. Biotech industries annually generate thousands of tons of waste fungal biomass residue from various processes, including the production of bread, beer, antibiotics, and other drugs, enzymes, and the cultivation of edible mushrooms. From the latter activity, for instance, around 50,000 tons of waste mushroom stalks and irregular fruit bodies are produced every year [32]. Additional 80,000 tons of waste biomass are derived annually from the cultivation of *A. niger* to produce organic acids, and more than 1 billion tons of fungal waste is generated annually from the penicillin production process [33]. These huge amounts of biomass usually are disposed of in landfills or incinerated. By exploiting them for chitin and chitosan production, waste from biotech industries could be turned into a profitable solution to waste management.

For the commercial production of chitin and chitosan by fermentation, several fungal species have been evaluated. Different results in terms of yields and polymer characteristics are obtained using different fungal growth mediums. Moreover, different forms of mycelia (e.g., aerial mycelia for sporulation and submerged mycelia for absorption of nutrients) with different chitin and chitosan content are produced by single filamentous fungi. Experimentation focuses mainly on zygomycetes because of the potential to extract chitin and chitosan directly. Analysis of the cell walls of several Zygomycetes conducted by Campos-Takaki and co-workers [34] in order to evaluate the most promising species for industrial production of these polymers showed an average chitin content of 10–16% of the cell wall and 26–28% chitosan. The species analyzed were *Gongronella butleri, Absidia blakesleeana, Mucor javanicus, Rhizopus arrhizus, Syncephalastrum racemosum*, and *Cunninghamella elegans. A. blakesleeana* and *R. arrhizus* had the highest content of chitin plus chitosan (43% of the cell wall). Chitosan content in Zygomycetes generally ranges from 1% to 10% of the cells' dry weight. However, fungal chitin and chitosan yields vary depending on fungus strain, type of cultivation (solid or submerged fermentation), growth rate and nutritional requirements of fungi, and extraction process.

Concerning the last category of fungal sources, existing myco-technological processes could be exploited to extract chitin and chitosan as an additional step after using fungal biomass for their primary production purpose. For instance, *L. edodes* is cultivated to produce lentinan, a polysaccharide with immune stimulation properties. After the extraction of lentinan, the fungal biomass can be used for chitosan purification. A further example is the side production of chitin and chitosan by *Mucor indicus* during its growing for fatty acids production [35].

Finally, it is worth mentioning that the production of fungal chitin and chitosan is not completely free of risk. Some genera of Zygomycetes used to extract these polymers (*Absidia* and *Rhizopus*) include pathogenic species to humans or animals. For example, *Rhizopus oryzae* can cause pulmonary mucormycosis in humans, and some species of *Absidia* cause abortion in domestic animals [36]. Therefore, specific safety measures should be adopted when handling these potentially pathogenic fungi to prevent their dispersal.

2.2.3 Insects

Insects are clothed with a protective exoskeleton structurally similar to that of crustaceans. Chitin contained in the two layers of the procuticle (endo- and exocuticle) can constitute up to 20–50% of its dry weight. Interest in insects as an alternative source of chitin and chitosan has grown alongside the rise of insect breeding facilities. Currently, large-scale insect farms are widespread worldwide, pursuing two primary purposes: producing insect-based protein feed for livestock and disposing of organic waste by bioconversion. The larval stages of reared insects are indeed able to feed and grow on a variety of decaying organic materials (e.g., manure, organic municipal waste, and agri-food scraps) by converting such waste into biomass of high biological value (the mature larvae) [19]. When development is complete, the larvae are destined for feed production. A portion of the larvae is kept until the completion of the life cycle to ensure the availability of adult breeding insects. Insect larvae can also be used for lipid extraction for the cosmetic and energetic sectors and for the isolation of molecules of pharmacological interest, such as antimicrobial peptides [37, 38]. Around 6 kilotons of insect meal for animal feed was produced in Europe in 2019 [39]. The insect farming industry is expected to grow further, given the increasing number of countries authorizing insects as human food. In 2023, the market for edible insects is estimated to exceed $522 million [40]. Beyond these valuable products, insect farms generate a huge amount of chitin-rich side streams, consisting of exoskeletons released by insect molts, which could be processed for chitin and chitosan purification.

The main insect species managed for feed production from larvae are the yellow mealworm (*Tenebrio molitor*), the lesser mealworm (*Alphitobius diaperinus*), the black soldier fly (*Hermetia illucens*), the wax moth (*Galleria mellonella*), and the silkworm (*Bombix mori*), as well as grasshoppers cricket species. Among them, *H. illucens* and *T. molitor* are the most relevant species produced in Europe [39]. During their life cycle, both these insect species go through different developmental stages: egg, larva, pupa, and adult. The larval stage is, in turn, divided into 5 or 15 to 17 larval instars, for *H. illucens* and *T. molitor*, respectively, during which the larvae grow and molt. At each molt, the larval exoskeleton is released. In the pupation stage, *H. illucens* synthetizes a chitinous puparium inside which it completes its development. Once adult, the insect emerges from the puparium, which is left as a waste material. Larval and pupal exuviae, as well as adult insects that die at the end of their life cycle, represent side streams of the insect breeding process that can be valorized for chitin and chitosan production [41].

Farmed insects are not subjected to seasonality, are easy to breed, and are promptly available, due to their high fertility and reproductive rate [19]. Generally, the whole insect body contains 5–25% chitin, 30–60% protein, 10–25% lipid, 2–10% minerals, and 5–10% catechols [14]. The yields that can be obtained from the extraction of chitin from insects on average range from 5% to 15% (chitin dry weight/insect dry weight) [14], with peaks up to 36% as reported for honey bee corpses and cicada sloughs [6]. These values are comparable to those reported on average for chitin purification from crustaceans, mainly crabs and shrimps (5–32% yield) [42]. Yields of chitosan from insects generally range from 2% to 8% of the insect's dry weight, just slightly lower than those average obtained from marine sources (4–15%) [43]. Insects can thus be a reliable sustainable source of these polymers.

Insect-based chitin and chitosan have already been found to be suitable for various applications. For instance, chitosan from *H. illucens* has been applied to fresh tomatoes as a preservative coating to extend their shelf life [13]. Chitin from *Pterophylla beltrani* showed excellent antifungal activity against entomopathogenic fungi [13], and chitin extracted from the molossus beetle (*Catharsius molossus*) has good mechanical properties suitable to produce biodegradable films. Chitosan derived from mealworms, silkworms, cicada sloughs, and grasshoppers has shown higher fat-binding and water-holding capacity than shrimp-derived chitosan [44].

2.3 Chitin Purification

Chitin extraction from the arthropods' cuticle involves the removal of minerals, proteins, lipids, and pigments contained therein. The two main types of chitin purification methods that can be performed are the chemical and the biotechnological processes. Due to its massive use as the traditional source of chitin on an industrial scale, the development and advancement of extraction techniques have been carried out on crustaceans' exoskeleton. Chitin purification from different sources still needs to be further deepened and optimized.

2.3.1 Chemical Method

The chemical process, involving the use of acids and alkalis, is the most applied method for the industrial production of chitin from crustacean shells for its rapidity and effectiveness in providing a high-purity polymer. The drawback is the large volume of waste chemical reagents residual from the purification reactions. The chitin chemical extraction consists of two main steps of mineral and protein removal, often followed by a decolorization reaction as the final treatment. Minerals are solubilized by acid treatment, while alkaline solutions are used to remove proteins. Bleaching is performed to improve chitin color and further remove residual pigments (Fig. 2.4) [45].

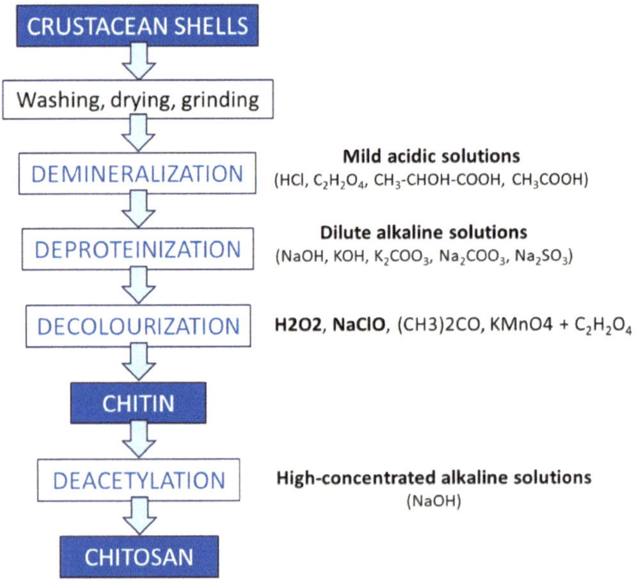

Figure 2.4 Chemical method for chitin extraction and deacetylation from crustacean waste.

Pre-treatment of the biomass can precede the extraction process. Crustacean shells are generally washed to remove coarse impurities, dried, and ground to powder to increase accessibility to chemical reagents. The optimal particle size for chitin purification has been found to be below 200–250 μm [46]. For chitin extraction from insects, an additional pre-treatment for fat removal may be necessary, especially for lipid-rich biomasses like larvae [47, 48].

2.3.1.1 Demineralization

Minerals in crustacean shells can amount to up to 50%, and they are mainly represented by calcium carbonate ($CaCO_3$) [49]. Demineralization implies the decomposition of $CaCO_3$ into its water-soluble salt (calcium chloride, $CaCl_2$) with the release of carbon dioxide and water by acid treatment, as shown in Eq. (2.1). Other minerals in the exoskeletons are simultaneously removed to give soluble salts.

$$2HCl + CaCO_3 \rightarrow CaCl_2 + H_2O + CO_2\uparrow \quad (2.1)$$

Dilute hydrochloric acid (HCl) (1–2 M) is the most common acid used for demineralization of crustacean shells, as well as insect cuticles [50]. This treatment is carried out under continuous stirring, applying different temperatures and times according to the manufacturer's protocol. Typically, the duration of treatment is 2–3 hours [10]. Longer times should not exceed 48 hours, however, to avoid chitin degradation. Room temperature is usually adequate to achieve demineralization, but higher temperatures are reported to promote penetration of the acid in the chitin matrix, accelerating the reaction. After mineral decomposition, the solubilized salts are removed by filtration and washing with deionized water of the solid residue.

Demineralization treatment can be influenced by several factors, including temperature and duration of the reaction, acid concentration, degree of mineralization of the chitin-bearing biomass, and particle size [50]. An additional crucial parameter is the solute/solvent ratio. To convert one molecule of $CaCOO_3$ into $CaCl_2$, two molecules of HCl are required. Thus, acid intake should be at least equal to the stoichiometric amount of minerals, but larger volumes are usually applied for higher efficacy.

Despite being the most widely used, HCl can have adverse effects on the structure and characteristics of chitin, lowering its molecular weight and degree of acetylation, especially when used in high concentrations, at high temperatures, and for long periods. Furthermore, HCl has a high environmental impact. For these reasons, HCl has been replaced by organic acids in many cases. Lactic acid and acetic acid, for instance, were found to be as effective as HCl in demineralizing shrimp shells [51]. Therefore, it would be worthwhile to study and optimize the treatment with organic acids to reduce the environmental impact and better maintain the characteristics of chitin.

In contrast to the exoskeletons of crustaceans and bivalve mollusks, squid pens contain negligible amounts of minerals (less than 0.05%). Demineralization treatment can therefore be omitted, saving time and resources and reducing pollutants. Fungal chitin has a much lower content of inorganic material, allowing to avoid the demineralization treatment.

For the demineralization of insect biomasses, the same procedures described for crustaceans are mainly applied [18]. Insects generally have a lower mineral content than marine sources (2–10% of the whole insect body) [52], so this step should require less harsh conditions.

2.3.1.2 Protein removal

Deproteinization of crustacean shells is commonly achieved by alkaline hydrolysis of the peptide bonds in the protein chains using alkaline solutions. A variety of base reagents has been proposed for chitin purification from proteins, including sodium hydroxide (NaOH), sodium carbonate, potassium carbonate, potassium hydroxide, calcium hydroxide, sodium sulfite, and sodium sulfide [50]. Like demineralization, deproteinization efficiency is affected by temperature and duration of the treatment, alkali concentration, and solid/solvent ratio. High deproteinization efficiencies in high temperatures are crucial, but they can lead to partial hydrolysis and deacetylation of chitin worsening its characteristics. This occurs especially for long treatment times [50].

NaOH is the preferred alkali for industrial chitin deproteinization [53]. Alternatively, Fu et al. [54] and Castillo et al. [55] proposed the use of potassium hydroxide as an eco-friendly reagent as the generated liquid waste can be used as a soil fertilizer. NaOH is typically applied in concentrations from 0.03 to 4 M, at 25–150 °C, with a solid/solvent ratio from 1:5 to 1:100 (w/v). Duration of the treatment can vary from hours to a few days, mainly ranging between 20 minutes and 96 hours [56]. As a rule of thumb, the harsher the treatment conditions (alkali concentration, temperature, time, and pH), the lower the molecular weight of the resulting chitin is expected to be [57].

Similar alkaline treatments are applied to fungi to remove proteins and alkali-soluble polysaccharides. Typical fungi deproteinization is reported to occur with 1 M NaOH at 100 °C for 3 hours [10].

Deproteinization of insect materials is generally carried out with similar conditions to those applied to crustaceans. NaOH is also the most used reagent for insects, applied in low concentrations (0.5–2 M), mainly at 80–100 °C for 16–20 hours [58–60].

The order of the demineralization and deproteinization treatments is not fixed and can be reversed without affecting the characteristics of the final crustacean-based chitin [61].

2.3.1.3 Decolorization

Chitin and chitosan must be as white as possible for the market. Demineralization and deproteinization already remove a small part of the pigments, but it is not sufficient. A step of decolorization is thus often performed to improve chitin color removing residual pigments. Sodium hypochlorite, hydrogen peroxide, acetone, and potassium permanganate are the most applied reagents for bleaching crustacean shells [53]. In addition to improving its appearance, the bleaching treatment can increase the degree of purity of the chitin obtained [41].

The effect of the bleaching treatment on the characteristics of chitin and the resulting chitosan is debated. The potentially detrimental effect of decolorization has particularly been investigated in the extraction of chitin from insects. For these organisms, decolorization is a highly required step, especially if the chitin is purified from pupal exuviae and adult insects. Indeed, during their development, sclerotization of the cuticle occurs, with the formation of cross-links with catechol compounds [62]. To remove catechols, the additional step of bleaching is practically mandatory. In contrast, this process may not be necessary for the purification of crustacean exoskeletons in which the reddish coloration is mainly due to non-covalent cuticle bonds with carotenes, not contaminated with catechols [63]. In insect-based chitin, a decrease in the crystallinity index after bleaching has been observed, suggesting a possible negative effect of this step on the crystalline structure of chitin [41]. Furthermore, the chitin decolorization treatment can lower the molecular weight and viscosity of the derived chitosan. It is conceivable that the decolorization causes the scission of some polysaccharide chains and that the absence of catechols makes the polymer more accessible to partial hydrolysis [41].

Concerning the purification of chitin from fungi, decolorization treatments with acetone [64], sodium hypochlorite, ethanol, or potassium permanganate combined with oxalic acid have been [65] reported for shiitake mushrooms. The most effective decolorization was achieved with the latter method. In other cases, decolorization of fungal chitin is considered unnecessary [28, 66].

2.3.2 Biotechnological Method

Chemical chitin extraction methods guarantee high yields and high purity of the polymer, but they have drawbacks. Strong acids and bases used in the treatments, often at high temperatures and for long times, pose a major environmental risk upon disposal. Effluents from the washing step after demineralization are rich in mineral acids, whose recycling is difficult and expensive, and each washing consumes a large volume of freshwater [57]. Furthermore, it is not unusual for severe purification conditions to impair the characteristics of the chitin. To counter these concerns, greener and more environment-friendly methods of chitin and chitosan production are increasingly gaining popularity, although they are not yet fully exploited on an industrial scale [67]. Biotechnological purification combines the chemical method with the biological process by application of microorganisms (Fig. 2.5).

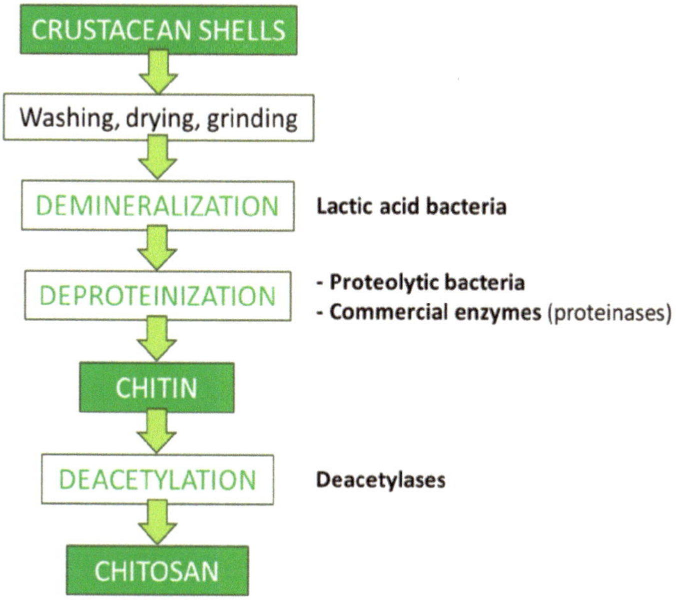

Figure 2.5 Biotechnological process for chitin and chitosan production from crustacean waste.

Demineralization is carried out by microbial fermentation, incubating the chitin-containing biomass with the microbial culture and the required nutrients [68]. Fermentation produces organic acids that react with minerals decomposing them into soluble salts. The latter can then be removed with a washing step. Microbial culture media are composed of sugars, fats, amino acids, magnesium, iron, sources of calcium, and other compounds specific to the cultivated microbial species [69]. *Lactobacillus* species, producing lactic acid, are the most used bacteria for microbial demineralization for their high efficiency and quality of the derived product [70]. However, the fermentation efficiency of lactic acid bacteria depends on the culture conditions, including the concentration of the carbon source, the quantity of inoculum, pH, and fermentation time [70].

Deproteinization can be performed by fermentation or enzymatic techniques. Fermentative processes exploit the production of digestive and microbial enzymes which consume organic material. The enzymatic method involves the use of proteases which, in addition to the action of protein removal, can give hydrolyzed proteins as valuable by-products [71]. A work by Synowiecki et al. [72] reported the application of proteolytic enzymes on shrimp shells after chemical demineralization to produce chitin and protein hydrolysate with good nutritional value. They use alcalase from *Bacillus licheniformis*, suitable to produce non-bitter hydrolysate, being specific for terminal hydrophobic amino acids. Digestion by alcalase yielded a hydrolysate that can be used as a food additive to supply essential amino acids. On the other hand, small peptides and amino acids resistant to enzymatic hydrolysis remain bound to the chitin chains, limiting the effectiveness of deproteinization. However, the low level of impurities residual in the chitin (approximately 4% protein and 1.5% ash) ensured sufficient purity for non-medical applications.

Biotechnological processes have also focused on the production of valuable by-products. Beyond protein hydrolysates, chitinases, carotenoids, lactic acid, calcium ions, and antioxidants are some of the bioactive compounds with nutraceutical potential that can be recovered making the chitin purification process more cost-effective and commercially viable [57].

The main disadvantages of the biotechnological purification method are the long time required for microbial fermentation, the high cost of proteolytic enzymes, and the generally lower chitin yields compared to the chemical process. Furthermore, the application of acid and alkaline reagents is often required to compensate for the incomplete efficiency of the biological process and to ensure adequate removal of minerals and proteins [10, 73].

The concurrent fermentation to produce chitin from crustaceans and fungi has been experimented with by placing crustacean shells in contact with the fermenting fungi. Teng et al. [74], cultivated strains of *A. niger* which produced proteolytic enzymes able to deproteinize the co-fermenting shrimp shells. The amino acids released by this reaction, in turn, acted as a nitrogen source for the fungal growth. They measured a residual protein content below 5% and between 10% and 15% in the chitin extracted from shrimps and fungi, respectively. Tan et al. [75] attempted to purify chitin from prawn shells and fungal chitosan in a concurrent fermentation with *Mucor circinelloides*, a producer of extracellular proteases. The fungus was grown on waste substrates, such as coffee waste, grape pomace, and avocado and durian seeds, and was able to successfully deproteinize the prawn shells. The acidic conditions resulting from this fermentation enable an adequate demineralization of the crustacean biomass. The final prawn-derived chitin and the fungal chitosan had a quality comparable to the commercial polymers, in terms of deacetylation degree and FTIR spectra. Fungal fermentation could thus be a cost-effective methodology to isolate chitin from both fungi and crustacean shells, avoiding the high cost of commercial proteases. However, this process needs further experimentation to confirm its effectiveness and to scale up from the laboratory to the industry.

2.4 Chitin Deacetylation

Chitin can be deacetylated to obtain chitosan. This process consists of the partial removal of acetyl groups from the chitin chain resulting in a polymer with a higher degree of free amino groups (NH_2) derived by the hydrolysis of the acetamide group, which are active sites for chemical reactions. Beyond a greater solubility, chitosan also has a high versatility lending itself to modification for several applications. The ratio of glucosamine to N-acetyl-glucosamine units defined the degree of deacetylation of chitosan. This is one of the most important characteristics of chitosan, affecting its chemical (e.g., solubility, viscosity, and conductivity) and biological properties (e.g., biodegradability, biocompatibility, antioxidant activity, and adsorption ability) [5, 76]. Chitin can be converted to chitosan mostly by alkaline hydrolysis or enzymatic deacetylation.

2.4.1 Chemical Method

Chemical deacetylation is the most used process for chitosan preparation on an industrial scale. It is a rapid, effective, and economical method for chitosan mass production. The chemical reaction is usually carried out by incubation of chitin in a solution of sodium hydroxide at a high concentration [50]. There are two main types of alkaline hydrolysis: heterogeneous and homogeneous reactions. Heterogeneous deacetylation involves the use of concentrated sodium hydroxide solutions at high temperatures for a few hours. In the homogeneous process, deacetylation occurs at room temperature for longer times, and it is followed by an additional step of dissolution in ice [50, 77]. These two methods result in a different distribution of the acetyl groups along the chitosan chains, which in turn can affect its physicochemical properties. In the heterogeneous chitosan, acetyl groups have a block-wise distribution, and the N-acetyl-glucosamine units are irregularly distributed along the polymer. The heterogeneous deacetylation acts more in the amorphous regions of the polymer and less in the crystalline ones. In the homogeneous chitosan, acetyl groups have a random distribution [78].

The deacetylation parameters can affect the quality and properties of chitosan. The degree of deacetylation especially depends on the alkali concentration, particle size, and previous treatment [79]. Chitin deacetylation inevitably leads to partial degradation of the polymer. Harsh conditions worsen this effect to the extent of damaging the chitosan structure and crystallinity. Alkaline solution with 75% (w/v) NaOH at 110 °C has been established as a heterogeneous deacetylation condition that causes a minimal degradation of the polymer [80], but even lower alkali concentrations (i.e., 50–60%) can yield good-quality chitosan with high solubility and around 80% deacetylation degree [81].

For the deacetylation of insect-derived chitin, the same methods reported for crustaceans are mainly used. Solutions of 40–60% sodium hydroxide are usually applied at 90–150 °C for 1–9 hours.

To produce fungal chitosan, a step of acid reflux is carried out, followed by precipitation of chitosan in alkali. Acid treatment for chitosan separation usually involves 2–10% hydrochloric or acetic acid at high temperature (around 95 °C) for a duration ranging from 3 to 14 hours. Acidic separation of chitosan generally occurs after alkaline deproteinization of the fungal biomass. For instance, 2% NaOH at 90 °C for 2 hours, followed by 10% acetic acid at 60 °C for 6 hours, has been applied to remove proteins and recover chitosan from *M. rouxii* by Synowiechi et al. [82].

2.4.2 Enzymatic Deacetylation

Enzymes (deacetylases) have received increasing interest as an alternative method to chemistry for converting chitin into chitosan while better preserving the polymer's properties and reducing the process's environmental impact [83]. However, the chitin insolubility, crystallinity, and high molecular weight pose challenges to enzymatic deacetylation [84]. For a long time, using deacetylases for chitin deacetylation was found to be unsuitable for chitosan production. Several strategies have been attempted to overcome these limitations and maximize deacetylation efficacy, such as chitin modification, supercritical technology, steam explosion, and chemical treatment. Chitin modification can significantly impact the efficiency of enzymatic deacetylation. One of the most successful treatments, capable of a 20-fold increase in the degree of deacetylation, has been dry-freezing a colloidal chitin suspension before incubation with fungal deacetylase. Another approach is a two-stage chemical and enzymatic method involving the improvement of chitin crystallinity, morphology, and thermal properties by chemical treatment, followed by the application of fungal deacetylase [84].

2.5 Conclusion and Future Prospects

Chitin and chitosan are among the natural polymers that have attracted the most research attention and are most widely used industrially for various applications. The traditional and most exploited source of chitin and chitosan is waste from the fishing industry, but this material has seasonal and geographical limitations that are becoming ever less sustainable considering the huge growth in market demand for chitin and chitosan. Therefore, alternative sources are being evaluated in recent years, of which mushrooms and insects have the best potential. Fungi are an advantageous source of chitin and chitosan due to the reduced use of chemical reagents required for extraction from these biomasses and their cultivability, making them available regardless of season and geographical factors. However, the amount of fungal biomass available to produce chitin and chitosan is much lower than for marine sources. The extraction process of these polymers from fungi has not yet been fully optimized, the yields obtained are generally lower than those from crustaceans, and therefore the production of fungal chitin and chitosan has not yet reached the industrial level. So far, the extraction of chitin and chitosan from insects has also remained limited to the laboratory scale, despite the increasing amounts of available biomass generated by breeding facilities for feed production and organic waste management. Better optimization of the extraction steps is also needed for this source, as the same procedures used for crustaceans are currently mainly applied.

It is desirable to focus more in the future on the quantitative evaluation of the extraction process from each different source, especially fungi, and insects, and on the efficiency of the respective demineralization, deproteinization, decolorization, and deacetylation treatments. This, and the proper evaluation of the degree of purification of the polymer obtained with the various extraction conditions, are crucial for accurately assessing the potential of alternative sources to produce chitin and chitosan and comparing them with the extraction process from crustaceans. Relevant applications of these polymers from alternative sources also need to be investigated.

An increasing need for more environmentally friendly extraction methods, based on reduced use of chemicals and the exploitation of enzymes and bacteria, is also becoming increasingly necessary. Biotechnological processes offer the opportunity to better preserve the useful characteristics of chitin and chitosan and, if properly optimized, reduce costs and safer waste disposal. It is conceivable that research into the implementation of biotechnological processes, currently limited to the laboratory scale, will leap forward in the near future, enabling the commercial-scale development of this promising technology for the sustainable production of chitin and chitosan.

References

1. Amor IB, Hemmami H, Laouini SE, Abdelaziz AG, Barhoum A (2023) Influence of chitosan source and degree of deacetylation on antibacterial activity and adsorption of AZO dye from water. *Biomass Conversion and Biorefinery*: 1–11.
2. Kaya M, Baublys V, Šatkauskienė I, Akyuz B, Bulut E, Tubelytë V (2015) First chitin extraction from Plumatella repens (Bryozoa) with comparison to chitins of insect and fungal origin. *International Journal of Biological Macromolecules* 79: 126–132.
3. Elieh-Ali-Komi D, Hamblin MR (2016) Chitin and chitosan: Production and application of versatile biomedical nanomaterials. *International Journal of Advanced Research* 4: 411.
4. Petrenko I, Bazhenov VV, Galli R, Wysokowski M, Fromont J, Schupp PJ, Stelling AL, Niederschlag E, Stöker H, Kutsova VZ (2017) Chitin of poriferan origin and the bioelectrometallurgy of copper/copper oxide. *International Journal of Biological Macromolecules* 104: 1626–1632.
5. Iber BT, Kasan NA, Torsabo D, Omuwa JW (2022) A review of various sources of chitin and chitosan in nature. *Journal of Renewable Materials* 10: 1097.
6. Sajomsang W, Gonil P (2010) Preparation and characterization of α-chitin from cicada sloughs. *Materials Science and Engineering: C* 30: 357–363.
7. Brasselet C, Pierre G, Dubessay P, Dols-Lafargue M, Coulon J, Maupeu J, Vallet-Courbin A, De Baynast H, Doco T, Michaud P (2019) Modification of chitosan for the generation of functional derivatives. *Applied Sciences* 9: 1321.
8. Ruiz GAM, Corrales HFZ (2017) Chitosan, chitosan derivatives and their biomedical applications. *Biological Activities and Application of Marine Polysaccharides* 87.
9. de Queiroz Antonino RSCM, Lia Fook BRP, de Oliveira Lima VA, de Farias Rached RÍ, Lima EPN, da Silva Lima RJ, Peniche Covas CA, Lia Fook MV (2017) Preparation and characterization of chitosan obtained from shells of shrimp (Litopenaeus vannamei Boone). *Marine Drugs* 15: 141.

10. Philibert T, Lee BH, Fabien N (2017) Current status and new perspectives on chitin and chitosan as functional biopolymers. *Applied Biochemistry and Biotechnology* 181: 1314–1337.

11. Goy RC, Morais ST, Assis OB (2016) Evaluation of the antimicrobial activity of chitosan and its quaternized derivative on E. coli and S. aureus growth. *Revista Brasileira de Farmacognosia* 26: 122–127.

12. Ong S-Y, Wu J, Moochhala SM, Tan M-H, Lu J (2008) Development of a chitosan-based wound dressing with improved hemostatic and antimicrobial properties. *Biomaterials* 29: 4323–4332.

13. Tafi E, Triunfo M, Guarnieri A, Ianniciello D, Salvia R, Scieuzo C, Ranieri A, Castagna A, Lepuri S, Hahn T (2023) Preliminary investigation on the effect of insect-based chitosan on preservation of coated fresh cherry tomatoes. *Scientific Reports* 13: 7030.

14. Hahn T, Roth A, Febel E, Fijalkowska M, Schmitt E, Arsiwalla T, Zibek S (2018) New methods for high-accuracy insect chitin measurement. *Journal of the Science of Food and Agriculture* 98: 5069–5073.

15. Hahn T, Bossog L, Hager T, Wunderlich W, Breier R, Stegmaier T, Zibek S (2019) Chitosan application in textile processing and fabric coating. *Chitin and Chitosan: Properties and Applications*: 395–428.

16. Arbia W, Arbia L, Adour L, Amrane A (2013) Chitin extraction from crustacean shells using biological methods–a review. *Food Technology and Biotechnology* 51: 12–25.

17. Pakizeh M, Moradi A, Ghassemi T (2021) Chemical extraction and modification of chitin and chitosan from shrimp shells. *European Polymer Journal* 159: 110709.

18. Hahn T, Tafi E, Paul A, Salvia R, Falabella P, Zibek S (2020) Current state of chitin purification and chitosan production from insects. *Journal of Chemical Technology & Biotechnology* 95: 2775–2795.

19. Jucker C, Lupi D, Moore CD, Leonardi MG, Savoldelli S (2020) Nutrient recapture from insect farm waste: Bioconversion with Hermetia illucens (L.)(Diptera: Stratiomyidae). *Sustainability* 12: 362.

20. Tacon AG (2018) Global trends in aquaculture and compound aquafeed production. *World Aquaculture* 49: 33–46.

21. Costello C, Cao L, Gelcich S, Cisneros-Mata MÁ, Free CM, Froehlich HE, Golden CD, Ishimura G, Maier J, Macadam-Somer I (2020) The future of food from the sea. *Nature* 588: 95–100.

22. Ni'mah YL, Harmami H, Ulfin I, Suprapto S, Saleh CW (2019) Water-soluble chitosan preparation from marine sources. *Malaysian Journal of Fundamental and Applied Sciences* 15: 159–163.

23. Hossain M, Iqbal A (2014) Production and characterization of chitosan from shrimp waste. *Journal of the Bangladesh Agricultural University* 12: 153–160.

24. Thirunavukkarasu N, Shanmugam A (2009) Extraction of chitin and chitosan from mud crab Scylla tranquebarica (Fabricius, 1798). *International Journal on Applied Bioengineering* 4: 31–33.

25. Nagasawa H (2012) The crustacean cuticle: Structure, composition and mineralization. *Frontiers in Bioscience* 4: 711–720.

26. Unterberger C, Luber F, Demmel A, Grünwald K, Huber I, Engel K-H, Busch U (2014) Simultaneous detection of allergenic fish, cephalopods and shellfish in food by multiplex ligation-dependent probe amplification. *European Food Research and Technology* 239: 559–566.

27. Yang F-C, Peters RD, Dies H, Rheinstädter MC (2014) Hierarchical, self-similar structure in native squid pen. *Soft Matter* 10: 5541–5549.

28. Crognale S, Russo C, Petruccioli M, D'annibale A (2022) Chitosan production by fungi: Current state of knowledge, future opportunities and constraints. *Fermentation* 8: 76.

29. Abo Elsoud MM, El Kady E (2019) Current trends in fungal biosynthesis of chitin and chitosan. *Bulletin of the National Research Centre* 43: 1–12.

30. Free SJ (2013) Fungal cell wall organization and biosynthesis. *Advances in Genetics* 81: 33–82.

31. Ghormade V, Pathan E, Deshpande M (2017) Can fungi compete with marine sources for chitosan production? *International Journal of Biological Macromolecules* 104: 1415–1421.

32. Di Mario F, Rapana P, Tomati U, Galli E (2008) Chitin and chitosan from Basidiomycetes. *International Journal of Biological Macromolecules* 43: 8–12.

33. Pareek N, Vivekanand V, Agarwal P, Saroj S, Singh RP (2013) Bioconversion to chitosan: A two stage process employing chitin deacetylase from Penicillium oxalicum SAEM-51. *Carbohydrate Polymers* 96: 417–425.

34. Rufino RD, de Luna JM, de Campos Takaki GM, Sarubbo LA (2014) Characterization and properties of the biosurfactant produced by Candida lipolytica UCP 0988. *Electronic Journal of Biotechnology* 17: 34–38.

35. Satari B, Karimi K, Zamani A (2016) Oil, chitosan, and ethanol production by dimorphic fungus Mucor indicus from different

lignocelluloses. *Journal of Chemical Technology & Biotechnology* 91: 1835–1843.

36. Pal M (2015) Growing role of fungi in mycotic abortion of domestic animal. *Journal of Bacteriology and Mycology* 2: 1009.
37. Manniello M, Moretta A, Salvia R, Scieuzo C, Lucchetti D, Vogel H, Sgambato A, Falabella P (2021) Insect antimicrobial peptides: Potential weapons to counteract the antibiotic resistance. *Cellular and Molecular Life Sciences* 78: 4259–4282.
38. Franco A, Salvia R, Scieuzo C, Schmitt E, Russo A, Falabella P (2021) Lipids from insects in cosmetics and for personal care products. *Insects* 13: 41.
39. Derrien C, Boccuni A (2018) Current status of the insect producing industry in Europe. *Edible Insects in Sustainable Food Systems*: 471–479.
40. Kim T-K, Yong HI, Kim Y-B, Kim H-W, Choi Y-S (2019) Edible insects as a protein source: A review of public perception, processing technology, and research trends. *Food Science of Animal Resources* 39: 521.
41. Triunfo M, Tafi E, Guarnieri A, Salvia R, Scieuzo C, Hahn T, Zibek S, Gagliardini A, Panariello L, Coltelli MB (2022) Characterization of chitin and chitosan derived from Hermetia illucens, a further step in a circular economy process. *Scientific Reports* 12: 6613.
42. Kaya M, Baublys V, Can E, Šatkauskienė I, Bitim B, Tubelytė V, Baran T (2014) Comparison of physicochemical properties of chitins isolated from an insect (Melolontha melolontha) and a crustacean species (Oniscus asellus). *Zoomorphology* 133: 285–293.
43. Luo Q, Wang Y, Han Q, Ji L, Zhang H, Fei Z, Wang Y (2019) Comparison of the physicochemical, rheological, and morphologic properties of chitosan from four insects. *Carbohydrate Polymers* 209: 266–275.
44. Mohan K, Ganesan AR, Muralisankar T, Jayakumar R, Sathishkumar P, Uthayakumar V, Chandirasekar R, Revathi N (2020) Recent insights into the extraction, characterization, and bioactivities of chitin and chitosan from insects. *Trends in Food Science & Technology* 105: 17–42.
45. Gopi S, Thomas S, Pius A (2020) *Handbook of Chitin and Chitosan*. Elsevier.
46. Reader AJ, Corda G, Mehranian A, da Costa-Luis C, Ellis S, Schnabel JA (2020) Deep learning for PET image reconstruction. *IEEE Transactions on Radiation and Plasma Medical Sciences* 5: 1–25.

47. Huet G, Hadad C, Husson E, Laclef S, Lambertyn V, Farias MA, Jamali A, Courty M, Alayoubi R, Gosselin I (2020) Straightforward extraction and selective bioconversion of high purity chitin from Bombyx eri larva: Toward an integrated insect biorefinery. *Carbohydrate Polymers* 228: 115382.

48. Khayrova A, Lopatin S, Shagdarova B, Sinitsyna O, Sinitsyn A, Varlamov V (2022) Evaluation of antibacterial and antifungal properties of low molecular weight chitosan extracted from Hermetia illucens relative to crab chitosan. *Molecules* 27: 577.

49. Aklog YF, Egusa M, Kaminaka H, Izawa H, Morimoto M, Saimoto H, Ifuku S (2016) Protein/$CaCO_3$/chitin nanofiber complex prepared from crab shells by simple mechanical treatment and its effect on plant growth. *International Journal of Molecular Sciences* 17: 1600.

50. Younes I, Rinaudo M (2015) Chitin and chitosan preparation from marine sources. Structure, properties and applications. *Marine Drugs* 13: 1133–1174.

51. Ameh AO, Isa MT, Abutu D, Danlami A (2014) Kinetic modelling of the demineralization of shrimp exoskeleton using citric acid. *Leonardo Electronic Journal of Practices and Technologies* 25: 99–108.

52. Finke MD (2013) Complete nutrient content of four species of feeder insects. *Zoo Biology* 32: 27–36.

53. Broquá J, Zanin B, Flach A, Mallmann C, Taborda F, Machado L, Alves S, Silva M, Dias R (2019) Methods of chitin production a short review. *American Journal of Biomedical Science & Research* 4: 307–314.

54. Fu X, Zhu L, Li L, Zhang T, Li M, Mou H (2019) Eco-friendly preparation of chitooligosaccharides with different degrees of deacetylation from shrimp shell waste and their effects on the germination of wheat seeds. *Marine Life Science & Technology* 1: 95–103.

55. Castillo V, Matos M, Müller AJ (2003) A new route towards the preparation of a biodegradable nanocomposite based on chitin whiskers from exoskeleton of shrimp and polycaprolacton. *Revista Latinoamericana de Metalurgia y Materiales* 23: 12–19.

56. Zainol Abidin NA, Kormin F, Zainol Abidin NA, Mohamed Anuar NAF, Abu Bakar MF (2020) The potential of insects as alternative sources of chitin: An overview on the chemical method of extraction from various sources. *International Journal of Molecular Sciences* 21: 4978.

57. Gortari MC, Hours RA (2013) Biotechnological processes for chitin recovery out of crustacean waste: A mini-review. *Electronic Journal of Biotechnology* 16: 14–14.

58. Mol A, Kaya M, Mujtaba M, Akyuz B (2018) Extraction of high thermally stable and nanofibrous chitin from Cicada (Cicadoidea). *Entomological Research* 48: 480–489.

59. Kaya M, Mujtaba M, Bulut E, Akyuz B, Zelencova L, Sofi K (2015) Fluctuation in physicochemical properties of chitins extracted from different body parts of honeybee. *Carbohydrate Polymers* 132: 9–16.

60. Soon CY, Tee YB, Tan CH, Rosnita AT, Khalina A (2018) Extraction and physicochemical characterization of chitin and chitosan from Zophobas morio larvae in varying sodium hydroxide concentration. *International Journal of Biological Macromolecules* 108: 135–142.

61. Teli M, Sheikh J (2012) Extraction of chitosan from shrimp shells waste and application in antibacterial finishing of bamboo rayon. *International Journal of Biological Macromolecules* 50: 1195–1200.

62. Mun S, Young Noh M, Dittmer NT, Muthukrishnan S, Kramer KJ, Kanost MR, Arakane Y (2015) Cuticular protein with a low complexity sequence becomes cross-linked during insect cuticle sclerotization and is required for the adult molt. *Scientific Reports* 5: 10484.

63. Vilasoa-Martínez M, Calaza-Ramos C, López-Hernández J, Lage-Yusty MA, Losada PP, de Quirós AR-B (2008) Determination of vitamin E and carotenoid pigments by high performance liquid chromatography in shell of Chionoecetes opilio. *Analytica Chimica Acta* 617: 225–229.

64. Żukiewicz-Sobczak W, Sobczak P, Zawiślak K, Zagórski J, Wojtyła-Buciora P, Wojtyła A (2015) Physical and chemical properties comparison of fungal and crustaceous chitosan. *Journal of Health Inequalities* 1: 7–14.

65. Yen M-T, Mau J-L (2006) Preparation of fungal chitin and chitosan from shiitake stipes. *Fungal Science* 21: 1–11.

66. Sebastian J, Rouissi T, Brar SK (2020) Fungal chitosan: Prospects and challenges. *Handbook of Chitin and Chitosan*: 419–452.

67. Kaur S, Dhillon GS (2015) Recent trends in biological extraction of chitin from marine shell wastes: A review. *Critical Reviews in Biotechnology* 35: 44–61.

68. Mao X, Zhang J, Kan F, Gao Y, Lan J, Zhang X, Hu Z, Li Y, Lin H (2013) Antioxidant production and chitin recovery from shrimp head fermentation with Streptococcus thermophilus. *Food Science and Biotechnology* 22: 1023–1032.

69. Aranday-García R, Guerrero AR, Ifuku S, Shirai K (2017) Successive inoculation of Lactobacillus brevis and Rhizopus oligosporus on shrimp wastes for recovery of chitin and added-value products. *Process Biochemistry* 58: 17–24.

70. Prameela K, Mohan CM, Smitha P, Hemalatha K (2010) Bioremediation of shrimp biowaste by using natural probiotic for chitin and carotenoid production an alternative method to hazardous chemical method. *International Journal of Applied Biology and Pharmaceutical Technology* 1(3): 903–910.

71. Mao X, Guo N, Sun J, Xue C (2017) Comprehensive utilization of shrimp waste based on biotechnological methods: A review. *Journal of Cleaner Production* 143: 814–823.

72. Synowiecki J, Al-Khateeb NAAQ (2000) The recovery of protein hydrolysate during enzymatic isolation of chitin from shrimp Crangon crangon processing discards. *Food Chemistry* 68: 147–152.

73. Hamed I, Özogul F, Regenstein JM (2016) Industrial applications of crustacean by-products (chitin, chitosan, and chitooligosaccharides): A review. *Trends in Food Science & Technology* 48: 40–50.

74. Teng WL, Khor E, Tan TK, Lim LY, Tan SC (2001) Concurrent production of chitin from shrimp shells and fungi. *Carbohydrate Research* 332: 305–316.

75. Tan YN, Lee PP, Chen WN (2020) Dual extraction of crustacean and fungal chitosan from a single Mucor circinelloides fermentation. *Fermentation* 6: 40.

76. Yeul VS, Rayalu SS (2013) Unprecedented chitin and chitosan: A chemical overview. *Journal of Polymers and the Environment* 21: 606–614.

77. Hahn T, Roth A, Ji R, Schmitt E, Zibek S (2020) Chitosan production with larval exoskeletons derived from the insect protein production. *Journal of Biotechnology* 310: 62–67.

78. Younes I, Sellimi S, Rinaudo M, Jellouli K, Nasri M (2014) Influence of acetylation degree and molecular weight of homogeneous chitosans on antibacterial and antifungal activities. *International Journal of Food Microbiology* 185: 57–63.

79. Bradić B, Bajec D, Pohar A, Novak U, Likozar B (2018) A reaction–diffusion kinetic model for the heterogeneous N-deacetylation step in chitin material conversion to chitosan in catalytic alkaline solutions. *Reaction Chemistry & Engineering* 3: 920–929.

80. Galed G, Diaz E, Goycoolea FM, Heras A (2008) Influence of N-deacetylation conditions on chitosan production from α-chitin. *Natural Product Communications* 3: 1934578X0800300414.

81. Toan NV (2009) Production of chitin and chitosan from partially autolyzed shrimp shell materials. T*he Open Biomaterials Journal* 1: 21–24.

82. Synowiecki J, Al-Khateeb NA (2003) Production, properties, and some new applications of chitin and its derivatives. *Critical Reviews in Food Science and Nutrition* 43(2): 145–171.

83. Kaczmarek MB, Struszczyk-Swita K, Li X, Szczęsna-Antczak M, Daroch M (2019) Enzymatic modifications of chitin, chitosan, and chitooligosaccharides. *Frontiers in Bioengineering and Biotechnology* 7: 243.

84. Pareek N, Vivekanand V, Saroj S, Sharma AK, Singh RP (2012) Purification and characterization of chitin deacetylase from Penicillium oxalicum SAEM-51. *Carbohydrate Polymers* 87: 1091–1097.

Chapter 3

Preparation of Chitin and Chitosan

Ali Alnazza Alhamad,[a] Ilham Ben Amor,[b,c] Soumeia Zeghoud,[b,c] Noura Gheraissa,[d] Salah Eddine Laouini,[b,e] Mohammed Fouad Ferhat,[b,f] Nani Sadok[g], and Shakeel Ahmed[h,i]

[a]*Department of Chemistry, Faculty of Science, University of Aleppo, Syria*
[b]*Department of Process Engineering and Petrochemical, Faculty of Technology, University of El Oued, El Oued, Algeria*
[c]*Renewable Energy Development Unit in Arid Zones (UDERZA), University of El Oued, El Oued, Algeria*
[d]*Faculty of Life and Natural Sciences, Department of Cellular and Molecular Biology, El Oued University, Algeria*
[e]*Laboratory of Biotechnology Biomaterials and Condensed Materials, Faculte de la Technologie, University of El Oued, El Oued, Algeria*
[f]*Laboratoire des Sciences et Techniques de l'Environnement et de la Valorisation, Faculty of Sciences and Technology, University Abdelhamid Ibn Badis-Mostaganem, Mostaganem, Algeria*
[g]*Laboratory of Valorization and Technology of Saharan Resources (VTRS), University of El Oued, El Oued, Algeria*
[h]*Postgraduate Department of Chemistry, Government Postgraduate College Rajouri, Jammu and Kashmir 185133, India*
[i]*University Centre for Research and Development, Chandigarh University, Mohali, Punjab, India*

che.alialnazaa@gmail.com

Chitin can be obtained sustainably from the shells of shrimp, crabs, etc. Research on extracting chitin from various sources is still ongoing, with a lot of focus on creating a sustainable method that produces high-quality chitin with little waste. To achieve this goal, chemical and biological techniques have been widely used.

Chitin and Chitosan: Physical and Chemical Properties
Edited by Shakeel Ahmed, Soumeia Zeghoud, Hadia Hemmami, and Ilham Ben Amor

ISBN 978-981-5129-39-7 (Hardcover), 978-1-003-58977-8 (eBook)
www.jennystanford.com

Recently, ionic liquid and eutectic solvent-based technologies have been investigated. Chitosan or nanochitene can be made from the removed chitin. Chitin can be used in many ways after it has been extracted and converted into the necessary form as a filling material, an ingredient in adsorbents, and an ingredient in biomaterials. This chapter deals with the chemistry of the procedures involved in chitosan synthesis, chitin extraction, and chitin formation. Additional topics covered include the relationship between the structural and biological properties of chitosan, several technologies for the production of chitin and chitosan, and their respective drawbacks.

3.1 Introduction

After cellulose, chitin is the second-most prevalent naturally occurring polysaccharide [1]. The polymer N-acetyl-D-glucosamine exhibits 1,4 glycosidic linkages [2]. There are three polymorphic types of chitins depending on how the polymer chains are arranged: β-chitin, which refers to chitin with parallel chains, α-chitin, which refers to chitin with antiparallel chains, and α-chitin, which is a combination of γ- and β-chitin with some parallel chains and antiparallel to each other. In addition to fungi, arthropods, and some mollusks, chitin may also be found in their cell walls and exoskeletons. The most common source of chitin in nature is crustacean shells [2, 3]. Chitin, despite its limited solubility in water and several organic solvents attributed to its high crystallinity and strong interchain hydrogen bonding, has versatile applications across multiple sectors such as textiles, papermaking, medicines, and treatment of wastewater [3–6].

Chitosan is a type of chitin that has been partly deacetylated, meaning that the N-acetyl-D-glucosamine moiety has been changed to D-glucosamine. The degree of deacetylation (DD) of chitosan is defined as the ratio of D-glucosamine to N-acetyl-D-glucosamine [7]. Chitin is neither entirely acetylated nor totally deacetylated in nature. Although the DD necessary to be deemed chitosan varies, it is often accepted that a DD of 50% or above qualifies. The only naturally occurring cationic polymer is chitosan [8]. Due to its numerous applications in industries, including medicine, agriculture, and the production of biodegradable polymers, chitosan is regarded as the most valuable chitin derivative. Chitosan's biocompatibility, biodegradability, antibacterial effects, and low environmental toxicity are well regarded [3, 9, 10]. In Fig. 3.1, chitin and chitosan are shown.

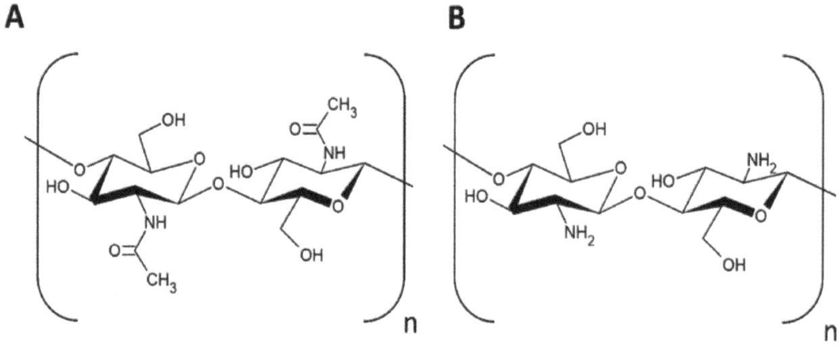

Figure 3.1 Structure of chitin (A) and chitosan (B), respectively.

In 2018, around 9.4 million tons of shellfish were caught in aquaculture. A further 6.3 million tons are wild-caught [11, 12]. According to estimates, the annual disposal of shellfish trash ranges from 6 to 8 million tons [13]. The primary source of chitin in shellfish is derived from the shells of prawns and crabs, while ongoing research is exploring the possibility of obtaining chitin from other species. Regardless of its origin, crustacean shell waste is widely recognized as a sustainable biomass source suitable for chitin extraction, which therefore enables the manufacturing of chitosan [3, 8, 14]. The development of chitin extraction from crab shells and chitosan manufacture is outlined in this chapter. Prior reviews in-depth studied the processes for obtaining chitin and associated components. Some describe a single set of chemical and biological methods, while others refer to the whole field [14–19]. The use of these materials has been covered in other reviews, including general studies with surveys with a variety of uses and more detailed reviews covering specific areas like energy, engineering with tissues, and drug delivery use [18, 20–23]. The purpose of this chapter is to confirm the information on the extraction and use of chitosan, nanochitin, and chitin, including new information that has emerged after these earlier evaluations were published.

3.2 Extraction Methods

The three primary components of crustacean shells are protein, calcium carbonate, and chitin [24]. Chitin must be separated from the remainder of the shell in order to be extracted. Chemical or biological approaches can be used to remove the materials. This section describes many chitin extraction techniques, including particular adaptations, and compares the benefits and drawbacks of each technique.

3.2.1 Chemical Techniques

Demineralization and deproteinization are typically the first two steps in the chemical extraction of chitin. In most circumstances, there is no need to follow this order. To expand the surface area for deproteinization, demineralization is frequently done before deproteinization. Deproteinization. Most of the time, it is irrelevant what sequence you complete these tasks. Often, demineralization and deproteinization are accomplished by exposing the shell to an alkali, referred to as the acid-base chitin extraction method [3, 14]. Because of the inexpensive cost of the chemicals and the lack of specialized equipment needed to carry out the reaction, the acid-base approach is the most used method in industrial extraction [25]. A hunt for alternatives has been sparked by worries about the discharge of hazardous acids and alkaline wastes into the environment. The extraction of green chitin has shown notable progress in recent times, with the adoption of innovative methodologies such as ionic liquids, deep eutectic solvents, and other emerging approaches [16, 26, 27].

3.2.1.1 Demineralization

The process of chemical demineralization often entails exposing shellfish to an acidic environment in order to dissolve the minerals present in their shells. These minerals are subsequently separated from the remaining solid shell components. Demineralization procedures generally involve the utilization of both mineral and organic acids. Hydrochloric acid (HCl) is the prevailing choice. However, the growing apprehension over environmental toxicity has prompted a transition towards organic acids, including lactic, citric, and acetic acids [14, 28]. The process of demineralization is commonly conducted at ambient temperature for a duration of 2–3 hours. However, research suggests that these time intervals may be excessive and offer minimal or no advantages compared to shorter durations [9, 28]. Because of the simplicity of chemical demineralization, the changes are usually related to acid concentration and selection. The acids should possess the capability to effectively dissolve calcium carbonate, which is the primary mineral constituent of the crust. Furthermore, it is desirable for these acids to generate salts soluble in water that can be readily separated from the remnant of the crust. The demineralization procedure was altered to incorporate several acid treatments. Citric acid was used in a two-stage demineralization process by Pohling et al. [29]. They also cleaned the crushed prawn shells in between treatments.

Table 3.1 Demineralization processes using chemicals

Technique source	Species	Concentration	Duration	Ash content (%)	Ref.
CO_2 (aq)	Gray shrimp	(Crangon) 10 atm	2 h	<1	[25]
HCl	Marine shrimp (Parapenaeopsis stylifera)	1 M	24 h	0.01	[28]
HCl	Northern shrimp (Pandalus borealis)	31.45%	2 h	0.29 ± 0.05	[29]
Citric acid	Northern shrimp (Pandalus borealis)	50%	35 min for 1 h	0.34 ± 0.12	[9]
HCl (12 hours in 0.28 M HCl)	White shrimp (Litopenaeus vannamei)	0.80 M	12 h	0.1	[30]
Lactic acid	Red crab (Chionoeetes japonicus)	5%	5 days	6.0 ± 0.2	[31]
HCl	Rock lobster (Jasis lalandii)	31%	2 h	<2	[32]

They were able to extract minerals from the shell to a degree equivalent to the conventional one-step demineralization of HCl despite calcium citrate's poor solubility [29]. Trung et al. [25] conducted a study by elevating the partial pressure of carbon dioxide, and a substantial extraction of minerals from the Earth's crust was achieved, resulting in the removal of almost 100% of these minerals. However, it is worth noting that a significant quantity of water was necessary for this process, mostly owing to the limited solubility of the calcium bicarbonate compound generated during the reaction. In Table 3.1, the demineralization procedures and outcomes are enumerated.

3.2.1.2 Deproteination

Chitin is isolated from the other organic parts of the shell by the process of chemical deproteinization. The most popular technique involves extracting proteins from chitin and solubilizing them in water using strong bases and high heat. Because it is inexpensive, sodium hydroxide (NaOH) is the most often used base for removing proteins [9, 14, 25]. While certain studies may conduct reactions at ambient temperature and others at temperatures beyond 100 °C, the customary range for reaction temperatures often falls between 65 °C and 100 °C [9, 14, 25, 29].

Researchers are looking for alternate bases with less harmful cations (such as KOH) or other deproteinization techniques due to concerns that excessive quantities of sodium ions might pollute the environment [14]. Yang et al. [25] substituted hot water in a pressure vessel for strong base reactions in order to mimic the way that pressure cookers may soften the proteins in meat. The molecular weight of the chitin generated using this process was found to be lower compared to that produced using the conventional acid–base method. However, the purity and shape of the chitin were found to be identical between the two methods [25]. Yayci et al. [33] conducted a study on the subject. The researchers chose to discontinue using the aqueous reaction medium and instead pursue a dielectric barrier discharge (DBD) plasma as an alternative method for deacetylation of chitin [26, 27]. The initial investigations with unprocessed prawn shells documented a notable decrease in protein content, reaching a maximum drop of 42% during a span of 6 minutes [26]. The subsequent phase of the study involved subjecting demineralized prawn shells to treatment using a DBD plasma device. The DBD plasma effectively eliminated leftover minerals and almost 90% of the proteins present in the shells [27]. The primary benefit associated with the DBD plasma technique in comparison to alternative deproteinization procedures is in the generation of gaseous by-products, including carbon monoxide, hydrogen, nitrogen, and carbon dioxide. Demineralization effluent was the sole solvent waste that was collected [27]. Table 3.2 summarizes the results of chemical deproteinization.

Table 3.2 Deproteinization processes involving chemicals

Technique	Source types	Concentration	Temperature/ duration	Residual protein/ chitin yield	Ref.
NaOH	Lagoon crab (Callinectes amnicola)	2.39 M	70 ° C for 2 h	19.36%/-	[34]
NaOH	Rock lobster (Jasis lalandii)	5%	80–85 °C for 2 × 30 min	24.0%/-	[32]
NaOH	White shrimp (Litopenaeus vannamei)	0.68 M	Ambient for 24 h	-/0.92–0.96%	[30]
boiling water under pressure	Gray shrimp (Crangon crangon)	-	180 °C for 1 h	-/4.7%	[25]

3.2.1.3 Deep eutectic acids and ionized liquids

Another method employed for the extraction of chitin involves the process of solvation using ionic liquids (ILs). ILs are a kind of salt that has a melting point of less than 100 °C. They are composed of a substantial organic cation and a comparatively minor organic or inorganic anion [35, 36]. Owing to their reduced vapor pressure [37], low flammability, thermal stability, and recyclability after use [16, 38], they are considered appealing as 'green' solvents.

Ionic liquids have garnered significant interest due to their capacity to solubilize biomass, such as polysaccharides with low solubility [16, 38]. The process of chitin extraction using ionic liquids involves the dissolution of shellfish in an ionic liquid, followed by the precipitation of chitin with the addition of an anti-solvent. The anti-solvent is a reagent that induces insolubility in the chitin.

Depending on the ionic liquid, temperatures above 100 °C can occur during dissolution [16, 38]. Typical anti-solvents are ethanol, methanol, and aqueous solutions. Aqueous acids have an inherent advantage in facilitating the process of demineralization and potentially producing chitin with fewer amounts of contamination [38–43].

Ionic liquids have the ability to dissolve chitin by means of delivering anions that cause disruption to the intricate network of hydrogen bonds present both between and within the chitin strands [43]. Several factors have been identified as influential in determining the solubility of chitin in ionic liquids. These factors include the molecular weight (Mw), crystallinity, and degree of deacetylation (DD) of chitin, as well as the specific identity of the ionic liquid [38, 43]. The solvation of the chitin sample is hindered by either its high crystallinity or its large molecular weight [43]. The solubility of chitin exhibits enhancement when DD increases, particularly when the DD reaches or exceeds 50% [43]. The dissolution of chitin is influenced by the capacity of anions to function as hydrogen bond acceptors, with the extent of dissolution also being contingent upon the interaction with the cation [38, 43]. In their study, Shimo et al. [44] discovered that ethylenediamine has the ability to penetrate chitin crystals and integrate into the existing hydrogen bonding network.

Chitin may dissolve in ionic liquids at room temperature when ethylenediamine is added, according to preliminary investigations [44]. An exception may be tetraalkylammonium hydroxides, as many of these have been shown to dissolve chitin at 25 °C [44, 45]. Ionic liquids based on dialkyl imidazole and tetraalkylammonium hydroxides have both it has been shown to dissolve chitin and may be used to extract it from biomass [16, 43, 45].

In their study, Tolessa et al. [41] discovered ionic liquids, namely ammonium diisopropylethylammonium acetate, diisopropylethylammonium propanoate, and dimethylbutylammonium acetate, which exhibit favorable properties for the dissolution of shrimp shells. Out of the options considered, it was observed that it exhibited the most favorable yield of chitin following its regeneration using an aqueous citric acid solution. In a study conducted by Mehmood et al. [46], chitin was extracted from prawn shells by the use of 1-ethyl-3-methylimidazolium acetate ([EMIM][OAc]) as a solvent for dissolving the shells, while deionized water was employed as an anti-solvent [40]. The researchers, Setoguchi et al. [42], expressed their interest in substituting the intricate synthesis of [EMIM][OAc] with a more straightforward method for synthesizing anionic halide ionic liquids. The process of chitin extraction from crab shells was undertaken utilizing 1-allyl-3-methylimidazolium bromide ([AMIM][Br]). The extraction process yielded positive results upon regeneration with citric acid. The designation of "green" attributed to ionic liquids has faced criticism on account of their toxicity and limited biodegradability [47]. Deep eutectic solvents (DES) have been formulated as potential substitutes for ionic liquids, with the added advantage of being moisture-insensitive and environmentally degradable, as highlighted in previous research [48].

DES consists of a pair of hydrogen bond donors and acceptors, allowing for solubility characteristics similar to those of ionic liquids. Early DES may have contained metal ions [48]. In order to attain reduced toxicity levels, researchers have devised natural deep eutectic solvents (NADESs). NADESs, or natural deep eutectic solvents, are a cost-effective and environmentally friendly alternative for extracting biomass. These solvents mostly consist of primary metabolites [47, 49, 50]. Combinations of choline chloride (ChCl) and organic acid have been extensively used as non-aqueous deep eutectic solvents for chitin extraction, primarily because of their little toxicity and capacity to dissolve calcium carbonate [16, 17, 51, 52]. Betaine is a well-recognized hydrogen bond acceptor, whereas organic compounds like ethylene glycol, glycerol, and urea serve as hydrogen bond donors. Deep eutectic solvents with choline chloride and numerous additional ingredients, including organic acids and alcohols, were evaluated by Saravanan et al. [53] for their capacity to remove chitin from Marsupenaeus japonicus prawn shells.

The optimal deep eutectic solvent for chitin extraction was found to be a mixture of choline chloride (ChCl) and malonic acid in a 1:2 ratio, based on the conducted studies. The use of NADES resulted in superior yields compared to the acid–base approach, notwithstanding the fact that the acid–base method showed more efficacy in the removal of proteins and minerals from the shell. To circumvent the relatively limited ability of NADES to remove minerals from the protein coat, Rodrigues et al. [51] After the extraction step, water was added to the NADES/chitin solution and stirred for 30 minutes. The chitin was precipitated from the NADES solution, resulting in the formation of an acidic environment that effectively promoted the extraction of minerals from the shells and the hydrolysis of proteins [51]. The addition of a hydrogen peroxide decolorization step resulted in chitin with a purity of up to 98% and a third of the losses associated with the acid–base technique [51]. The application of microwave irradiation has demonstrated enhanced capabilities in the demineralization and deproteinization processes of deep eutectic solvents (DES). Huang et al. [54] conducted a study revealing that microwave irradiation for a duration of 9 minutes resulted in the removal of over 99% of minerals from shrimp shells. Furthermore, the deproteinization efficiency improved as the ratio of non-aqueous DES to shell material increased, reaching a maximum ratio of 20:1.

Zhao et al. [55] found similar results. However, they did not notice any statistically significant variation in chitin product following a 7-minute microwave irradiation period. In this work, a citric acid demineralization step was incorporated before DES solvation in order to generate chitin with comparable purity to the acid–base technique but with a greater yield [55]. Nanoparticle-based delivery systems have the potential to incorporate a greater number of components beyond the conventional two-component systems. In their study, Wang et al. [37] conducted experiments to evaluate the efficacy of several binary and ternary NADES in extracting chitin from snow crab shells. These NADES consisted of chitin-like compounds, including

D-gluconic and N-acetyl-D-glucosamine. It was discovered that the ratio of ChCl/N-acetyl-D-glucosamine/formic acid for the tested NADES was determined to be 1:0.6:1. A ratio of 4 resulted in chitin with higher purity and yield, which was found to be equivalent to the acid–base approach [34]. Additional experiments were conducted on NADES demonstrated a confluence of diminished productivity and reduced quality in comparison to the acid–base test [37].

The authors of the study chose the chitin that was created with the highest quality, as determined by NADES, and had a molecular weight that was more than three times more than the chitin synthesized using the acid–base approach [33]. Bisht et al. [56] conducted a study that provided empirical evidence supporting the recyclability of deep eutectic solvents (DES). The findings indicated that chitin experienced a weight loss of around 5% with each subsequent usage of the DES [48]. The discoloration of DES was seen after three extractions, despite the cleaning of protein and mineral debris using solvent and pure ethanol in between each extraction [52]. It has been observed that DES has the potential for recycling without the need for the removal of pollutants using other solvents. However, it should be noted that the presence of accumulated protein and mineral debris can result in increased viscosity of DES, hence limiting its effectiveness as a solvent [54]. The extraction of chitin on a large scale is challenging due to the high viscosity of DES, even in the absence of waste contamination.

3.2.2 Biological Processes

Biological chitin extraction techniques can be classified as fermentation techniques or enzymatic techniques. As the name implies, the former employs enzymes to crack open crustacean shells, while the latter employs microorganisms to break down the shells until only chitin is left [3, 14, 15]. Chitin with low DD and high molecular weight is often produced by biological techniques, although complete demineralization and deproteinization are challenging [14, 31].

3.2.2.1 Enzymatic processes

Proteinases are used in enzymatic procedures to digest coat proteins and isolate chitin. The proteinases required for chitin extraction are frequently derived from bacteria or the intestinal tracts of aquatic organisms [14, 57]. Enzymatic techniques eliminate proteins with little deacetylation and chitin strand degradation by using the specificity of enzymes and moderate reaction conditions (typically 25–59 °C) [14, 17, 58]. Similar to chemical deproteinization, reactions take between 2 and 8 hours to complete. However, some studies allow for up to 24 hours [17]. The lack of active sites to which enzymes may attach when chitin proteins are removed is the reason why deproteinization by enzymatic techniques seldom exceeds 90%, even with extended reaction durations [14, 17, 57–59]. This restriction still exists, according to Hamdi et al. [57], even when maximal enzyme performance is attained in alkaline conditions (pH 8.0). When contemplating increasing chitin synthesis, the cost is a limiting issue as well since proteinases are more expensive than the bases employed in chemical deproteinization [14, 58]. Despite being expensive, the supernatant created by the enzymatic extraction of chitin is important as a dietary resource since it includes amino acids [15, 57].

3.2.2.2 Fermentation processes

By using bacterial cultures instead of enzymes, fermentation technologies boost reactions while decreasing expenses. Contrary to enzymatic extraction, demineralization can occur during fermentation procedures using bacteria that produce acid. Lactic acid-producing microorganisms are commonly employed for this objective [14, 15, 17]. For full chitin extraction, lactic acid fermentation may be sufficient [60]. Protease-producing cultures can be added to the fermentation process after lactic acid-producing bacteria to increase the purity of the chitin product [61–63]. Protease-producing bacteria seldom produce sufficient demineralization on their own and frequently need an extra step [62–65]. The provision of a medium to enable culture development is necessary for chitin extraction fermentation procedures.

To produce the best chitin extraction, the medium may be a complex combination of reagents [65]. The bacteria must have a carbon source to convert to acid in all conditions, regardless of optimization [15, 17]. Glucose is commonly utilized as a viable choice for doing studies on a smaller scale. Tan et al. [63] experimented with utilizing autoclaved peel and pulp debris from 10 different fruits as a carbon source for fermentation to cut expenses and utilize numerous waste streams. Red grape pulp, out of all the fruit remnants examined, emerged as the most effective carbon source for chitin extraction, yielding 12.2% of the mass of the original prawn shells. It generates less protein- and mineral-rich chitin than other fruit remnants [63].

Similar to enzymatic techniques, the liquid portion obtained after fermentation includes amino acids of significant value [65]. The biggest disadvantage of fermentation techniques is the long fermentation times. The tests often run three days or longer, and one week of fermentation is not uncommon [15, 17].

3.2.3 Comparing Extraction Techniques

The acid/alkaline extraction of chitin is one of the most affordable processes, due to the generally low prices of bases and acids. Alkali/acid extraction effectively eliminates proteins and minerals to generate high-purity chitin with quick reaction times without the need for highly specialized reactors [14, 17]. The process of deproteinization, even when utilizing low-toxicity organic acids for demineralization, leads to the formation of environmentally harmful alkaline residue. Additionally, the use of intense reaction conditions can inadvertently cause deacetylation and chain breakdown, resulting in reduced yield and molecular weight of the produced chitin [14, 17, 66]. Ionic liquids (IOLs) provide a more expensive solution to the low molecular weight and undesirable deacetylation issues, while part of the expense is offset by IOLs' capacity to be recycled. Once discharged into the environment, ionic liquids continue to be dangerous [16, 38].

Deep eutectic solvents have the same advantages as ionic liquids in being less expensive and less polluting, although the cost is higher than acid/base extraction, and the high viscosity of DES frustrates attempts to scale up chitin extraction [16, 66]. High molecular weight products and little environmental toxicity are advantages of biological techniques, although they are much more expensive than the acid-base method and have difficulties in the complete demineralization and deproteinization of chitin [14, 31]. Due to the high cost of the enzymes, enzymatic approaches have limited scalability and need an additional demineralization phase that is frequently done using the same chemicals as the acid-base method [14, 58]. Fermentation solves the fouling problems associated with enzymatic procedures but requires special reactors and reaction media preparation to get the most out of the cultures [14, 15, 17, 31]. When trying to drift bacterial strains, biosecurity can also be a problem [14]. Table 3.3 and Fig 3.2 illustrate how chitin and chitosan are extracted chemically and biologically.

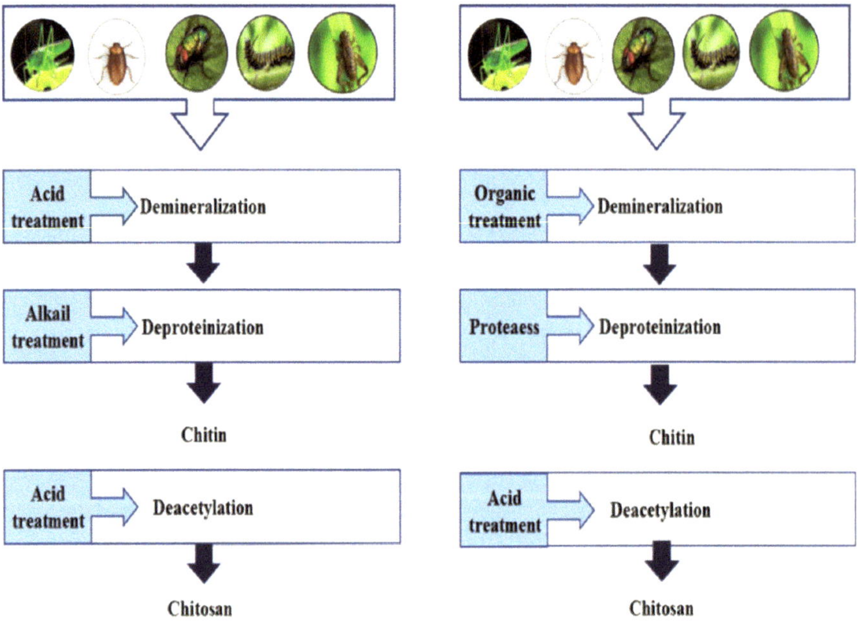

Figure 3.2 Schematic representation of the production of chitosan and chitin from insects by biological and chemical processes.

Table 3.3 Comparison of biological and chemical approaches for extracting chitosan and chitin [57, 58]

Method	Techniques for treatment	Positive and negative aspects
Chemical	Demineralization: HNO_3, HCl, and H_2SO_4 acid treatment. Deproteinization: Using KOH or NaOH to treat an alkaline solution. Organic solvents or acetone can cause discoloration. Deacetylation: A process using strong KOH or NaOH alkalines.	Shortened processing time; high-end product DD%; widely used; all organic salts are eliminated. Not very environmentally friendly; Because they are not biodegradable, minerals and proteins that have been dissolved in a solvent cannot be given to either people or animals.
Biological	Demineralization: Bacterial lactic acid is used in the demineralization process. Deproteinization: To deproteinize the culture, proteases from the fermentation media are released into the medium. Acetone or other organic solvents work well as discoloration agents. Chitin is deacetylated by chitin deacetylase, which is made by bacteria.	A superb product; environmentally safe; minerals and proteins that are extracted from the food supply can be fed to people and animals. Processing is only appropriate for laboratory testing since it takes a lengthy time (a few days).

3.3 Deacetylation

By deacetylating chitin and changing the N-acetyl groups into amines, chitin may be transformed into chitosan. Chitin is often exposed to concentrated alkaline solutions at temperatures above 100 °C as part of the conventional chemical method [14, 67]. NaOH is a popular option for deacetylation, similar to deproteinization, but at considerably larger concentrations; it is common for sodium hydroxide concentrations to range up to 50%, and concentrations as high as 70% have been utilized in certain publications [9, 14]. The products' molecular weight may decrease due to the severe circumstances of chemical deacetylation [14, 67]. Deacetylation reactions can last up to 24 hours, although most last between 30 minutes and 5 hours [9, 14, 68]. Increased deacetylation occurs at the expense of molecular weight at higher temperatures, longer reaction periods, and greater alkali concentrations [9, 68]. The enzymatic techniques of chitin extraction are comparable in performance and limit the biological process of deacetylation, which is carried out by the fungal enzyme chitin deacetylase [14, 61]. Due to the lack of commercially available enzymes and the poor activity of enzymes isolated from biomass, this approach is uncommon [45, 69, 70].

The activity of chitin deacetylase is increased by a variety of ionic liquids. Twelve different ionic liquids were examined by Aspras et al. to see how they might affect chitin deacetylase; of those, 1-allyl-3-methylimidazolium chloride ([Amim][Cl]) had the greatest influence on deacetylase activity [69]. The activity of enzymes was reduced in other ionic solutions that included the chloride anion [69]. The bromide anion, which is present in all of the examined ionic liquids, boosted enzyme activity [69]. Later studies utilizing butyl-3-methylimidazolium bromide ([Bmim][Br]) revealed that the ionic liquid enhanced deacetylase activity at low concentrations and increased dosages that inhibited activity (when the concentration of the solution is more than 1 mg/mL, in the presence of chitin deacetylase) [70]. Tetrabutylammonium hydroxide was also observed to boost deacetylase activity by Ma et al. [45], indicating that the encouraging effect is not exclusive to ionic liquids based on imidazolium.

Additionally, it has been discovered that ionic liquids can increase the deacetylation of chitosan. Chitosan was dissolved in 1-butyl-3-methylimidazolium acetate (BMIMOAc) and heated at 100 °C for two hours while under a nitrogen environment by Ishi et al. [71]. With acetic acid as the only by-product, this enhanced the DD in chitosan to 77.3% to 86.7% [71]. Previous studies have demonstrated that tetraalkylammonium hydroxide solutions possess the ability to deacetylate dissolved chitin under ambient conditions. However, the research conducted using Shimo et al. [44] revealed that this process needed a duration of two weeks. BMIMOAc was recovered, suggesting that the solvent might be recycled for subsequent use [71]. Table 3.4 summarizes the results of the deacetylation of alkaline and ionic liquids.

Table 3.4 Deacetylation of chitin: Procedures and outcomes

Method	Reagent	Temperature/ duration	DD	Chitosan yield	Ref.
Alkali	NaOH 50%	84.46 °C for 187 min	84.2%	-	[36]
Alkali	NaOH 40%	100 °C for 12 h	93%	-	[41]
IL	[N2,2,2,2] [OH]/ water (1:7.5)	IL Ambient for 2 weeks	91%	-	[44]
Alkaline	NaOH 50%	120 °C for 4 h	71.9%	-	[59]
Alkaline	NaOH 50%	121 °C for 30 min	80%	50%	[67]
Alkaline	KOH50%, ethanol, monoethyleneg-lycol	120 °C for 24 h	77% (*Homarus vulgaris*) 75% (*Maia squinado*)	97% 90%	[68]
IL	BMIMO[Ac]	100 °C for 2 h	-	86%	[71]

3.4 Nanochitin Synthesis

The word "nanochitin" describes chitin formations at the nanoscale. The morphology of chitin nanostructures exhibits variability contingent upon the source and extraction process employed. These nanostructures are frequently categorized as nanoparticles, nanocrystals, and nanofibers based on their respective morphologies [18]. Nanocrystals and nanofibers both feature long, crystalline rods with high aspect ratios as their morphology. Nanocrystals can reach lengths of several hundred nanometers, but nanofibers can reach lengths of up to a micron [19]. Due to differing production techniques, nanoparticles are nanostructures that lack the crystallinity and aspect ratios of the remaining shapes [18]. Chitin was initially converted into nanocrystals by Revol and Marches Sault in 1993 by exposing it to 3 M HCl for up to 6 hours [72]. This procedure has been used as the model for the acid hydrolysis technique of nanochitin production, one of the most popular techniques [18, 19, 73]. Strong acids with concentrations reaching 3 M are used in acid hydrolysis to dissolve and leave behind amorphous areas of the chitin crystalline fibers [18, 19, 72]. Amorphous portions can dissolve more easily at reaction temperatures near 100 °C, and sonication is sometimes used after to spread the fibers more evenly [18, 19].

The acid concentration impacts the shape of the synthesized nanochitin; at sufficiently low concentrations, certain amorphous areas can be left intact to make chitin nanofibers rather than nanocrystals [74]. Lower concentrations result in bigger particle sizes. To reduce waste, acidic strong eutectic solvents can take the position of strong acids with a recyclable substitute [75]. Nanochitin has also been synthesized using oxidation mechanisms facilitated using the 2,2,6,6-tetramethylpiperidine-1-oxyl radical (TEMPO). In these methods, the C6 hydroxyl group of chitins is subjected to selective oxidation utilizing TEMPO, sodium bromide (NaBr), and a co-oxidant such as sodium hypochlorite (NaClO) under alkaline circumstances with a pH range of 10–11. This oxidation reaction occurs within a time frame of 48 hours, resulting in the solubility of amorphous sections of chitin in water [76]. Chitin nanocrystals make up the insoluble crystalline components. Shorter crystal lengths are the result of more co-oxidants being present in the process, which causes more sites on the chitin chain to be oxidized and dissolved [18, 19, 76]. Room temperature TEMPO-mediated oxidation processes are possible, although they take substantially longer to complete than acid hydrolysis [18, 19, 76].

The growing usage of acid hydrolysis for nanochitin production is attributed to the lengthy reaction durations and relatively expensive prices. Fernández-Marn et al. [77] synthesized chitin nanocrystals and nanofibers using a microwave-assisted extraction approach to shorten the reaction time and energy requirements of nanochitin extraction [78]. Chitin was submerged in 1–3 M HCl and heated for 10–30 minutes to generate a reaction [78]. Despite faster reaction durations, supported microwave yields and morphology extraction have been compared to TEMPO oxidation, conventional HCl, and mechanical techniques [78]. The biggest drawback of this technology was the requirement for a microwave reactor. Chitin nanofibers, as opposed to nanocrystals, are often generated during mechanical disintegration [19]. Only mildly acidic conditions are needed for mechanical procedures, which can include blenders, mills, homogenizers, or microfluidizers [19, 79]. Under neutral pH settings, ultrasound can produce chitin nanofibers, albeit the process is aided by surface ionization techniques such as oxidation or partial deacetylation [80]. These delicate techniques cannot completely split apart amorphous areas to produce nanocrystals [19, 79]. Because deacetylated, chitin is more soluble and cationic charges on the crystal surface cause electrostatic repulsion, partial deacetylation of chitin before mechanical breakdown produced nanocrystals instead of nanofibers, according to Fan et al. [81] Since chitin-nanoparticles need to be crosslinked using amine groups, which are absent from acetylated chitin, they also need a sufficient amount of deacetylated chitin for synthesis [18]. Chitin that has been chemically or partly deacetylated typically dissolves when an ionic crosslinking agent (such as tripolyphosphate or ferric chloride) is present [18, 82]. The amorphous nanoparticles are expelled from the solution after crosslinking [82]. Table 3.5 provides an overview of the nanochitin production techniques.

Table 3.5 Synthesis of nanochitin: processes and outcomes

Technique	Product	Yield (%)	Dimensions (diameter/length) (nm)	Ref.
Acid hydrolysis	Nanocrystals	65	6–8; 50–300	[72]
Acid hydrolysis	Nanocrystals	87.5	42–49; 257–670	[75]
TEMPO-mediated oxidation	Nanocrystals	90	8; 340	[76]
Mechanical disintegration	Nanofibers	-	3.6–3.9; 1000–1500	[79]
Deacetylation	Nanocrystals	85–90	6.2 ± 1.1; 250 ± 140	[80]
Decompose via link	Nanoparticles	-	237–429 (diameter)	[82]
Acid decomposition	Nanocrystals	86	18–40; 200–560	[83]
Acid decomposition	Nanocrystals	55–60	6–8; 100–200	[84]
Acid decomposition	Nanocrystals	40	20; 300	[85]

3.5 Applications

Table 3.6 The different uses of chitin and chitosan

Area of applications	Example	Ref.
Pharmaceutical	Absorbable material with controlled release of active ingredients; formation of microcapsule gels with anionic polymers; excipients; drug delivery systems; vaccine adjuvants; drug encapsulation drug release	[86, 87]
Medical	Artificial blood vessels bandage; antitumor agent; anti-infective, antibacterial; antifungal agent; anticoagulant	[88, 89]
Water and wastewater treatment	An agent that coagulates or flocculates, removing heavy metal ions	[90, 91]
Biotechnology	Enzyme immobilization; gene delivery	[92, 93]
Cosmetics and toiletries	Skincare	[94, 95]
Textile	Fibers for textile	[96, 97]
Food industry	Antimicrobial agent; bactericidal; fungicidal; film industry; controlled release of antioxidants, nutrients, flavors, and medicines; controlled release of antimicrobial compounds; lowering the partial pressure of oxygen; controlled respiratory rate; temperature regulation controlled fruit enzymatic browning; reverse osmosis membrane additive; fruit and beverage clarification and deacidification; natural flavor enhancer; texture-regulating substance food imitation color stabilization; a thickening and stabilizing agent nutrient content; fiber in food; influence on cholesterol levels; addition for fish and livestock feed; lowering the absorption of lipids; single-cell protein production; a gastric ulcer treatment; infant food component; affinity flocculation, agar fractionation, enzyme immobilization, and nutraceutical encapsulation	[98–100]
Agriculture	Plant defense mechanisms; plant growth stimulation; seed coating, frost protection; time-release fertilizers Earth with nutrients	[101, 102]

Chitin exhibits inertness inside the mammalian digestive system, possesses low levels of toxicity, and demonstrates biodegradability. Because of these properties, chitin finds applications in various fields, like cosmetics, papermaking, wound dressings, tissue engineering, and wastewater treatment [8]. Chitosan has been extensively investigated for its potential applications in analogous domains. The enhanced solubility of chitosan, as opposed to chitin, has facilitated its use in the realm of plastics [8, 10]. Table 3.6 lists the benefits of chitosan and chitin.

3.6 Conclusion and Future Prospects

A sustainable supply of chitin can be found in the substantial quantities of crustacean shell waste left over after mussel fishing. Large-scale chitin extraction procedures are known to be expensive, slow to react, or produce hazardous waste. Each of the above approaches has seen some improvement: Deep eutectic solvents have the same issues as ionic liquids, albeit at a lower cost. In order to justify the increased costs and longer reaction times, it is imperative to enhance the efficacy of biological techniques in eliminating remaining minerals and proteins. The primary objective of chemical methodologies should be the mitigation of hazardous waste by-products. One approach to achieve this objective involves the substitution of chemicals with lower toxicity levels, as well as the optimization of concentration and volume parameters to minimize the potential for unintended release. Deep eutectic solvents, on the other hand, must deal with considerable viscosity, which prevents efficient chitin extraction. Because reactions at higher temperatures or longer reaction periods need larger concentrations of alkali, enzymes with costly ionic liquids, or low activity, chitin deacetylation exacerbates many of the same issues as deproteination. In order to enhance the economics of chitosan manufacturing and lessen the environmental effect, greater energy and detector costs should be reduced. Similar to this, the frequent acid hydrolysis during nanochitin production indicates demineralization issues that are made worse by increased mineral acid concentrations that prohibit organic acids from replacing them. Intense conditions are needed to eliminate amorphous chitin in the creation of nanocrystals.

Substituting deep eutectic solvents for mineral acids, and reducing energy costs with substituting or microwave radiation techniques like TEMPO oxidation for acid hydrolysis results in reduced environmental impact, although each of these measures requires more expensive reagents or special reactors. The enhancement of cost-efficiency in relation to these ecologically conscious techniques would significantly bolster the long-term viability and sustainability of nanochitin manufacturing. Numerous products, such as biodegradable plastics and biomedical materials, are made from chitin and its derivatives. As the body of knowledge grows, new applications for chitin and its derivatives are found. In particular, efforts are being made to overcome shortcomings like chitin's insolubility and brittleness while incorporating the favorable features of chitin into materials. More research will be done to create chitin-containing materials and other renewable materials with functionality that can replace less environmentally friendly materials.

References

1. Amor IB, Hemmami H, Laouini SE, Temam HB, Zaoui H, Barhoum A (2023) Biosynthesis MgO and ZnO nanoparticles using chitosan extracted from Pimelia Payraudi Latreille for antibacterial applications. *World Journal of Microbiology and Biotechnology* 39: 19.
2. Meyers MA, Chen P-Y (2014) *Biological Materials Science: Biological Materials, Bioinspired Materials and Biomaterials*. Cambridge University Press.
3. Hossin MA, Al Shaqsi NHK, Al Touby SSJ, Al Sibani MA (2021) A review of polymeric chitin extraction, characterization, and applications. *Arabian Journal of Geosciences* 14: 1870.
4. Shamshina JL, Zavgorodnya O, Rogers RD (2019) Advances in processing chitin as a promising biomaterial from ionic liquids. *Application of Ionic Liquids in Biotechnology*: 177–198.
5. Amor IB, Hemmami H, Laouini SE, Abdelaziz AG, Barhoum A (2023) Influence of chitosan source and degree of deacetylation on antibacterial activity and adsorption of AZO dye from water. *Biomass Conversion and Biorefinery*: 1–11.

6. Ben Amor I, Hemmami H, Laouini SE, Zeghoud S, Benzina M, Achour S, Naseef A, Alsalme A, Barhoum A (2023) Use of insect-derived chitosan for the removal of methylene blue dye from wastewater: process optimization using a central composite Design. *Materials* 16: 5049.
7. Ben Amor I, Hemmami H, Laouini SE, Mahboub MS, Barhoum A (2022) Sol-gel synthesis of ZnO nanoparticles using different chitosan sources: Effects on antibacterial activity and photocatalytic degradation of AZO Dye. *Catalysts* 12: 1611.
8. Rinaudo M (2006) Chitin and chitosan: Properties and applications. *Progress in Polymer Science* 31: 603–632.
9. No HK, Meyers SP (1995) Preparation and characterization of chitin and chitosan—a review. *Journal of Aquatic Food Product Technology* 4: 27–52.
10. Ji J, Wang L, Yu H, Chen Y, Zhao Y, Zhang H, Amer WA, Sun Y, Huang L, Saleem M (2014) Chemical modifications of chitosan and its applications. *Polymer-Plastics Technology and Engineering* 53: 1494–1505.
11. Bianchi M, Chopin F, Farme T, Franz N, Fuentevilla C, Garibaldi L, Laurenti A (2014) *FAO: The State of World Fisheries and Aquaculture. Food and Agriculture Organization of the United Nations*: Rome, Italy: 1–230.
12. Kozma M, Acharya B, Bissessur R (2022) Chitin, chitosan, and nanochitin: Extraction, synthesis, and applications. *Polymers* 14: 3989.
13. Yan N, Chen X (2015) Sustainability: Don't waste seafood waste. *Nature* 524: 155–157.
14. Kou SG, Peters LM, Mucalo MR (2021) Chitosan: A review of sources and preparation methods. *International Journal of Biological Macromolecules* 169: 85–94.
15. Arbia W, Arbia L, Adour L, Amrane A (2013) Chitin extraction from crustacean shells using biological methods–a review. *Food Technology and Biotechnology* 51: 12–25.
16. Morais ES, Lopes AMdC, Freire MG, Freire CS, Coutinho JA, Silvestre AJ (2020) Use of ionic liquids and deep eutectic solvents in polysaccharides dissolution and extraction processes towards sustainable biomass valorization. *Molecules* 25: 3652.
17. Mohan K, Ganesan AR, Ezhilarasi P, Kondamareddy KK, Rajan DK, Sathishkumar P, Rajarajeswaran J, Conterno L (2022) Green and

eco-friendly approaches for the extraction of chitin and chitosan: A review. *Carbohydrate Polymers* 287: 119349.

18. Jin T, Liu T, Lam E, Moores A (2021) Chitin and chitosan on the nanoscale. *Nanoscale Horizons* 6: 505–542.
19. Salaberria AM, Labidi J, Fernandes SC (2015) Different routes to turn chitin into stunning nano-objects. *European Polymer Journal* 68: 503–515.
20. Parhi R (2020) Drug delivery applications of chitin and chitosan: A review. *Environmental Chemistry Letters* 18: 577–594.
21. Ali A, Ahmed S (2018) A review on chitosan and its nanocomposites in drug delivery. *International Journal of Biological Macromolecules* 109: 273–286.
22. Tao F, Cheng Y, Shi X, Zheng H, Du Y, Xiang W, Deng H (2020) Applications of chitin and chitosan nanofibers in bone regenerative engineering. *Carbohydrate Polymers* 230: 115658.
23. Ikram R, Mohamed Jan B, Abdul Qadir M, Sidek A, Stylianakis MM, Kenanakis G (2021) Recent advances in chitin and chitosan/graphene-based bio-nanocomposites for energetic applications. *Polymers* 13: 3266.
24. Nagasawa H (2012) The crustacean cuticle: Structure, composition and mineralization. *Frontiers in Bioscience* 4: 711–720.
25. Yang H, Gözaydýn Gk, Nasaruddin RR, Har JRG, Chen X, Wang X, Yan N (2019) Toward the shell biorefinery: Processing crustacean shell waste using hot water and carbonic acid. *ACS Sustainable Chemistry & Engineering* 7: 5532–5542.
26. Borić M, Puliyalil H, Novak U, Likozar B (2018) An intensified atmospheric plasma-based process for the isolation of the chitin biopolymer from waste crustacean biomass. *Green Chemistry* 20: 1199–1204.
27. Borić M, Vicente FA, Jurković DL, Novak U, Likozar B (2020) Chitin isolation from crustacean waste using a hybrid demineralization/DBD plasma process. *Carbohydrate Polymers* 246: 116648.
28. Percot A, Viton C, Domard A (2003) Optimization of chitin extraction from shrimp shells. *Biomacromolecules* 4: 12–18.
29. Pohling J, Dave D, Liu Y, Murphy W, Trenholm S (2022) Two-step demineralization of shrimp (Pandalus Borealis) shells using citric acid: An environmentally friendly, safe and cost-effective alternative to the traditional approach. *Green Chemistry* 24: 1141–1151.

30. Trung TS, Van Tan N, Van Hoa N, Minh NC, Loc PT, Stevens WF (2020) Improved method for production of chitin and chitosan from shrimp shells. *Carbohydrate Research* 489: 107913.
31. Jung W-J, Jo G-H, Kuk J-H, Kim K-Y, Park R-D (2005) Demineralization of crab shells by chemical and biological treatments. *Biotechnology and Bioprocess Engineering* 10: 67–72.
32. Blumberg R, Southall C, Van Rensburg N, Volckman O (1951) South african fish products. XXXII.—The rock lobster: A study of chitin production from processing wastes. *Journal of the Science of Food and Agriculture* 2: 571–576.
33. Yayci A, Dirks T, Kogelheide F, Alcalde M, Hollmann F, Awakowicz P, Bandow JE (2020) Protection strategies for biocatalytic proteins under plasma treatment. *Journal of Physics D: Applied Physics* 54: 035204.
34. Olafadehan O, Ajayi T, Amoo K (2020) Optimum conditions for extraction of chitin and chitosan from Callinectes amnicola shell waste. *Theoretical Foundations of Chemical Engineering* 54: 1173–1194.
35. Abbasi NM, Farooq MQ, Anderson JL (2021) Investigating the variation in solvation interactions of choline chloride-based deep eutectic solvents formed using different hydrogen bond donors. *ACS Sustainable Chemistry & Engineering* 9: 11970–11980.
36. Berton P, Shamshina JL, Ostadjoo S, King CA, Rogers RD (2018) Enzymatic hydrolysis of ionic liquid-extracted chitin. *Carbohydrate Polymers* 199: 228–235.
37. Wang Y, Yang Y, Wang R, Zhu Y, Yang P, Lin Z, Wang Z, Cong W (2022) Efficient extraction of chitin from crustacean waste via a novel ternary natural deep eutectic solvents. *Carbohydrate Polymers* 286: 119281.
38. Roy JC, Salaün F, Giraud S, Ferri A, Chen G, Guan J (2017) Solubility of chitin: Solvents, solution behaviors and their related mechanisms. *Solubility of Polysaccharides* 3: 20–60.
39. Tan X, Wang G, Zhong L, Xie F, Lan P, Chi B (2021) Regeneration behavior of chitosan from ionic liquid using water and alcohols as anti-solvents. *International Journal of Biological Macromolecules* 166: 940–947.
40. Wineinger HB, Kelly A, Shamshina JL, Rogers RD (2020) Farmed Jumbo shrimp molts: An ionic liquid strategy to increase chitin yield per animal while controlling molecular weight. *Green Chemistry* 22: 6001–6007.

41. Tolesa LD, Gupta BS, Lee M-J (2019) Chitin and chitosan production from shrimp shells using ammonium-based ionic liquids. *International Journal of Biological Macromolecules* 130: 818–826.
42. Setoguchi T, Kato T, Yamamoto K, Kadokawa J-I (2012) Facile production of chitin from crab shells using ionic liquid and citric acid. *International Journal of Biological Macromolecules* 50: 861–864.
43. Wang W-T, Zhu J, Wang X-L, Huang Y, Wang Y-Z (2010) Dissolution behavior of chitin in ionic liquids. *Journal of Macromolecular Science, Part B* 49: 528–541.
44. Shimo M, Abe M, Ohno H (2016) Functional comparison of polar ionic liquids and onium hydroxides for chitin dissolution and deacetylation to chitosan. *ACS Sustainable Chemistry & Engineering* 4: 3722–3727.
45. Ma Q, Gao X, Bi X, Han Q, Tu L, Yang Y, Shen Y, Wang M (2020) Dissolution and deacetylation of chitin in ionic liquid tetrabutylammonium hydroxide and its cascade reaction in enzyme treatment for chitin recycling. *Carbohydrate Polymers* 230: 115605.
46. Mehmood N, Husson E, Jacquard C, Wewetzer S, Büchs J, Sarazin C, Gosselin I (2015) Impact of two ionic liquids, 1-ethyl-3-methylimidazolium acetate and 1-ethyl-3-methylimidazolium methylphosphonate, on Saccharomyces cerevisiae: Metabolic, physiologic, and morphological investigations. *Biotechnology for Biofuels* 8: 1–14.
47. Smith EL, Abbott AP, Ryder KS (2014) Deep eutectic solvents (DESs) and their applications. *Chemical Reviews* 114: 11060–11082.
48. Abbott AP, Capper G, Davies DL, Munro HL, Rasheed RK, Tambyrajah V (2001) Preparation of novel, moisture-stable, Lewis-acidic ionic liquids containing quaternary ammonium salts with functional side chainsElectronic supplementary information (ESI) available: plot of conductivity vs. temperature for the ionic liquid formed from zinc chloride and choline chloride (2:1). See http://www.rsc.org/suppdata/cc/b1/b106357j. *Chemical Communications* 2010–2011.
49. Paiva A, Craveiro R, Aroso I, Martins M, Reis RL, Duarte ARC (2014) Natural deep eutectic solvents–solvents for the 21st century. *ACS Sustainable Chemistry & Engineering* 2: 1063–1071.
50. Dai Y, van Spronsen J, Witkamp G-J, Verpoorte R, Choi YH (2013) Natural deep eutectic solvents as new potential media for green technology. *Analytica Chimica Acta* 766: 61–68.
51. Rodrigues LA, Radojčić Redovniković I, Duarte ARC, Matias AA, Paiva A (2021) Low-phytotoxic deep eutectic systems as alternative

extraction media for the recovery of chitin from brown crab shells. *ACS Omega* 6: 28729–28741.

52. Bisht M, Macário IsP, Neves MC, Pereira JL, Pandey S, Rogers RD, Coutinho JoA, Ventura SP (2021) Enhanced dissolution of chitin using acidic deep eutectic solvents: A Sustainable and simple approach to extract chitin from crayfish shell wastes as alternative feedstocks. *ACS Sustainable Chemistry & Engineering* 9: 16073–16081.
53. Saravana PS, Ho TC, Chae S-J, Cho Y-J, Park J-S, Lee H-J, Chun B-S (2018) Deep eutectic solvent-based extraction and fabrication of chitin films from crustacean waste. *Carbohydrate Polymers* 195: 622–630.
54. Huang W-C, Zhao D, Guo N, Xue C, Mao X (2018) Green and facile production of chitin from crustacean shells using a natural deep eutectic solvent. *Journal of Agricultural and Food Chemistry* 66: 11897–11901.
55. Li Z, Li M-C, Liu C, Liu X, Lu Y, Zhou G, Liu C, Mei C (2023) Microwave-assisted deep eutectic solvent extraction of chitin from crayfish shell wastes for 3D printable inks. *Industrial Crops and Products* 194: 116325.
56. Bisht S, Pandey P, Bhargava B, Sharma S, Kumar V, Sharma KD (2015) Bioremediation of polyaromatic hydrocarbons (PAHs) using rhizosphere technology. *Brazilian Journal of Microbiology* 46: 7–21.
57. Hamdi M, Hammami A, Hajji S, Jridi M, Nasri M, Nasri R (2017) Chitin extraction from blue crab (Portunus segnis) and shrimp (Penaeus kerathurus) shells using digestive alkaline proteases from P. segnis viscera. *International Journal of Biological Macromolecules* 101: 455–463.
58. Hongkulsup C, Khutoryanskiy VV, Niranjan K (2016) Enzyme assisted extraction of chitin from shrimp shells (Litopenaeus vannamei). *Journal of Chemical Technology & Biotechnology* 91: 1250–1256.
59. Sayari N, Sila A, Abdelmalek BE, Abdallah RB, Ellouz-Chaabouni S, Bougatef A, Balti R (2016) Chitin and chitosan from the Norway lobster by-products: Antimicrobial and anti-proliferative activities. *International Journal of Biological Macromolecules* 87: 163–171.
60. Adour L, Arbia W, Amrane A, Mameri N (2008) Combined use of waste materials—recovery of chitin from shrimp shells by lactic acid fermentation supplemented with date juice waste or glucose. *Journal of Chemical Technology & Biotechnology: International Research in Process, Environmental & Clean Technology* 83: 1664–1669.

61. Tan JS, Abbasiliasi S, Lee CK, Phapugrangkul P (2020) Chitin extraction from shrimp wastes by single step fermentation with Lactobacillus acidophilus FTDC3871 using response surface methodology. *Journal of Food Processing and Preservation* 44: e14895.

62. Liu Y, Xing R, Yang H, Liu S, Qin Y, Li K, Yu H, Li P (2020) Chitin extraction from shrimp (Litopenaeus vannamei) shells by successive two-step fermentation with Lactobacillus rhamnoides and Bacillus amyloliquefaciens. *International Journal of Biological Macromolecules* 148: 424–433.

63. Tan YN, Lee PP, Chen WN (2020) Microbial extraction of chitin from seafood waste using sugars derived from fruit waste-stream. *AMB Express* 10: 1–11.

64. Doan CT, Tran TN, Vo TPK, Nguyen AD, Wang S-L (2019) Chitin extraction from shrimp waste by liquid fermentation using an alkaline protease-producing strain, Brevibacillus parabrevis. *International Journal of Biological Macromolecules* 131: 706–715.

65. Cahyaningtyas HAA, Suyotha W, Cheirsilp B, Prihanto AA, Yano S, Wakayama M (2022) Optimization of protease production by Bacillus cereus HMRSC30 for simultaneous extraction of chitin from shrimp shell with value-added recovered products. *Environmental Science and Pollution Research*: 1–16.

66. Gamal RF, El-Tayeb TS, Raffat EI, Ibrahim HM, Bashandy A (2016) Optimization of chitin yield from shrimp shell waste by Bacillus subtilis and impact of gamma irradiation on production of low molecular weight chitosan. *International Journal of Biological Macromolecules* 91: 598–608.

67. Agarwal M, Agarwal MK, Shrivastav N, Pandey S, Gaur P (2018) A simple and effective method for preparation of chitosan from chitin. *International Journal of Life Sciences Research* 41: 721–1728.

68. Rhazi M, Desbrieres J, Tolaimate A, Alagui A, Vottero P (2000) Investigation of different natural sources of chitin: Influence of the source and deacetylation process on the physicochemical characteristics of chitosan. *Polymer International* 49: 337–344.

69. Aspras I, Kamińska M, Karzyński K, Kawka M, Jaworska MM (2016) The influence of selected ionic liquids on activity of chitin deacetylase. *Chemical and Process Engineering* 37: 77–82.

70. Aspras I, Jaworska MM, Górak A (2017) Kinetics of chitin deacetylase activation by the ionic liquid [Bmim][Br]. *Journal of Biotechnology* 251: 94–98.

71. Ishii D, Ohashi C, Hayashi H (2014) Facile enhancement of the deacetylation degree of chitosan by hydrothermal treatment in an imidazolium-based ionic liquid. *Green Chemistry* 16: 1764–1767.
72. Revol J-F, Marchessault R (1993) In vitro chiral nematic ordering of chitin crystallites. *International Journal of Biological Macromolecules* 15: 329–335.
73. Kadokawa Ji (2013) Preparation and applications of chitin nanofibers/nanowhiskers. *Biopolymer Nanocomposites: Processing, Properties, and Applications* 131–151.
74. Joseph B, Mavelil Sam R, Balakrishnan P, Maria HJ, Gopi S, Volova T, CM Fernandes S, Thomas S (2020) Extraction of nanochitin from marine resources and fabrication of polymer nanocomposites: Recent advances. *Polymers* 12: 1664.
75. Yuan Y, Hong S, Lian H, Zhang K, Liimatainen H (2020) Comparison of acidic deep eutectic solvents in production of chitin nanocrystals. *Carbohydrate Polymers* 236: 116095.
76. Fan Y, Saito T, Isogai A (2008) Chitin nanocrystals prepared by TEMPO-mediated oxidation of α-chitin. *Biomacromolecules* 9: 192–198.
77. Ardakani AG, Nezhadhaghighi MG (2015) Controlling Anderson localization in disordered heterostrctures with Levy-type distribution. *Journal of Optics* 17: 105601.
78. Fernández-Marín R, Hernández-Ramos F, Salaberria AM, Andrés MÁ, Labidi J, Fernandes SC (2021) Eco-friendly isolation and characterization of nanochitin from different origins by microwave irradiation: Optimization using response surface methodology. *International Journal of Biological Macromolecules* 186: 218–226.
79. Mushi NE, Butchosa N, Salajkova M, Zhou Q, Berglund LA (2014) Nanostructured membranes based on native chitin nanofibers prepared by mild process. *Carbohydrate Polymers* 112: 255–263.
80. Bai L, Liu L, Esquivel M, Tardy BL, Huan S, Niu X, Liu S, Yang G, Fan Y, Rojas OJ (2022) Nanochitin: Chemistry, structure, assembly, and applications. *Chemical Reviews* 122: 11604–11674.
81. Fan Y, Saito T, Isogai A (2010) Individual chitin nano-whiskers prepared from partially deacetylated α-chitin by fibril surface cationization. *Carbohydrate Polymers* 79: 1046–1051.
82. Geetha P, Sivaram AJ, Jayakumar R, Mohan CG (2016) Integration of in silico modeling, prediction by binding energy and experimental approach to study the amorphous chitin nanocarriers for cancer drug delivery. *Carbohydrate Polymers* 142: 240–249.

83. Phongying S, Aiba S-I, Chirachanchai S (2007) Direct chitosan nanoscaffold formation via chitin whiskers. *Polymer* 48: 393–400.

84. Araki J, Yamanaka Y, Ohkawa K (2012) Chitin-chitosan nanocomposite gels: Reinforcement of chitosan hydrogels with rod-like chitin nanowhiskers. *Polymer Journal* 44: 713–717.

85. Ma B, Qin A, Li X, Zhao X, He C (2014) Structure and properties of chitin whisker reinforced chitosan membranes. *International Journal of Biological Macromolecules* 64: 341–346.

86. Abdelhak M (2019) A review: Application of biopolymers in the pharmaceutical formulation. *Pharmacology, Pharmaceutics & Pharmacovigilance* 1: 15–25.

87. Gopi S, Amalraj A, Sukumaran NP, Haponiuk JT, Thomas S (2018) Biopolymers and their composites for drug delivery: A brief review. In: *Macromolecular Symposia*. Wiley Online Library, p. 1800114.

88. Fatullayeva S, Tagiyev D, Zeynalov N, Mammadova S, Aliyeva E (2022) Recent advances of chitosan-based polymers in biomedical applications and environmental protection. *Journal of Polymer Research* 29: 259.

89. Murugesan S, Scheibel T (2021) Chitosan-based nanocomposites for medical applications. *Journal of Polymer Science* 59: 1610–1642.

90. Kanmani P, Aravind J, Kamaraj M, Sureshbabu P, Karthikeyan S (2017) Environmental applications of chitosan and cellulosic biopolymers: A comprehensive outlook. *Bioresource Technology* 242: 295–303.

91. Nechita P (2017) Applications of chitosan in wastewater treatment. *Biological Activities and Application of Marine Polysaccharides* 1: 209–228.

92. Verma ML, Kumar S, Das A, Randhawa JS, Chamundeeswari M (2020) Chitin and chitosan-based support materials for enzyme immobilization and biotechnological applications. *Environmental Chemistry Letters* 18: 315–323.

93. Al-Jbour ND, Beg MD, Gimbun J, Alam AM (2019) An overview of chitosan nanofibers and their applications in the drug delivery process. *Current Drug Delivery* 16: 272–294.

94. Madera-Santana TJ, Herrera-Méndez CH, Rodríguez-Núñez JR (2018) An overview of the chemical modifications of chitosan and their advantages. *Green Materials* 6: 131–142.

95. Gaspar AL, Gaspar AB, Contini LR, Silva MF, Chagas EG, Bahú JO, Concha VO, Carvalho RA, Severino P, Souto EB (2022) Lemongrass

(Cymbopogon citratus)-incorporated chitosan bioactive films for potential skincare applications. *International Journal of Pharmaceutics* 628: 122301.

96. Huang L, Xiao L, Yang G (2018) Chitosan application in textile processing. *Current Trends in Fashion Technology & Textile Engineering* 4: 32–34.
97. Elamri A, Zdiri K, Hamdaoui M, Harzallah O (2023) Chitosan: A biopolymer for textile processes and products. *Textile Research Journal* 93: 1456–1484.
98. Maleki G, Woltering EJ, Mozafari M (2022) Applications of chitosan-based carrier as an encapsulating agent in food industry. *Trends in Food Science & Technology* 120: 88–99.
99. Gutiérrez TJ (2017) Chitosan applications for the food industry. *Chitosan: Derivatives, Composites and Applications* 183–232.
100. Manigandan V, Karthik R, Ramachandran S, Rajagopal S (2018) Chitosan applications in food industry. *Biopolymers for Food Design*. Elsevier, pp. 469–491.
101. Xing K, Zhu X, Peng X, Qin S (2015) Chitosan antimicrobial and eliciting properties for pest control in agriculture: A review. *Agronomy for Sustainable Development* 35: 569–588.
102. Pandey P, Verma MK, De N (2018) Chitosan in agricultural context-A review. *Bulletin of Environment, Pharmacology and Life Sciences* 7: 87–96.

Chapter 4

Physical Properties of Chitin and Chitosan

Ahmed Barhoum,[a,b] **Ilham Ben Amor,**[c,d] **and Shakeel Ahmed**[e,f]

[a]*NanoStruc Research Group, Chemistry Department, Faculty of Science, Helwan University, Cairo, Egypt*
[b]*School of Chemical Sciences, Dublin City University, Dublin, Ireland*
[c]*Department of Process Engineering and Petrochemical, Faculty of Technology, University of El Oued, El Oued, Algeria*
[d]*Renewable Energy Development unit in Arid Zones (UDERZA), University of El Oued, El Oued, Algeria*
[e]*Postgraduate Department of Chemistry, Government Postgraduate College Rajouri, Jammu and Kashmir 185133, India*
[f]*University Centre for Research and Development, Chandigarh University, Mohali, Punjab, India*

ilhambenamor97@gmail.com

Next to cellulose, chitin is the second most prevalent organic amino polysaccharide in the world. It is deacetylated to produce chitosan. Researchers and industry scientists are looking at chitosan as a resource that is underutilized and as a novel functional material with great promise in many industries. The biopolymers chitosan and chitin have been produced using a variety of techniques over the years. The preparation methods and study of the physical-chemical characteristics of these biopolymers are discussed in terms of the many types of crustaceans and methodologies for characterizing them. These characteristics affect biopolymer functioning differently, contingent on the kind

Chitin and Chitosan: Physical and Chemical Properties
Edited by Shakeel Ahmed, Soumeia Zeghoud, Hadia Hemmami, and Ilham Ben Amor

ISBN 978-981-5129-39-7 (Hardcover), 978-1-003-58977-8 (eBook)
www.jennystanford.com

of crustacean and the process of preparation. This chapter's goal is to provide a summary of the physical properties of chitosan.

4.1 Introduction

Chitin is a very prevalent amino polysaccharide that is often found in nature, second only to cellulose. It is derived from several sources, such as the exoskeletons of crustaceans including crabs, prawns, crawfish, and Oniscus asellus [1]. Growing interest in chitin and its derivative chitosan as beneficial biopolymers. Chitin and chitosan, which compete with starch and cellulose for the title of a most abundant naturally occurring organic compound, are thought to be materials with enormous potential for modification of structure to give desired characteristics and functions and to improve their myriad uses in different areas of science. Additionally, the degree of N-acetylation significantly affects how insoluble chitin is, and compared to cellulose, it has fewer water-swelling characteristics [2, 3]. Chitin's highly basic polysaccharide property that makes it water-insoluble is most likely, due to the hydrogen links both intramolecular and intermolecular associations with nearby –OH or –NH function groups that the oxygen of the acetamido group forms [4]. Because of these factors, scientists have converted these complicated compounds into chitooligosaccharides, low molecular weight (Mw) oligosaccharides with the least viscosity, and very small molecular dimensions that make them easily absorbed in vivo and water soluble [5].

However, biotechnology fermentation techniques, such as deacetylation by chitin deacetylase, demineralization, and deproteination using organic acid bacteria and protease, were effective and functional ways to extract chitin and their resources. Due to their high cost, variable Mw and degree of acetylation (DD), and hostile environmental practices, traditional techniques for commercially preparing chitosan from their resources have some limitations [6, 7]. Due to their numerous biological applications in the fields of agriculture, pharmaceuticals, food, medicine, products for personal care, and the environment, chitosan and oligosaccharides have become more lucrative [8]. In this chapter, we have tried to summarize the most important physicochemical properties that distinguish both chitin and chitosan.

4.2 Molecular Weight (Mw)

There are several applications for chitosan and its derivatives. However, it has been shown that the crystallinity, DD, and Mw of these materials affect how effective they are [9]. Temperature, reagent concentrations, alkaline step repetition, duration, and ambient conditions during the deacetylation all significantly affect the Mw and DD of chitosan [10]. Enzymatic and chemical techniques are used to manufacture chitosan [11]. Poor structural order (glucose ring) and low product yields are just two of the disadvantages of the chemical process. Additionally, the chemical process's use of acids and alkalis may have an adverse effect on the environment [12]. An alternate, more environmentally friendly method of producing chitosan is using the enzymatic process. Instead of using acidic treatment, lactic acid-producing bacteria have been employed throughout the enzymatic process to demineralize crustacean shells. The resulting calcium lactate may be precipitated and eliminated after the calcium carbonate and lactic acid reaction. Proteases from bacteria (*Bacillus subtilis, Serratia marcescens*, and *Pseudomonas aeruginosa*) are employed to deproteinize crustacean shells. *Bacillus sp.* and *Serratia sp.* are bacteria that may be utilized to make chitosan since they also produce chitin deacetylase [13]. The efficient deacetylation is accomplished by sometimes rinsing the intermediate byproduct using water through the alkaline treatment. Chitosan has a B500 kDa average Mw and 100% DD. When treating alkaline during the first hours (50% NaOH) at 100 °C, the DD rises quickly to around 68% and then gradually rises over time [14].

To calculate Mw, a different set of techniques are used, including capillary viscometry, gel permeation chromatography (GPC), and light scattering. Measurement of capillary viscosity is the quickest and most used technique for determining chitosan's Mw. Through the study, a capillary viscometer of the Ubbelohde type measures the passage time of fluid flow during capillaries at 25 °C. In 0.2 M NaCl and 0.1 M acetic acid solutions, different viscosity solutions of chitosan are employed at concentrations that span from 0.00125% to 0.15%. The specimen is introduced into the capillary viscometer, thereafter undergoing calibration within a water bath maintained at a temperature of 25 °C. The solvent and solution flow times are calculated in order to establish relative viscosity.

The relationship between intrinsic viscosity (η) and *Mw* is provided by the Mark Houwink Sakurada equation, which is presented below (Eq. 4.1) [15]:

$$[\eta] = k \times [Mw]^{\alpha} = 3.04 \times 10^{-5} \times [Mw]^{1.26}. \quad (4.1)$$

Where *Mw* is the polymer's molecular weight as measured by viscosity, and *a, k,* and temperature are constants that vary with the polymer, temperature, and solvent system. The macromolecule's hydrodynamic volume affects how intrinsic viscosity is measured. The hydrodynamic volume is determined by the Mw conformation properties and the interaction of the polymer and solvent. In solution, the polymer molecules frequently take the form of coils rather than an extended straight chain [16]. These clumps or coils provide mobility resistance and obstruct the movement of molecules that cause viscosity. Various solvent and chitosan combinations can be used to find the chitosan's Mw [1]. For a certain polymer-solvent system, the viscosity and molecular weight are nearly correlated.

Chitosan with a few Mw (LMWC; 50 kDa); chitosan with an average Mw (50–250 kDa); and chitosan with a big Mw (HMWC; > 250 kDa) are the three main forms of chitosan that are categorized based on Mw. According to some publications, LMWC has improved antibacterial, antifungal, intestinal disaccharidase, lipid metabolism, anticancer, and mucoadhesive characteristics [17]. In addition, the Mw is crucial for the rheological characteristics of the biopolymer. It has an immediate effect on the creation of chitosan-based biomaterials [18]. LMWC is used as an anticreasing agent to offer a finishing agent when treating cotton materials for anticreasing. HMWC is used in quaternized chitosan films, due to its free radical scavenging and water solubility properties. It is also used in high-performance cells and affects polymer aggregation and phase separation.

4.3 Viscosity

Due to one of its special qualities, the polymorphic character it takes on when dissolved in an acidic solution, chitosan is the most well-known deacetylated chitin derivative. Because of this, this biopolymer is advantageous for a wide variety of industrial and biological uses. A viscous solution is produced when chitosan is dispersed in an acidic solution. The temperature, Mw, pH, concentration, DD, and viscosity of the chitosan solution are all connected to each other. At the molecular level, the size of the molecules of polymer is directly proportional to the viscosity of the chitosan solution, which is affected by the size of the molecular or chain length.

A capillary viscometer is frequently utilized to test the viscosity of polymer solutions, and the results are employed to calculate the mean Mw of polymer by the Mark Houwink equation (Eq. 4.1). The following equation (Eq. 4.2) can be used to calculate the intrinsic viscosity $[\eta]$:

$$[\eta]=\frac{(\eta-\eta s)}{(\eta sC)}. \tag{4.2}$$

Where C is the solution concentration, s is the reference solvent viscosity, and η (dL/g) is the viscosity of the solution. When $(\eta - \eta s)/(\eta sC)$ decreased viscosity (η_{red}) is plotted versus concentration (C), and the intercept agrees to $[\eta]$. The $[\eta]$ value is directly applied to calculate the Mw of various chitosan grades. Table 4.1 lists some of the usual chitosan viscosities for various Mws [19].

Table 4.1 Different degrees of molecular weight and intrinsic viscosity for chitosan

Chitosan	Intrinsic viscosity $[\eta]$ (dL/g)	Mw
Chitosan (1)	9.4	654.127
Chitosan (2)	4.72	285.231
Chitosan (3)	2.55	135.839
Chitosan (4)	1.5	71.676
Chitosan (5)	0.535	20.698

The Arrhenius equation (Eq. 4.3) is used to analyze the flow activation energy (*Ef*) of the viscous chitosan specimen.

$$[\eta] = Ae^{Ef/RT}, \tag{4.3}$$

where *A* and *R* stand for absolute temperature (in Kelvin) and the preexponential factor, respectively. A study [20] revealed that the *Ef* of two various grades of chitosan, 70% DD and 91% DD, was 15 and 25 kJ/mol, respectively. This indicates that due to the nature of the polymer, chain entanglement grows when chitosan's DD increases.

Chitosan's concentrated solutions with various levels of deacetylation differ in their viscosity and flow characteristics. As the DD of chitosan increases, there is a corresponding increase in the viscosities of the solutions and the manifestation of non-Newtonian flow characteristics. Conversely, the addition of salt decreases the viscosity and alters the non-Newtonian flow properties of the chitosan solution [21].

4.4 Degree of Deacetylation (DD) and Chitosan Dissociation Constant

DD is one of the structural elements that affect the biological and physical properties of chitosan. Depending on the degree of deacetylation, the biopolymer is identified as either chitosan or chitin. The production process for chitin and chitosan has a big effect on the DD of chitosan. Chitosan requires DD to be greater than 50%.

4.5 Techniques for Estimating Deacetylation of Degree

Since it affects their characteristics, the DD must be established in order to utilize the chitosan samples for various industrial and biological purposes. For chitosan DD determination, many techniques are used. Potentiometric titration, 1H-NMR, FTIR, and proton nuclear magnetic resonance spectroscopy are among approximately that are often utilized [22].

4.5.1 Fourier Transforms Infrared Spectroscopy (FTIR)

Two absorption bands are used in the commonly employed FTIR technique for determining the DD of chitosan: A characteristic band that reflects a group that is found in both the N-acetyl-d-glucosamine and d-glucosamine monomers and a distinctive band (probe band) that corresponding to the amide group of the N-acetyl-d-glucosamine remains [22]. The peaks at 3450 and 1650 cm^{-1}, which stand for the hydroxyl groups and amide I of chitosan, respectively, are the most frequently employed probe and reference bands (Fig. 4.1) [23]. The proportion of DD is estimated using these absorption bands using the formula in Eq. (4.4).

$$DD\% = 100 - \frac{A_{1652}}{A_{3450}} \times 115 \tag{4.4}$$

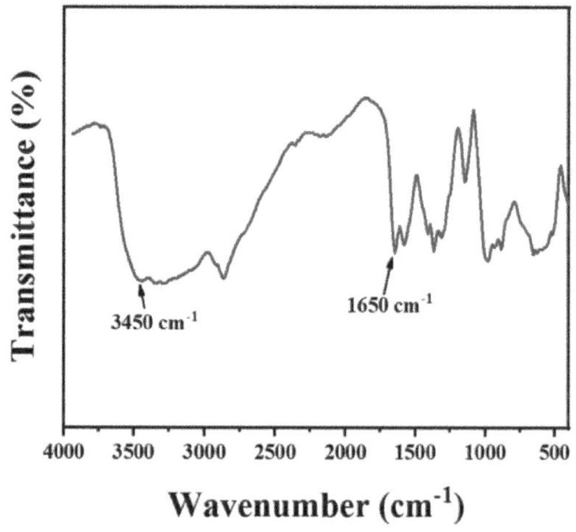

Figure 4.1 Chitosan's typical infrared Fourier transform spectrum (reproduced from Ilham et al. [3]).

According to the aforementioned equation, the DD of chitosan material increases as the amide I group absorption decreases [24].

This absorbance is also shown to decrease when the deacetylation process is prolonged. In addition to this calibration equation, numerous studies have presented additional calibration curves and used various peaks to calculate the percentage of DD. The FTIR approach for calculating DD has a number of benefits and drawbacks.

4.5.1.1 Drawbacks

- Prior to analysis, synthetically N-acetylated chitosan must first have its ester groups dissolved in ethanolic KOH
- This approach is only applicable to chitosan with DD values between 0 and 55%
- Poorly resolved spectra for materials with low DD
- Issues with OH absorption from polysaccharides and water. Samples must thus be cleaned and well-dried
- The margin of error for DD values produced using IR is relatively considerable

4.5.1.2 Advantages

- Spectrometers for infrared light are commonly accessible
- This method is the best option when just cost and analysis time are taken into account
- This method might be used to determine the DD of insoluble substances

4.5.2 ^{1}H-NMR Spectroscopy

The most effective approach for calculating the proportion of DD is ^{1}H-NMR spectroscopy. This technique involves dissolving an established amount sample of chitosan in deuterated aqueous solutions of DCl/D_2O and measuring the spectra between 0 and 6 ppm [25]. The subsequent (Eq. 4.5) is used to compute the percentage of DD and takes into account the integrals (I) of the N-acetyl group peak regions at 2.05 ppm (CH_3) and 3.15 ppm (H_2).

$$DDA\% = \left\{1 - \left(\frac{I_{CH_3}/3}{I_{H_2}}\right)\right\} \times 100 \tag{4.5}$$

Other proton signals that can be used to compute the DD are included in Table 4.2.

Table 4.2 Proton chemical changes in a solution of CD_3COOD/D_2O or DCl/D_2O at 65 °C [26]

Proton type	Position (δ, ppm)
H-Ac or $HN{-}COCH_3$	1.95–2.09
H1 (H-1(A) or H1 of GluNAc)	4.62–4.85
H1 (H-1(D) or H1 of $GluNH_2$)	4.85–4.97
H2 (H-2(D) or H2 of $GluNH_2$)	3.18–3.24
H2 (H-2(A) or H2 of GluNAc)	3.38–3.65
H3 (H-3(D) or H3 of $GluNH_2$)	3.52–3.87
H3 (H-3(A) or H3 of GluNAc)	3.52–3.65
H6′, H6, H5, H4, H3	3.74–4.34

To determine the percentage of DD, the three protons of the acetyl group's peak (H-Ac) are taken into account [27].

$$DD\% = \left\{1 - \left(\frac{H1 - D}{H1 - D + H - Ac/3}\right)\right\} \times 100 \tag{4.6}$$

The proportion of DD is computed utilizing the signal resulting in protons H6′, H6, H5, H4, H3 (H2–6) of both the peak of acetyl group and the monomers (H-Ac).

$$DD\% = \left\{1 - \left(\frac{H - Ac/3}{H2 - 6/6}\right)\right\} \times 100 \tag{4.7}$$

For DD less than 90%, the percentage of DD is determined by utilizing the peaks of protons H1 of both acetylated monomer and deacetylated (*H1-A*, *H1-D*).

$$DD\% = \left(\frac{H1 - D}{H1 - D + H1 - A}\right) \times 100 \tag{4.8}$$

In their study evaluating several techniques to ascertain the DD, Czechowska et al. [25] found that NMR is the most dependable, quick, easy, and precise methodology when compared to other techniques. The price of the equipment and chemicals needed, the need for skilled people to utilize deuterated solvents, and the low solubility of chitosan make this process potentially unfavorable.

4.5.3 UV-Vis Spectroscopy

N-acetyl glucosamine absorbs in the UV spectrum between 199 and 203 nm and has linear concentration dependence that is unaffected by solvent presence; DD of chitosan may also be estimated using UV-vis spectroscopy. Utilizing N-acetyl glucosamine in 0.001 mol/l for HCl and altering the concentration (0.01–0.05 mg/mL), a standard solution is first made according to this procedure [27]. Using HCl as a reference, the absorbance of the standard solutions is calculated at 199 nm. Then, an accurately calculated amount of chitosan has been digested in an HCl solution, and absorbance will be determined at 199 nm.

The formula $DD = 100\% - C_1/C$, where C is the concentration of the sample, and C_1 is the concentration of acetyl in the sample. This can be used to figure out the concentration of the acetyl group in chitosan.

As illustrated in Fig. 4.2, Wu et al. [28] suggested an additional technique for calculating the proportion of DD. The proportion of DD is computed using Eq. (4.9) in this manner.

$$DD\% = 100 - \left(\frac{\frac{m_1}{203.21}}{\frac{m_1}{203.21} + \frac{m_1}{203.17}} \right) \times 100 \tag{4.9}$$

Where m_1 is the amount of acetyl glucosamine in 1 mL of chitosan/chitin solution, determined through the calibration curve by the associated H_2O_3 (UV value of the first derivative); m_2 is the amount of glucosamine in 1 mL of chitosan/chitin solution, determined as $m_2 = M - m_1$. The mass of chitin/chitosan (M) in the 1 mL solution was calculated as $M = (M_1 \times M_3)/(M_1 + M_2)$, where M_1 is a mass of solid chitin/chitosan sample obtained for measurement (100 ± 10 mg); M_2 is a mass of 20 ml for phosphoric acid solution (85%); and M_3 is mass of 1 mL chitosan/chitin solution in focused phosphoric acid. First derivative UV methods and UV spectrophotometric techniques both offer a quick, easy way to calculate the percentage of DD. However, when compared to other methodologies, these methods are less accurate.

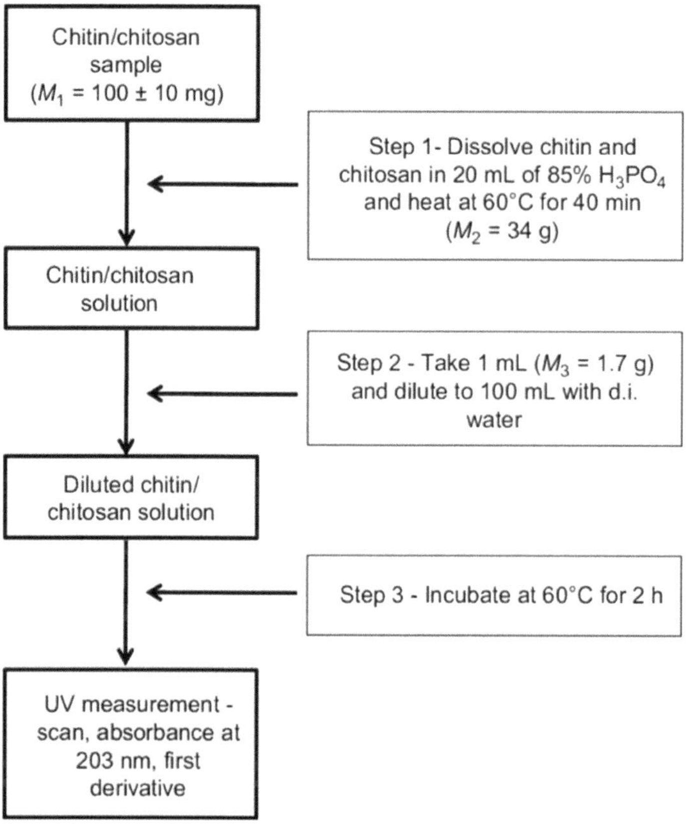

Figure 4.2 Use the first derivative of UV spectrophotometry to calculate the percentage of DD (reproduced with permission from Qiao et al. [28]).

4.5.4 Potentiometric Titration

Among non-NMR approaches, potentiometric titration is regarded as the technique that is most reliable and consistent

[25]. However, several drawbacks include the need for solutions with precisely known concentrations and the labor-intensive and time-consuming nature of measurements [22]. A known quantity of chitosan sample will dissolve in a known volume of standard HCl solution for potentiometric titration, and the result is compared to the same volume of standard NaOH solution. For each indicated transition, the inflection points of the pH against NaOH titration volume titration curve are obtained. The amine content of the sample is determined by employing Eq. (4.10) and measuring the quantity at each inflection point, NaOH was used.

$$NH_2\% = 16.1 \times (Y - X)/M \tag{4.10}$$

Where X(mL) is the first point of inflection on the pH measurement graph, the second inflection point at Y (mL) is the titration volume, and M(g) is the mass of chitosan utilized. The following calculation is then used to determine DD based on the hypothetical $NH_2\%$ content (9.94%).

$$DD\% = (NH_2\%/9.94\%) \times 100 \tag{4.11}$$

The following formula can be used to determine the percentage of DD when using a different titration technique.

$$DD\% = 100 - \left[2.03\left(\frac{(V_2 - V_1)}{m + 0.0042\,(V_2 - V_1)}\right)\right] \tag{4.12}$$

Where the sample weight, m (g), is given. The volumes of 0.1 M NaOH solution V_1, V_2 (mL) correspond to the deflection points. The free amine content is determined and then employed in the estimation of DD based on theoretically free amine in the preceding computation based on volumes of solution at points of inflection; it should be highlighted.

4.5.5 Raman Spectroscopy

According to Zajac et al. [29], Raman spectroscopy is a better option than FTIR for determining DD because IR can only be used to roughly estimate DD for chitosan. This method is quick and does not call for sample purification or solvent dissolution. The researchers conducted calculations and performed a comparative analysis of DD for six chitosan samples with different DD. This analysis was carried out using FTIR and Fourier transform–Raman spectroscopy techniques. A discernible alteration in the intensity was observed in several bands upon comparing the FT-Raman spectra of the six samples with the corresponding variation in DD percentage. The spectral peaks observed at wavenumbers 1591, 1654, and 936 cm^{-1} correspond to the partial acetylation of NH_2 groups within the chitosan polymer. These peaks can be attributed to the vibrational modes δ (NH_2), ν (CN), and ν (C=O), as illustrated in Fig. 4.3. The following equations illustrate the correlation between the DD% of the chitosan samples and the integral intensities:

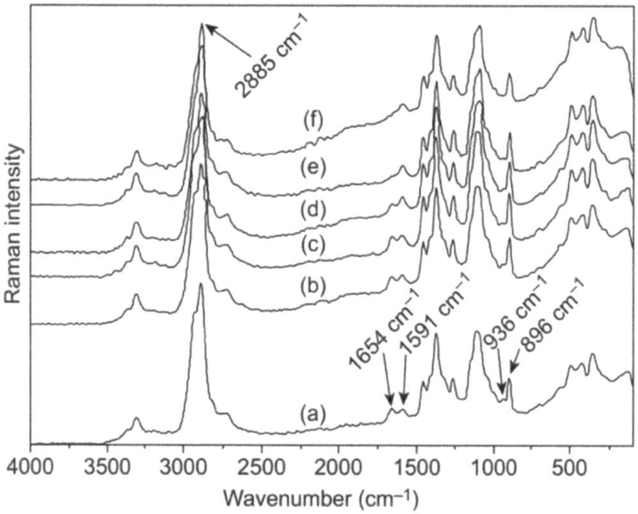

Figure 4.3 Chitosan's Raman spectra with DD values of 70, 75, 80, 85, 90, and 95% (reproduced with permission from [29]).

$$A_{1654} = -0.021DDA + 2.68\ (R^2 = 0.9468) \text{ for the band at } 1654\ \text{cm}^{-1} \tag{4.13}$$

$$A_{1591} = 0.016DDA + 0.92\ (R^2 = 0.9787) \text{ for the band at } 1591\ \text{cm}^{-1} \tag{4.14}$$

A_{936} = −0.075DDA + 0.71 (R^2 = 0.9848) for the band at 936 cm^{-1} (4.15)

4.5.5.1 Elemental analysis

The determination of DD of chitosan can be achieved by utilizing an elemental analyzer to perform an elemental analysis of nitrogen and carbon. Chitosan, which has been totally deacetylated, has a carbon-to-nitrogen ratio (C/N) of 5.145; whereas chitin, a polymer that has been completely N-acetylated, has a C/N of 6.861. In this method, Eq. (4.16) may be used to determine DD based on the C/N ratio [30].

$$DD\% = \left(1 - \frac{C/N - 5.145}{6.861 - 5.145}\right) \times 100 \tag{4.16}$$

In comparison to FTIR analysis, it was discovered that this procedure is quicker. To determine the proportion of DD [24], employed the below formulas to derive the weight percentage of nitrogen (N) or carbon (C) based on elemental analysis:

$$DD\% = \frac{9600}{(364 \times W_C + 2400)} \times 100 \tag{4.17}$$

$$DD\% = \frac{1400}{(364 \times W_N)} \times 100, \tag{4.18}$$

where the weight percentages of nitrogen and carbon in the chitosan samples are denoted as W_C and W_N, respectively.

4.5.5.2 Method of capillary zone electrophoresis

A technique for measuring DD using capillary zone electrophoresis was devised by Wu et al. [31] (Fig. 4.4). By applying optimum capillary zone electrophoresis conditions, the electrophoretic mobilities of the chitosan samples are assessed in this manner. For the capillary zone electrophoresis method, a calibration curve is made with the help of samples with ^{1}H-NMR values for DD that run from 55.3 to 96.2%. The calibration curve was employed to determine the percentage of DD of the chitosan peaks observed in the electropherograms. The results gathered indicated extremely high agreement with the samples' ^{1}H-NMR values. Capillary zone electrophoresis's great resolving power enables it to discern the amount of DD in mixed chitosan samples, which ^{1}H-NMR was unable to achieve. For the identification of compounds with a wide range of DD inside a specific chitosan polymer sample, this technique is, therefore, highly beneficial.

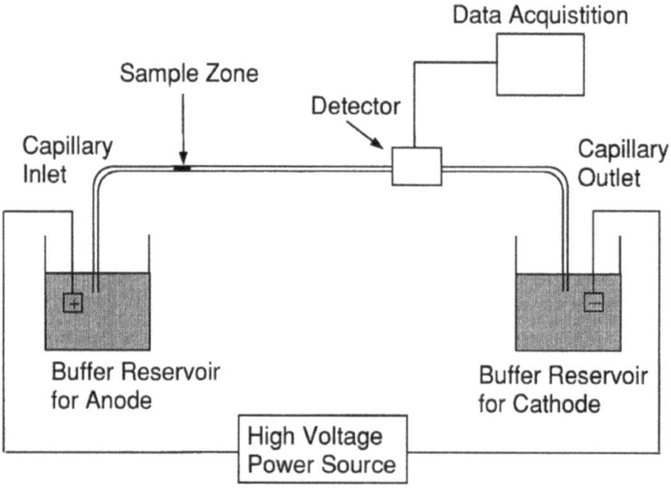

Figure 4.4 Capillary electrophoresis setup schematic diagram.

4.5.5.3 Base-acid titration

Results from acid-base titration are equivalent to those from FTIR and ^{1}H-NMR. Due to its simplicity, reproducibility, affordability, and portability, acid-base titration is preferable to potentiometric and spectroscopic approaches [32].

4.6 Degradation of the Polymer or Stability

(1) A small quantity of methyl orange indicator is put into a solution consisting of 30 mL of 0.1 M hydrochloric acid (HCl) at 25 °C, along with 0.3–0.5 g of chitosan to facilitate its dissolution.

(2) The chitosan solution is titrated using 0.1 M NaOH solution. The color transitions from pink to orange-yellow at the titration point.

(3) Chitosan undergoes (W_1) a heating process at a temperature of 105 °C until it reaches a stable weight (W_2) for the purpose of determining its moisture content. Chitosan's moisture content percentage is computed using W_2 and W_1 using the formula $[(W_1 W_2)/W_1]100$.

The formula below can be used to determine the fraction of the free NH_2 group:

$$NH_2\% = [(C_1V_1 - C_2V_2) \times 0.016]/[G\,(100 - W)] \times 100 \qquad [4.19]$$

where

C_1 is the concentration of hydrochloric acid (M);
C_2 is NaOH concentration (M);
V_1 is the quantity of added HCl (mL);
V_2 is NaOH volume (mL) applied during titration;
G is sample weight (g);
W is moisture content (%) in a sample; and chitosan's theoretical NH_2 concentration (in grams) in 1 mL of 1M HCl is 0.016.

Stability is another important property of chitosan biopolymers. How stable chitosan is under specific conditions depends on how much of its polymeric chain degrades. Chitosan typically undergoes degradation by the cleavage of its polymeric chain during acid hydrolysis under conditions of low pH. In order to depolymerize and then deacetylate, the cleavage process separates 1,4-glycosidic connections and N-acetylglucosamine linkages, respectively. The process of depolymerization results in the generation of free radical species within the solution, which subsequently initiate oxidation reactions.

In contrast, increased deacetylation leads to the formation of molecules with less weight. Following their rupture, it has been shown that they exhibit significant mutual interaction, leading to the formation of interchain cross-links. This phenomenon can induce alterations in their overall structure and attributes that are dependent on their structure. The primary factors influencing the degradation of chitosan encompass its molecular weight, temperature, moisture content, polydispersity, and level of purity. Additionally, chitosan may be broken down in vivo by a number of enzymes, including the protease lysozyme found in human cells, to create safe oligosaccharides that may be added to glycoproteins and glycosaminoglycans. The regulated in vitro degradation of chitosan may be achieved by several chemical processes, such as oxidation and enzymatic hydrolysis, which leads to the production of chitosan with reduced Mw [33].

4.7 Thermal Properties

The chitosan solution can undergo changes in viscosity, solubility, structural characteristics, and other physicochemical qualities when exposed to high temperatures. The application of heat to a polymeric solution frequently leads to deterioration and a decrease in stability. Chitosan has a number of noteworthy traits when it comes to its thermal properties, including:

Thermal stability: Chitosan has relatively good thermal stability, with its decomposition temperature typically ranging between 200 °C and 300 °C. The exact decomposition temperature may vary depending on factors like the degree of deacetylation and molecular weight [34].

Temperature of glass transition (Tg): Chitosan exhibits a glass transition temperature, denoting the critical temperature at which it undergoes a shift from a hard and brittle state to a more pliable and elastomeric one. The Tg of chitosan is subject to many influences, including its DD, Mw, and the presence of plasticizers. Chitosan often demonstrates a Tg within the temperature interval of 50–100 °C [35, 36].

Melting point: Chitosan does not have a well-defined melting point like some crystalline materials, but it undergoes gradual degradation and decomposition upon heating. However, it is important to note that chitosan can be blended or modified with other substances to alter its thermal behavior and create materials with specific melting or degradation temperatures.

Thermal conductivity: Chitosan has a comparatively low thermal conductivity, rendering it well-suited for scenarios that necessitate thermal insulation. This property can be advantageous in areas such as packaging materials or wound dressings, where thermal insulation and moisture management are important [37].

Thermal expansion: Chitosan typically exhibits a positive coefficient of thermal expansion, meaning it expands when heated and contracts when cooled. The magnitude of thermal expansion depends on the specific chitosan sample and its composition.

It is noteworthy to emphasize that the thermal characteristics of chitosan may be altered by several parameters, including its DD, Mw, processing techniques, and the incorporation of other chemicals. These modifications can be utilized to tailor chitosan for specific applications requiring different thermal characteristics.

4.8 Swelling Capacity and Hygroscopicity Chitosan and Chitin

Both chitin and chitosan are known to be hygroscopic substances, which means they have a predilection for moisture naturally and may take in water from their surroundings [38]. The presence of hydroxyl (-OH) and amino ($-NH_2$) groups, which have the ability to establish hydrogen bonds with water molecules, is accountable for this characteristic. Chitin, a polymer found in the cellular structures of fungi and the exoskeletons of crustaceans, serves as the precursor of chitosan. It has a little hygroscopicity and can take in moisture from the surroundings. Chitin has a far higher level of acetylation than chitosan, which limits its capacity to swell and absorb water. On the other side, chitosan is created when chitin is converted to deacetylated chitosan. Deacetylation causes the acetyl groups in chitin to be removed, which results in the production of chitosan. Because there are more free amino groups in chitosan than in chitin, it is more hydrophilic. Chitosan has much better hygroscopicity and swelling ability than chitin, thanks to its improved hydrophilicity. Chitosan's capacity for swelling is especially notable and frequently used in a variety of applications [39]. Chitosan has rapid water absorption and retention capabilities upon exposure to water or aqueous solutions, resulting in the formation of a gelatinous matrix. Hydrogen connections are formed between the amino groups of chitosan and water molecules due to this characteristic. Several factors influence the swelling behavior of a material, including but not limited to DD, Mw, and the inclusion of additional chemicals or crosslinking agents. Chitosan and chitin are useful for applications such as moisture-absorbing films, wound dressings, medication delivery systems, and food packaging, where moisture control and water absorption are sought due to their hygroscopic and swelling qualities [40]. Moreover, they exhibit utility in water treatment protocols, particularly in the adsorption of heavy metals or dyes from aqueous solutions, owing to their affinity for water molecules.

Once a stable value for the weighted sample has been acquired for a certain time period, usually at a temperature of 37 °C, the swelling ratio may be estimated using the subsequent procedure.

$$[SR] = \frac{w_f - w_i}{w_i} \tag{4.20}$$

Where w_i denotes the dosage form's beginning weight, *SR* denotes the swelling ratio, and w_f denotes the dosage form's ultimate weight before and after swelling, respectively.

4.9 Conclusions and Future Prospects

The major physical characteristics of chitin and chitosan are discussed in this chapter. The most significant aspect is that we can create biomaterials with good performance from these polysaccharides because of their excellent film-forming properties. Additionally, they can be employed in the fabrication of composites for fibers or gels, wherein their antibacterial, biodegradable, biocompatible, or hydrophilic properties are investigated. Chitin, a notable natural polysaccharide, exhibits the ability to undergo specific reactions at the NH_2 site located at the C2 position following deacetylation. After undergoing deacetylation, it assumes the role of the only pseudo-natural cationic polysaccharide at a pH of 6.

References

1. Beier S, Bertilsson S (2013) Bacterial chitin degradation—mechanisms and ecophysiological strategies. *Frontiers in Microbiology* 4: 149.

2. Amor IB, Hemmami H, Laouini SE, Abdelaziz AG, Barhoum A (2023) Influence of chitosan source and degree of deacetylation on antibacterial activity and adsorption of AZO dye from water. *Biomass Conversion and Biorefinery*: 1–11.
3. Amor IB, Hemmami H, Laouini SE, Temam HB, Zaoui H, Barhoum A (2023) Biosynthesis MgO and ZnO nanoparticles using chitosan extracted from Pimelia Payraudi Latreille for antibacterial applications. *World Journal of Microbiology and Biotechnology* 39: 19.
4. Silvestre J, Delattre C, Michaud P, de Baynast H (2021) Optimization of Chitosan properties with the aim of a water resistant adhesive development. *Polymers* 13: 4031.
5. Sinha S, Chand S, Tripathi P (2016) Recent progress in chitosanase production of monomer-free chitooligosaccharides: bioprocess strategies and future applications. *Applied Biochemistry and Biotechnology* 180: 883–899.
6. Kaur S, Dhillon GS (2015) Recent trends in biological extraction of chitin from marine shell wastes: a review. *Critical Reviews in Biotechnology* 35: 44–61.
7. Francisco FC, Simora RMC, Nuñal SN (2015) Deproteination and demineralization of shrimp waste using lactic acid bacteria for the production of crude chitin and chitosan. *Aquaculture, Aquarium, Conservation & Legislation* 8: 107–115.
8. Katiyar D, Hemantaranjan A, Singh B, Bhanu AN (2014) A future perspective in crop protection: Chitosan and its oligosaccharides. *Advances in Plants & Agriculture Research* 1: 1–8.
9. Zhao D, Yu S, Sun B, Gao S, Guo S, Zhao K (2018) Biomedical applications of chitosan and its derivative nanoparticles. *Polymers* 10: 462.
10. Younes I, Rinaudo M (2015) Chitin and chitosan preparation from marine sources. Structure, properties and applications. *Marine Drugs* 13: 1133–1174.
11. Abdel-Rahman RM, Hrdina R, Abdel-Mohsen A, Fouda MM, Soliman A, Mohamed F, Mohsin K, Pinto TD (2015) Chitin and chitosan from Brazilian Atlantic Coast: Isolation, characterization and antibacterial activity. *International Journal of Biological Macromolecules* 80: 107–120.
12. Dash M, Chiellini F, Ottenbrite RM, Chiellini E (2011) Chitosan—A versatile semi-synthetic polymer in biomedical applications. *Progress in Polymer Science* 36: 981–1014.

13. Hamed I, Özogul F, Regenstein JM (2016) Industrial applications of crustacean by-products (chitin, chitosan, and chitooligosaccharides): A review. *Trends in Food Science & Technology* 48: 40–50.

14. Hoffmann K, Daum G, Köster M, Kulicke W-M, Meyer-Rammes H, Bisping B, Meinhardt F (2010) Genetic improvement of Bacillus licheniformis strains for efficient deproteinization of shrimp shells and production of high-molecular-mass chitin and chitosan. *Applied and Environmental Microbiology* 76: 8211–8221.

15. Al-Jbour ND, Beg MD, Gimbun J, Alam AM (2019) An overview of chitosan nanofibers and their applications in the drug delivery process. *Current Drug Delivery* 16: 272–294.

16. Ladchumananandasivam R, da Rocha BG, Belarmino DD, Galv AO (2012) The use of exoskeletons of shrimp (Litopenaeus vanammei) and crab (Ucides cordatus) for the extraction of chitosan and production of nanomembrane. *Materials Sciences and Applications* 3: 495–508.

17. Collado-González M, Montalbán MG, Peña-García J, Pérez-Sánchez H, Víllora G, Baños FGD (2017) Chitosan as stabilizing agent for negatively charged nanoparticles. *Carbohydrate Polymers* 161: 63–70.

18. Kaya M, Baran T, Erdoðan S, Menteþ A, Özüsaðlam MA, Çakmak YS (2014) Physicochemical comparison of chitin and chitosan obtained from larvae and adult Colorado potato beetle (Leptinotarsa decemlineata). *Materials Science and Engineering: C* 45: 72–81.

19. Kaya M, Baran T, Karaarslan M (2015) A new method for fast chitin extraction from shells of crab, crayfish and shrimp. *Natural Product Research* 29: 1477–1480.

20. Kumari S, Kishor R (2020) Chitin and chitosan: origin, properties, and applications. *Handbook of Chitin and Chitosan*. Elsevier, pp. 1–33.

21. Cho J, Heuzey M-C, Bégin A, Carreau PJ (2006) Viscoelastic properties of chitosan solutions: Effect of concentration and ionic strength. *Journal of Food Engineering* 74: 500–515.

22. Dimzon IKD, Knepper TP (2015) Degree of deacetylation of chitosan by infrared spectroscopy and partial least squares. *International Journal of Biological Macromolecules* 72: 939–945.

23. Wijesena RN, Tissera N, Kannangara YY, Lin Y, Amaratunga GA, de Silva KN (2015) A method for top down preparation of chitosan nanoparticles and nanofibers. *Carbohydrate Polymers* 117: 731–738.

24. Hussain MR, Iman M, Maji TK (2013) Determination of degree of deacetylation of chitosan and their effect on the release behavior

of essential oil from chitosan and chitosan-gelatin complex microcapsules. *International Journal of Advanced Engineering Applications* 6: 4–12.

25. Czechowska-Biskup R, Jarosińska D, Rokita B, Ulański P, Rosiak JM (2012) Determination of degree of deacetylation of chitosan-comparision of methods. *Progress on Chemistry and Application of Chitin and its Derivatives*: 5–20.
26. Kasaai MR (2010) Determination of the degree of N-acetylation for chitin and chitosan by various NMR spectroscopy techniques: A review. *Carbohydrate Polymers* 79: 801–810.
27. Sivashankari P, Prabaharan M (2017) Deacetylation modification techniques of chitin and chitosan. *Chitosan Based Biomaterials* Volume 1. Elsevier, pp. 117–133.
28. Wu T, Zivanovic S (2008) Determination of the degree of acetylation (DA) of chitin and chitosan by an improved first derivative UV method. *Carbohydrate Polymers* 73: 248–253.
29. Zając A, Hanuza J, Wandas M, Dymińska L (2015) Determination of N-acetylation degree in chitosan using Raman spectroscopy. *Spectrochimica Acta Part A: Molecular and Biomolecular Spectroscopy* 134: 114–120.
30. Abdulkarim A, Isa MT, Abdulsalam S, Muhammad AJ, Ameh AO (2013) Extraction and characterisation of chitin and chitosan from mussel shell. *Extraction* 3: 108–114.
31. Wu C, Kao CY, Tseng S-Y, Chen KC, Chen S-F (2014) Determination of the degree of deacetylation of chitosan by capillary zone electrophoresis. *Carbohydrate Polymers* 111: 236–244.
32. Yuan Y, Chesnutt BM, Haggard WO, Bumgardner JD (2011) Deacetylation of chitosan: Material characterization and in vitro evaluation via albumin adsorption and pre-osteoblastic cell cultures. *Materials* 4: 1399–1416.
33. Benchamas G, Huang G, Huang S, Huang H (2021) Preparation and biological activities of chitosan oligosaccharides. *Trends in Food Science & Technology* 107: 38–44.
34. Yodkhum K, Phaechamud T (2012) Thermal analysis of chitosan-lactate and chitosan-aluminum monostearate composite system. *Advanced Materials Research* 506: 278–281.
35. Dong Y, Ruan Y, Wang H, Zhao Y, Bi D (2004) Studies on glass transition temperature of chitosan with four techniques. *Journal of Applied Polymer Science* 93: 1553–1558.

36. Qiao C, Ma X, Zhang J, Yao J (2019) Effect of hydration on water state, glass transition dynamics and crystalline structure in chitosan films. *Carbohydrate Polymers* 206: 602–608.

37. Ye H, Zheng G, Zuo S, Yu Q, Xia C, Sheng Y, Shi Y, Wang D, Li J, Ge S (2023) Lightweight, bacteriostatic and thermally conductive wood plastic composite prepared by chitosan modified biointerfaces. *Applied Surface Science* 615: 156313.

38. Klein M, Poverenov E (2020) Natural biopolymer-based hydrogels for use in food and agriculture. *Journal of the Science of Food and Agriculture* 100: 2337–2347.

39. Ahmed S, Ikram S (2016) Chitosan based scaffolds and their applications in wound healing. *Achievements in the Life Sciences* 10: 27–37.

40. Nesic AR, Seslija SI (2017) The influence of nanofillers on physical–chemical properties of polysaccharide-based film intended for food packaging. *Food Packaging*. Elsevier, pp. 637–697.

Chapter 5

Chemical Properties of Chitin and Chitosan

Olfa Ouled Ltaief, Wiem Hamza, Mabrouk Eloussaief, and Mourad Benzina

Water, Energy, and Environment Laboratory, National School of Engineers of Sfax, University of Sfax, Sfax, Tunisia

olfa.ol@live.fr

A common aminated polysaccharide found in nature is chitosan. Chitosan has a special set of functional features because of its chemical makeup. The primary properties of chitosan encompass its molecular weight and level of acetylation. These factors have a substantial impact on the solubility, material creation capacity, biodegradability, and variety of bioactive qualities of chitosan, along with other functional characteristics. Chitosan has many different applications, and its immense potential has long been understood. Chitosan is very well suited for purposes in agricultural, culinary, and environmental engineering because of its biologically linked qualities. The link between chitosan's chemical properties and its primary functional qualities is examined in this chapter.

Chitin and Chitosan: Physical and Chemical Properties
Edited by Shakeel Ahmed, Soumeia Zeghoud, Hadia Hemmami, and Ilham Ben Amor

ISBN 978-981-5129-39-7 (Hardcover), 978-1-003-58977-8 (eBook)
www.jennystanford.com

5.1 Introduction

The majority of the time, discarded crab, prawns, and lobster shells are used to extract chitin from seafood waste. It is renewable and has a variety of sources. It is the most prevalent polysaccharide in nature [1, 2].

One of the most significant chemical raw ingredients is sugar. The result of deacetylating chitin is chitosan [3]. It is simple to convert chitin and chitosan into a variety of materials, including membranes [4], gels [5], microspheres [6], and nanofibers [7]. The aforementioned materials possess several uses within the biological fields, including but not limited to medication delivery, tissue engineering, and wound healing [8].

In other words, the chitin crystalline regions are regular, straight, and tightly organized in three-dimensional space. The natural chitin structure is created by the staggered arrangement of ordered crystalline areas and disordered amorphous regions. Loose molecular chains that coil randomly and interact with one another make up the amorphous area.

Twisted, chitin forms a highly organized and crystalline microfibril network structure as a result of its intramolecular and intermolecular hydrogen bond interactions, which makes it challenging to deacetylate chitosan [9]. Chitin undergoes deacetylation under elevated temperatures in a concentrated sodium hydroxide (NaOH) solution, resulting in the formation of chitosan [1, 10]. However, the acetamide group is less accessible due to the semi-crystalline structure of chitin, meaning that it is more difficult for –OH to access the acetamide group, which results in poor reaction efficiency and partial deacetylation.

Chitin may now be dissolved by a relatively restricted number of solvents, including LiCl/DMAc, $CaCl_2/CH_3OH$, ionic liquid, hexafluoroisopropanol, and alkali/urea [11]. Alkali/urea is a brand-new kind of environmentally friendly solvent created recently. The tiny molecules of NaOH and urea in the solvent first come together to form hydrates at low temperatures. These hydrates then break the initial intermolecular force between the macromolecular chains of chitin and create new hydrogen bonds with them. The molecular chains of chitin are finally coated by sodium hydroxide/urea hydrate to create a tubular clathrate, which causes chitin to dissolve in the solvent [12, 13]. In light of the preceding discussion, this chapter will enumerate the foremost chemical attributes of chitin and chitosan.

5.2 Solubility

5.2.1 Chitin's Solubility

The limited availability of data about the physical characteristics of chitin in solution might be ascribed to the inherent difficulties involved with its dissolution. One of the often-employed systems is as follows: The complexation of chitin with LiCl, at a concentration of 5% weight/weight, leads to the coordination of lithium with the carbonyl group of acetylated units. Previous research has indicated that this compound has a solubility in dimethylacetamide and N-methyl-2-pyrrolidone [12–16]. The solvents that were tested in this study included a saturated solution of formic acid [17], lithium thiocyanate [17], a combination of formic acid with dichloroacetic and trichloroacetic acids [12], phosphoric acid [18], and several alkali solvents [19–22]. Chitin exhibits solubility in aqueous NaOH solutions, specifically at concentrations of 16% w/w [23] or a mass fraction of 0.1 [19, 21]. This solubility may be achieved at low temperatures, below 0 °C, with the presence of broken ice. Consequently, this solubility property enables the formation of films that can be regenerated in water at 25 °C or in methanol. The events seen under these circumstances encompass deacetylation, depolymerization, and destructuration of the crystalline substance, which are dependent on the length of interaction between chitin and alkali [21, 22]. In comparison to chitosan, possessing a lower degree of acetylation (DA) and chitin, the resulting chitin exhibits an amorphous structure and, under certain conditions, has the ability to be solubilized in water. According to scientific researchers, this phenomenon is attributed to the occurrence of deacetylation as well as a reduction in molecular weight (Mw) in alkaline conditions. It has been established that achieving a DD value near 0.5 is essential for optimizing water solubility.

Furthermore, in order to impede the packing of chains and the disruption of secondary structure in extremely alkaline settings, it is likely that a uniform distribution of acetyl groups along the chain is necessary [22, 24]. The level of chitin deacetylation was recorded as 0.55 following 77 hours of alkaline treatment, and it subsequently increased to 0.88 after 580 hours [22]. Following an extended period of interaction, the authors assert that they have successfully synthesized a water-soluble regenerated chitin, essentially chitosan, with a degree of acetylation (DA) of 0.5. A recent study conducted an investigation that integrated rheology, turbidimetry, and fluorescence techniques to demonstrate that alkali chitin functions as a polymeric system with a lower critical solution temperature (LCST). The critical temperature of this system was found to be about 30 °C [20]. Following the removal of NaOH by washing, a chitin gel was generated, and further analysis demonstrated its sensitivity to changes in temperature and pH [23].

5.2.1.1 NaOH/urea aqueous solution

In their study, Zhang et al. [14] employed a novel environmentally friendly solvent, NaOH/urea aqueous solution, to effectively dissolve cellulose at reduced temperatures. In their study, Hu et al. [15] utilized a NaOH/urea aqueous solution as a solvent to dissolve chitin. This choice was motivated by the better performance of this solution in dissolving biomacromolecules, as previously demonstrated [15]. They discovered that a concentration of 8 weight percent (wt%) NaOH and a temperature of 20 wt% urea were both satisfactory. Furthermore, there was a prevailing belief that urea exerted a substantial influence on the stability of chitin in an aqueous solution. In contrast, the aqueous solution containing chitin exhibited temperature sensitivity, resulting in gelation as the temperature increased.

In their study, Hu et al. [15] utilized a NaOH/urea aqueous solution to dissolve several forms of natural chitin produced from crabs, prawns, fly shells, and silkworm chrysalis. Subsequently, they utilized laser light scattering and viscometry techniques to investigate the behavior of the chitin in diluted solutions [16]. The results indicate that chitin molecules in the NaOH/urea aqueous solution exhibit an unexpected chain shape, with increasing chain flexibility seen from C1 to C4.

Zhang et al. [17] created the freezing/thawing method to successfully produce a transparent chitin solution without the need for derivatization. This was achieved by completely dissolving chitin in an aqueous solution containing 8 wt% NaOH/4 wt% urea. The chitin solution was employed in the direct fabrication of a hydrogel, enabling the preservation of the unique structure, and the characteristics of chitin are being examined for the first time.

Cai et al. [18] conducted pretreatment of the raw chitin powder using a 0.3% $NaClO_2$ solution and a 1 M NaOH solution. The requisite amount of the refined chitin powder was subsequently mixed with a solution comprising NaOH, urea, and water in a weight ratio of 11:4:85 to produce a suspension with 38 wt%. The suspension was next subjected to freezing at 20 °C for the duration of one night. It was then thawed at a temperature of 5 °C while being agitated, resulting in the formation of a chitin solution that exhibited clarity and viscosity [19]. The results indicated that the amounts of urea and NaOH, together with the dissolving temperature, were the key factors that significantly influenced the enhancement of solubility. The chitin solution with the highest quality was produced using an aqueous solution containing 11 wt% NaOH/4 wt% urea.

5.2.1.2 $CaCl_2$/solvent

The dissolution of chitin in a methanol solution containing calcium chloride dihydrate resulted in a very effective dissolution process, leading to the formation of a chitin solution with a high viscosity. The use of a mild solvent solution effectively inhibited the degradation of chitin. The dissolution of chitin in $CaCl_2$ dihydrate-saturated methanol was shown to be influenced significantly by the concentration of water and calcium ions, as observed by Tokura et al. [20]. Furthermore, the solubility of chitin was influenced by its molecular weight (Mw) and degree of N-acetylation. The concept posited that calcium ions interacted with the amide bond inside the chitin side chains, leading to the disruption of the rigid crystalline structure during the dissolution process.

Chitin powder and nanofiber were dissolved in methanol saturated with calcium chloride and dehydrated under favorable circumstances in order to generate chitin hydrogels. The investigation revealed the application of the therapy-induced nanofibrillation of the powder.

X-ray powder diffraction (XRD) analyses were conducted on chitin samples derived from squid pens, which had undergone treatment with a calcium chloride dihydrate/methanol solvent solution under varying conditions. The purpose of these studies was to evaluate the crystalline structure of chitin. The findings were reported in a previous study [21]. The experimental results indicate that the $CaCl_2$/methanol solvent had a particular impact on the (010) plane of the chitin structure.

Kadokawa et al. [22] devised a unique solvent system employing a calcium chloride dihydrate/methanol solvent. The researchers in their work employed the substitution of calcium chloride dihydrate with calcium bromide dihydrate as a means to successfully dissolve chitin for the purpose of gel formation [19]. The research utilized a regeneration methodology to transform the resultant gel into a chitin substance characterized by a porous architecture. The results further showed that the porosities and mechanical characteristics of the porous chitin were significantly influenced by the quantities of calcium bromide dihydrate inside the gel.

5.2.1.3 N,N-dimethylacetamide/lithium chloride solvent

In 1976, for the first time, researchers utilized lithium chloride (LiC) and N,N-dimethylacetamide (DMAc) as a solvent to effectively dissolve chitin under mild conditions [23]. The amount of chitin in the solvent was affected by the Mw, and solutions containing up to 15% chitin were produced.

A 13 wt% chitin concentration was achieved using N,N-dimethylacetamide, which also included 5 wt% lithium chloride, in order to ascertain the mechanism by which chitin dissolves in the solvent [24]. The ^1H NMR analysis revealed that the dissolution of chitin in DMA/LiCl may be attributed to the significant contact between LiCl molecules and the acetamide groups, as well as the intermolecular hydrogen bonding involving hydroxyl groups. By binding one LiCl molecule to each labile proton, this interaction perturbed the intermolecular hydrogen bonding, leading to the expansion of chitin and its subsequent dissolution in the solvent.

Furthermore, it demonstrated good efficacy in the dissolving of synthetic polyamide, protein, and cellulose, as shown in reference [25]. The use of the solvent system was often utilized in the examination of the polymer product using a range of methods, including gel permeation chromatography (GPC), Fourier-transform infrared spectroscopy (FT-IR), and nuclear magnetic resonance (NMR), owing to its favorable characteristics. As a result, it garnered a substantial reputation as a solvent in the field of polymer analysis. In the context of the acetylation of chitin, the substance under consideration has been recognized as a viable reaction medium with significant potential for application.

5.2.1.4 Ionic liquid

Ionic liquids (ILs) are new and promising media that provide a variety of positive benefits, including nearly zero vapor pressure, great thermal stability, a large electrochemical window, low flammability, and programmable characteristics. Due to these particular attributes, these substances have been identified as eco-friendly benign catalysts and solvents and have been extensively utilized across diverse domains such as electrochemistry, biomass, separations, and gas absorption [26, 27].

Zhang et al. [28] conducted a study on the ionic liquid 1-butyl-3-methyl-imidazolium chloride ([Bmim][Cl]), wherein they successfully produced a 10% chitin/IL solution with a clear and viscous consistency within a 5-hour timeframe. This was achieved by employing an oil bath at a temperature of 110 °C. The ionic liquid demonstrated remarkable efficacy in dissolving chitin and chitosan, as previously reported.

In the study conducted by Kadokawa et al. [29], the dissolution of chitin was achieved by the utilization of 1-allyl-3-methyl-imidazolium bromide ([Amim][Br]). Following a 48-hour heating period at 100 °C, a transparent solution consisting of 7% (w/w) and 5% (w/w) chitin with the ionic liquid (IL) was successfully obtained. It has been proven that the Mw of chitin typically remains stable or does not undergo degradation or reduction during the process of producing a transparent liquid. The researchers conducted an investigation into the chemical modification of chitin in an ionic liquid (IL) by dissolving chitin in a 2% (w/w) solution of [Amim][Br] for a duration of 24 hours at 100 °C. This experiment demonstrated that [Amim][Br] exhibited high efficacy as a solvent for the modification of chitin [30]. Chitin and [Amim][Br] were subjected to a reaction at a temperature of 100 °C for a duration of 24 hours, resulting in the formation of a chitin solution with a weight percentage of 10%. This chitin solution was subsequently employed in the production of a hybrid gel consisting of cellulose chitin. To utilize chitin from different sources and different molecular weights the ionic biomass fluid 1-butyl-3-methylimidazolium acetate ([Bmim][OAc]) is used [31]. It was shown that the solubility of chitin in [Bmim][OAc] is dependent on the sources and molecular weights of the chitin. Specifically, it was observed that chitin may dissolve in [Bmim][OAc] at elevated temperatures above 85 °C. Additionally, at appropriate concentrations, transparent viscous solutions of chitin could be generated at a temperature of 110 °C. The results indicated that there was a strong correlation between the dissolution of chitin in [Bmim][OAc] and its molecular weight when the impact of the crystal morphologies of the chitin samples was not taken into consideration. The ionic liquid known as 1-butyl-3-methylimidazolium acetate ([Bmim] [OAc]) exhibited the capacity to dissolve 6 wt% of the chitin sample characterized by a lower molecular weight (-chitin-L). However, it was found that only 3 wt% of the chitin sample with a higher molecular weight (-chitin-H) could be dissolved.

Simone et al. [32] achieved the highest chitin concentration by preparing a chitin solution using [Bmim] [OAc] at temperatures ranging from 90 °C to 95 °C for a duration of 5 hours with continuous stirring. They discovered that as the polymer content was raised, the chitin/IL solution's viscosity and dissolving time also rose.

5.2.2 Chitosan's Solubility in a Solution

Understanding how chitosan reacts in various solvents is essential for optimal chitosan utilization. Solvents like inorganic and organic acids, water, and bases are essential in the biomedical sector [33]. Chitin may be readily dissolved by a diverse range of inorganic acids, bases, and salts. These substances function as solvents, effectively dissolving both chitin and chitosan. Chitosan was discovered to be soluble in 4 wt% urea and 8 wt% NaOH without any previous preparation. Chitosan has a high solubility in diluted organic acidic solutions when the pH (5 to 6.5) is below the pKa. The rationale behind this phenomenon is attributed to the presence of amino groups within the chitosan structure, which facilitates ionization in acidic conditions. This ionization leads to the generation of two NH_3^+ ions, hence enhancing the solubility of chitosan. Chitosan exhibits solubility in acidic solutions characterized by a pKa value below 6.2, due to the full protonation of its amino groups, which possess pKa values ranging from 6.2 to 7.0, in such acidic environments. Nevertheless, the polymer undergoes precipitation at elevated pH values. Both the protonation of the chitosan molecule by the medium and the solubility of the polyelectrolytes are essential factors for achieving the solubility of chitosan in organic acids. Chitosan exhibits solubility with several acids, such as acetic acid, lactic acid, formic acid, succinic acid, and L-glutamic acid.

During the depolymerization process of chitosan in the presence of organic acids, the glycosidic link experiences premature hydrolysis, which is then followed by the subsequent deacetylation of the N-acetyl groups [34]. The solubility of chitosan is affected by the process of protonation that occurs when chitosan is exposed to hydrochloric acid and acetic acid. The occurrence of this phenomenon is contingent upon the pKa and pH values of the acid [35]. The prevalent approach for assessing the solubility of chitosan involves its dissolution in acetic acid solutions with concentrations of 1% or 0.1 M. The quantity of acid required for the dissolution of chitosan is contingent upon the quantity of chitosan that necessitates dissolution. The quantity of NH_2 units should correspond to the intended proton concentration.

5.3 Mucoadhesion

The composition of reactive groups and the chemical makeup of polymers have an impact on the mucoadhesive property, which is demonstrated by a number of natural and synthesized polymers. A number of reactive groups, including $-NH_2$, $-SO_3H$, $-COOH$, and $-OH$, are present in mucoadhesive polymers. The phenomenon of mucoadhesion occurs when these functional groups establish non-covalent interactions with mucin molecules, resulting in their adherence to the mucosal surface.

Chitosan's mucoadhesion is based on the fact that it is cationic. The mucus gel layer's anionic functions include sialic acid and sulfonic acid. The occurrence of mucoadhesion may be observed through the electrostatic attraction between the positively charged chitosan groups and the negatively charged constituents found in the mucus gel layer [36]. However, at greater cross-linking levels, mucoadhesion of chitosan drops while its degree of deacetylation (DD) and Mw both rise. The inherent property of chitosan facilitates its adherence to the upper respiratory tract and gastrointestinal system, hence exhibiting the drug's sustained release [37]. A significant cationic charge is produced by trimethylating chitosan. The high mucoadhesive characteristics of trimethyl chitosan can be attributed to its cationic charge. The solubility of trimethyl chitosan is seen at basic and neutral pH conditions, which can be attributed to the alkylation of the amino group with methyl groups. Amino carboxymethyl chitosan demonstrates mucoadhesive properties at both acidic and basic pH levels as a result of the presence of carboxyl groups [37]. Chitosan's mucoadhesive properties are improved by a strong contact between its thiol groups and the cysteine portion of mucus glycoprotein [38].

5.4 Forming Properties of Films

One of chitosan's most significant features is its capacity to create films. It is feasible to create membranes or solid/gel films with intriguing chemical, mechanical, and textural properties because of this feature. These films can also be blended with other substances to enhance their properties and functions, such as essential oils, proteins, and other substances [39]. Chitosan's mucoadhesive and gelling qualities are thought to be responsible for its capacity to form films. These characteristics of chitosan are conferred by its polycationic nature and the potential for hydrogen and hydrophobic connections between its polymeric chains [40, 41].

5.5 Biocompatibility and Biodegradability

Biocompatibility is characterized as "compatibility with living tissue or a living system by not being toxic, injurious, physiologically reactive, and not causing immunological rejection" [42]. Chitosan's biocompatibility is one of its most vital qualities. Chitosan has the ability to be used in wound care, tissue engineering, bone regeneration, and other related areas because of its feature, making it a potential biomaterial. Both in vitro and in vivo methodologies may be employed to evaluate the biocompatibility of chitosan and chitosan-derived products [43]. The cytotoxicity test is the most used in vitro assessment. The characterization for in vivo testing depends on the application. It is feasible to do tests for genetic toxicity, intramuscular implantation, and others.

Numerous industries, including agriculture, medicine, packaging, medication release, and others, use biodegradable materials [44]. Biodegradation refers to processes occurring naturally that include living things or living tissues that involve the breakdown, mineralization, and redistribution of various chemicals through elemental cycles like the nitrogen or carbon cycles. Since microorganisms are essential to these processes, biodegradation can only take place in the biosphere [45].

Chitobiases, chitinases, and several other enzymes with the ability to degrade chitosan are widely distributed throughout a diverse range of taxa, including prokaryotes, higher mammals, yeasts, and plants [46]. Chitosan is often broken down by enzymes to liberate amino sugars, which can then be used in the metabolism of glycosaminoglycans and glycoproteins or eliminated as harmless substances [47]. Under the right circumstances of ionic strength, temperature, and pH, other enzymes like lysozyme and a number of proteinases might potentially break down chitosan. For non-specific enzymes like chitinases, the reaction rates often rise as acetylation levels grow [48]. In comparison to chitinases, chitosanases are a class of enzymes that exhibit selective hydrolysis of chitosan. However, it is worth noting that the occurrence of chitosanases is considerably less prevalent. Chitosanases, which are totally lacking in mammals, are thought to have originated in soil-dwelling Actinomycetes, particularly the Streptomyces species, such as *Vibrio sp.* and *Bacillus sp.* [49].

5.6 Reactivity

Chitosan is a biopolymer with a linear chain and a reactive backbone made up of functional groups $-NH_2$ and $-OH$ [50]. The amount to which certain functional groups, particularly amino groups, are protonated results in chitosan being more reactive. The polymeric chain's protonated $-NH_2$ group's reactivity and consequent capacity to connect with poisonous metal ions as well as form chelates with transition metal ions increase as more of the group is present. Chitosan's chelating properties highlight their value in the water and air cleaning industries.

The primary motivations for chemically modifying chitosan are frequently described as the need to boost its solubility under neutral or alkaline conditions, enable its solubility in organic solvents, and improve certain functional attributes. Typically, chitosan undergoes similar derivatization procedures as chitin. But, since chitosan's amino groups are accessible, various kinds of changes are possible. Derivatization of chitosan is typically accomplished chemically. The enzymatic method of modifying chitosan, however, is intriguing due to its selectivity and very minimal environmental effect [51].

Chitosan has two major groups that are amenable to chemical modification: hydroxyl and amino. In order to achieve the desired derivatization on hydroxyl groups, it is necessary to protect the amino group due to its higher susceptibility to derivatization. The reaction between chitosan and phthalate anhydride results in the formation of chitosan phthalate. The protection of amino groups by the use of a derivative that is soluble in organic solvents is a common approach when the objective is to derivatize hydroxyl groups. Chitosan's amino groups might be restored by removing the phthaloyl group using hydrazine treatment [52].

Various chemical reactions, such as the generation of quaternary amino salts, Schiff base formation, acylation, alkylation, copolymerization, and others, have the potential to impact the amino group. The predominant form of N-acylation seen in chitosan is n-acetylation. In order to generate a water-soluble, homogenous form of half-acetylated chitosan, the following process is employed. Schiff base generation is a crucial factor in several chitosan derivatizations. The amino groups that are unbound in chitosan have the capacity to engage in a chemical reaction with ketones and aldehydes, leading to the creation of the matching Schiff base. The amine group that is formed can undergo reduction to yield N-alkyl derivatives and exhibit stability under neutral or alkaline conditions [52]. The possible quaternization of amino groups can be accomplished by exposing chitosan to methyl iodide treatment in an alkaline setting. The process of synthesizing N-methylene phosphonic chitosan, which is an anionic derivative, entails the interaction between chitosan and phosphoric acid, as well as formaldehyde, under specific conditions. These conditions include a temperature of 70 °C and a reaction period of 7 hours [53]. Numerous methodologies have been documented for conducting graft copolymerization on chitosan. These methodologies frequently include the utilization of cerium ions, γ radiation, 2,2′-azobisisobutyronitrile, or redox systems to facilitate the generation of radicals, which are imperative for initiating the graft polymerization process. Chitosan has been subjected to copolymerization with several vinyl monomers, like vinyl acetate, acrylonitrile, and methyl acrylate. Various chemicals have been successfully attached to chitosan by grafting techniques. These compounds include dendrimers, monomers, and a few amino acids and sugars, ethylene oxide, lactic acid, aniline, and poly(ethylene glycol) [54].

5.7 Bioactivity

Chitosan has garnered significant interest due to its many bioactive properties. The various interactions of chitosan with proteins, lipids, genetic material, anionic molecules, and other biological components can be attributed to its polycationic nature. Chitosan has a wide range of biological activities that have been noted, which have been observed to include the stimulation of plant growth and resistance mechanisms, as well as the manifestation of antiviral, antifungal, and antibacterial properties. Additionally, chitosan demonstrates various biomedically significant effects, such as its hypocholesterolemic effect, ability to facilitate the transfection of genetic material, and promotion of wound healing, among others [55, 56].

Chitosan serves as a structural component in the cell walls of certain plant fungal infections. Most plants have enzymes, lectins, and proteins that can recognize and break down chitin and chitosan as part of their resistance mechanism. Chitosan elicits phytoalexins without altering the permeability of the plant cell membrane, according to studies [57, 58]. Due to this, the formation of these chemicals as well as other interesting metabolites (such as isoflavonoids, taxol, etc.), has been induced or enhanced using chitosan and its derived oligosaccharides. It appears that chitosan's elicitor activity and antibacterial potential work together to boost plant growth impacts [59].

Chitosan has been shown to have antimicrobial efficacy against a variety of bacterial, fungal, and viral illnesses [60]. The physicochemical characteristics of chitosan have a significant role in determining the extent of its effectiveness. The inhibitory effect is enhanced by the increase in the molecular weight of chitosan [61]. The antibacterial efficiency of chitosan is influenced by pH, with lower pH values demonstrating enhanced action against bacteria. The bioactivity of chitosan and its derivatives is linked to several phenomena, such as membrane effects, biodegradation, the induction of innate immunity, and direct antibacterial activity [59]. There is a considerable body of evidence documenting the resistance of several pathogens, encompassing both Gram (+) and Gram (–) bacteria, as well as food deterioration agents, to the antibacterial effects of chitosan. Evidently, chitosan activity affects Gram-positive bacteria more readily. The need for a more in-depth understanding of the mechanism by which chitosan acts against various types of microorganisms is made evident by some results, reports of homogeneous water-soluble chitosan's antibacterial activity being selectively inhibited or lacking altogether but having a significant impact when combined with amylose [62]. Regarding antiviral function, it is believed that chitosan does not affect virus development but instead suppresses infection, perhaps by preventing replication. It is still unclear, nevertheless, which step of replication chitosan is targeting [63].

Chitosan has been observed to exert various effects on higher animals. The effects observed are significantly influenced by the physicochemical characteristics of chitosan, as well as the specific type and interaction terms. Chitosan has a broad spectrum of present and prospective applications due to its compatibility with and physiological impacts on both human and animal systems. Consequently, extensive research has been conducted on these factors associated with chitosan, and a multitude of evaluations pertaining to pertinent subjects are readily available [64].

For a number of species and people, the nutritional effects of chitosan have been studied. It has been shown [65] how chitosan-supplemented feeds affect the diets of agricultural and lab animals, including rabbits, chickens, mice, fish, and rats. The majority of studies show a pattern of development that is normal or increased and a decrease in the absorption of dietary fat. Due to alterations in the digestive microbiota and lack of appetite, several writers have suggested that chitosan-supplemented diets may have long-term negative consequences.

5.8 Adsorption Capacity

The process of adsorption is influenced by many types of intermolecular forces, namely ionic interactions, hydrophobic interactions, Van der Waals forces, and hydrogen bonding. Functional groups in chitosan may cause these interactions with many sorts of molecules. Numerous investigations have been conducted to examine the phenomenon of adsorption, including proteins, metals, lipids, as well as other substances like dyes and aromatic compounds onto chitosan. The adsorption capacity of chitosan is influenced by several aspects, including its physicochemical qualities, such as DD and molecular weight, as well as the conditions of contact (such as pH and ionic strength). Additionally, the type and features of the adsorbate also play a significant role in determining the adsorption capacity of chitosan [66].

The pH, which controls how much both chemicals are ionized, has a significant impact on how chitosan interacts with proteins. For the purpose of generating electrostatic contacts, it is necessary for the isoelectric point of the protein to be lower than the pKa of chitosan.

Chitosan-protein interactions are typically weak; therefore, under the right circumstances, proteins can be desorbed [66].

Chitosan's functional qualities, such as cell or mucoadhesion and microbial flocculation, show protein adsorption. At the moment, chitosan's adsorption ability is crucial to its primary commercial applications, such as those involving food or wastewater flocculation [67].

Figure 5.1 illustrates three early techniques of reaction used by Ilham et al. [68] to demonstrate the adsorption process of the methylene blue dye by chitosan.

(1) Beta-hydrophobic sections of dye molecules, such as CH_3 groups, can interact with hydrophobic contacts. Hydrophobic interactions cause hydrophobic sections of chitosan chains (such as CH_3 groups) to become denatured. Methylene blue molecules may bind or adsorb to the surface of the chitosan or to its internal structure as a result of this process. Additionally, the aromatic rings of methylene blue and chitosan may engage in stacking interactions.

(2) Electrostatic interactions occur between chitosan (-COCHa) and methylene blue due to the presence of negatively charged groups in chitosan that may interact with the positively charged (S^+) groups in methylene blue. This results in the formation of electrostatic attractions between the two molecules.

(3) Hydrogen bonds: The structure of the methylene blue dye includes nitrogen atoms, which can serve as acceptors in hydrogen bonds. In contrast, chitosan possesses hydroxyl (-OH) and amino (-NH) groups, which may function as

both hydrogen bond donors and acceptors. Together, these three types of electrostatic pulls, interactions hydrophobic interactions, and hydrogen bonds explain how methylene blue dye bonds to and is adsorbed on chitosan. It is significant to note that several elements, including pH, temperature, dye concentration, and particular chitosan characteristics, might have an impact on the actual adsorption process [68].

Chitosan's ability to adsorb lipids has been investigated. However, the interaction processes are still not fully understood [69]. According to some experts, chitosan's cationic nature causes it to interact strongly with lipids that have an opposite charge. However, independent of the DD, it has been suggested that the creation of hydrophobic aggregates is a cause of lipid-chitosan interactions [70]. When ingested by both people and animals, chitosan's ability to absorb lipids is associated with a reduction in cholesterol levels.

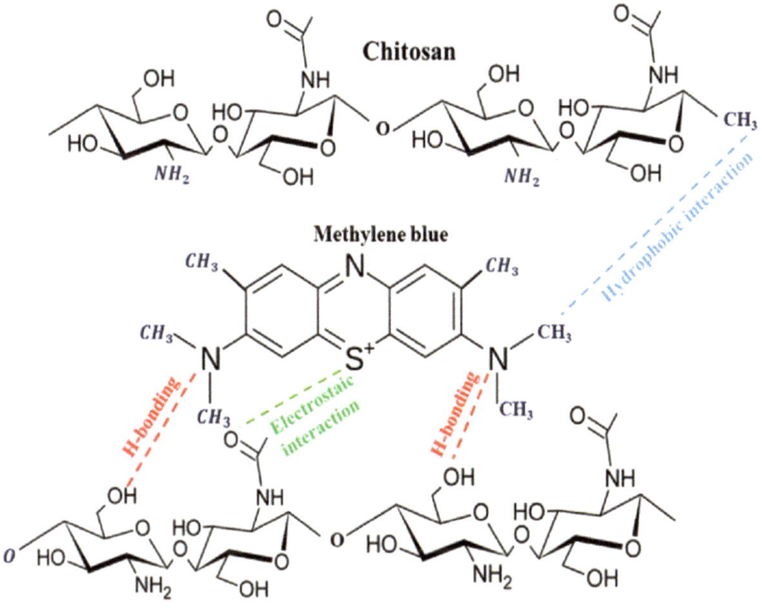

Figure 5.1 Diagram showing the hydrogen bonds, hydrophobic interactions, and electrostatic interactions between chitosan and the dye methylene blue (reproduced with permission from Ben Amor et al. [7]).

5.9 Conclusion

In conclusion, the two polysaccharides, chitin and chitosan, have unique chemical properties that have led to numerous applications in different fields. They also offer a practical, environmentally responsible substitute for synthetic materials because they are renewable and biodegradable. Due to its unique chemical features, it has the potential to be utilized across a diverse array of domains, encompassing food and agriculture, pharmaceutical applications, and environmental uses. As long as scholars keep looking into their potential, chitin and chitosan are expected to make major contributions to the development of sustainable technology.

References

1. Amor IB, Hemmami H, Laouini SE, Abdelaziz AG, Barhoum A (2023) Influence of chitosan source and degree of deacetylation on antibacterial activity and adsorption of AZO dye from water. *Biomass Conversion and Biorefinery*: 1–11.
2. Ben Amor I, Hemmami H, Laouini SE, Mahboub MS, Barhoum A (2022) Sol-gel synthesis of ZnO nanoparticles using different chitosan sources: effects on antibacterial activity and photocatalytic degradation of AZO Dye. *Catalysts* 12: 1611.
3. Amor IB, Hemmami H, Laouini SE, Temam HB, Zaoui H, Barhoum A (2023) Biosynthesis MgO and ZnO nanoparticles using chitosan extracted from Pimelia Payraudi Latreille for antibacterial applications. *World Journal of Microbiology and Biotechnology* 39: 19.
4. Fang Y, Zhang R, Duan B, Liu M, Lu A, Zhang L (2017) Recyclable universal solvents for chitin to chitosan with various degrees of acetylation and construction of robust hydrogels. *ACS Sustainable Chemistry & Engineering* 5: 2725–2733.
5. Xu D, Huang J, Zhao D, Ding B, Zhang L, Cai J (2016) High-flexibility, high-toughness double-cross-linked chitin hydrogels by sequential chemical and physical cross-linkings. *Advanced Materials* 28: 5844–5849.
6. Duan B, Shou K, Su X, Niu Y, Zheng G, Huang Y, Yu A, Zhang Y, Xia H, Zhang L (2017) Hierarchical microspheres constructed from chitin nanofibers penetrated hydroxyapatite crystals for bone regeneration. *Biomacromolecules* 18: 2080–2089.

7. Wu J, Meredith JC (2014) Assembly of chitin nanofibers into porous biomimetic structures via freeze drying. *ACS Macro Letters* 3: 185–190.
8. Duan B, Huang Y, Lu A, Zhang L (2018) Recent advances in chitin based materials constructed via physical methods. *Progress in Polymer Science* 82: 1–33.
9. Deringer VL, Englert U, Dronskowski R (2016) Nature, strength, and cooperativity of the hydrogen-bonding network in α-chitin. *Biomacromolecules* 17: 996–1003.
10. Alhamad AA, Zeghoud S, Amor IB, Hemmami H (2023) Chitosan-based hydrogels for wound healing: correspondence. *International Journal of Surgery* 109: 1821–1822.
11. Stevens CV (2019) *Chitin and Chitosan: Properties and Applications.*
12. Jiang Z, Fang Y, Xiang J, Ma Y, Lu A, Kang H, Huang Y, Guo H, Liu R, Zhang L (2014) Intermolecular interactions and 3D structure in cellulose–NaOH–urea aqueous system. *The Journal of Physical Chemistry B* 118: 10250–10257.
13. Fang Y, Duan B, Lu A, Liu M, Liu H, Xu X, Zhang L (2015) Intermolecular interaction and the extended wormlike chain conformation of chitin in NaOH/urea aqueous solution. *Biomacromolecules* 16: 1410–1417.
14. Zhou J, Zhang L (2000) Solubility of cellulose in NaOH/urea aqueous solution. *Polymer Journal* 32: 866–870.
15. Hu X, Du Y, Tang Y, Wang Q, Feng T, Yang J, Kennedy JF (2007) Solubility and property of chitin in NaOH/urea aqueous solution. *Carbohydrate Polymers* 70: 451–458.
16. Li G, Du Y, Tao Y, Liu Y, Li S, Hu X, Yang J (2010) Dilute solution properties of four natural chitin in NaOH/urea aqueous system. *Carbohydrate Polymers* 80: 970–976.
17. Chang C, Chen S, Zhang L (2011) Novel hydrogels prepared via direct dissolution of chitin at low temperature: structure and biocompatibility. *Journal of Materials Chemistry* 21: 3865–3871.
18. Cai J, Yang J, Du Y, Fan L, Qiu Y, Li J, Kennedy JF (2006) Enzymatic preparation of chitosan from the waste Aspergillus niger mycelium of citric acid production plant. *Carbohydrate Polymers* 64: 151–157.
19. Ding B, Cai J, Huang J, Zhang L, Chen Y, Shi X, Du Y, Kuga S (2012) Facile preparation of robust and biocompatible chitin aerogels. *Journal of Materials Chemistry* 22: 5801–5809.

20. Tamura H, Tsuruta Y, Tokura S (2002) Preparation of chitosan-coated alginate filament. *Materials Science and Engineering: C* 20: 143–147.
21. Nagahama H, Higuchi T, Jayakumar R, Furuike T, Tamura H (2008) XRD studies of β-chitin from squid pen with calcium solvent. *International Journal of Biological Macromolecules* 42: 309–313.
22. Tajiri R, Mihata A, Yamamoto K, Kadokawa J-I (2014) Facile preparation of chitin gels with calcium bromide dihydrate/methanol media and their efficient conversion into porous chitins. *RSC Advances* 4: 5542–5546.
23. Dupont A-L (2003) Cellulose in lithium chloride/N,N-dimethylacetamide, optimisation of a dissolution method using paper substrates and stability of the solutions. *Polymer* 44: 4117–4126.
24. Younes I, Rinaudo M (2015) Chitin and chitosan preparation from marine sources. Structure, properties and applications. *Marine Drugs* 13: 1133–1174.
25. Taniguchi M, Belfort G (2004) Low protein fouling synthetic membranes by UV-assisted surface grafting modification: varying monomer type. *Journal of Membrane Science* 231: 147–157.
26. Rogers RD, Voth GA (2007) Ionic liquids. *Accounts of Chemical Research* 40: 1077–1078.
27. Greer AJ, Jacquemin J, Hardacre C (2020) Industrial applications of ionic liquids. *Molecules* 25: 5207.
28. Xie H, Zhang S, Li S (2006) Chitin and chitosan dissolved in ionic liquids as reversible sorbents of CO_2. *Green Chemistry* 8: 630–633.
29. Prasad K, Murakami M-A, Kaneko Y, Takada A, Nakamura Y, Kadokawa J-I (2009) Weak gel of chitin with ionic liquid, 1-allyl-3-methylimidazolium bromide. *International Journal of Biological Macromolecules* 45: 221–225.
30. Mine S, Izawa H, Kaneko Y, Kadokawa J-i (2009) Acetylation of α-chitin in ionic liquids. *Carbohydrate Research* 344: 2263–2265.
31. Wu Y, Sasaki T, Irie S, Sakurai K (2008) A novel biomass-ionic liquid platform for the utilization of native chitin. *Polymer* 49: 2321–2327.
32. Silva SS, Duarte ARC, Carvalho AP, Mano JF, Reis RL (2011) Green processing of porous chitin structures for biomedical applications combining ionic liquids and supercritical fluid technology. *Acta Biomaterialia* 7: 1166–1172.

33. Roy JC, Salaün F, Giraud S, Ferri A, Chen G, Guan J (2017) Solubility of chitin: solvents, solution behaviors and their related mechanisms. *Solubility of Polysaccharides* 3: 20–60.

34. Kasaai MR, Arul J, Charlet G (2013) Fragmentation of chitosan by acids. *The Scientific World Journal* 2013.

35. Pardo-Castaño C, Bolaños G (2019) Solubility of chitosan in aqueous acetic acid and pressurized carbon dioxide-water: experimental equilibrium and solubilization kinetics. *The Journal of Supercritical Fluids* 151: 63–74.

36. Ariful Islam M, Park T-E, Reesor E, Cherukula K, Hasan A, Firdous J, Singh B, Kang S-K, Choi Y-J, Park I-K (2015) Mucoadhesive chitosan derivatives as novel drug carriers. *Current Pharmaceutical Design* 21: 4285–4309.

37. M. Ways TM, Lau WM, Khutoryanskiy VV (2018) Chitosan and its derivatives for application in mucoadhesive drug delivery systems. *Polymers* 10: 267.

38. Ta LP, Bujna E, Kun S, Charalampopoulos D, Khutoryanskiy VV (2021) Electrosprayed mucoadhesive alginate-chitosan microcapsules for gastrointestinal delivery of probiotics. *International Journal of Pharmaceutics* 597: 120342.

39. Kumar S, Mukherjee A, Dutta J (2020) Chitosan based nanocomposite films and coatings: emerging antimicrobial food packaging alternatives. *Trends in Food Science & Technology* 97: 196–209.

40. Wang H, Ding F, Ma L, Zhang Y (2021) Edible films from chitosan-gelatin: physical properties and food packaging application. *Food Bioscience* 40: 100871.

41. Gao H-X, He Z, Sun Q, He Q, Zeng W-C (2019) A functional polysaccharide film forming by pectin, chitosan, and tea polyphenols. *Carbohydrate Polymers* 215: 1–7.

42. Sharma A, Sood K, Kaur J, Khatri M (2019) Agrochemical loaded biocompatible chitosan nanoparticles for insect pest management. *Biocatalysis and Agricultural Biotechnology* 18: 101079.

43. Canciani B, Semeraro F, Herrera Millar VR, Gervaso F, Polini A, Stanzione A, Peretti GM, Di Giancamillo A, Mangiavini L (2023) In vitro and in vivo biocompatibility assessment of a thermosensitive injectable chitosan-based hydrogel for musculoskeletal tissue engineering. *International Journal of Molecular Sciences* 24: 10446.

44. Morin-Crini N, Lichtfouse E, Torri G, Crini G (2019) Applications of chitosan in food, pharmaceuticals, medicine, cosmetics, agriculture,

textiles, pulp and paper, biotechnology, and environmental chemistry. *Environmental Chemistry Letters* 17: 1667–1692.

45. Meena M, Sonigra P, Yadav G (2021) Biological-based methods for the removal of volatile organic compounds (VOCs) and heavy metals. *Environmental Science and Pollution Research* 28: 2485–2508.
46. Kumari R, Kumar M, Vivekanand V, Pareek N (2023) Chitin biorefinery: a narrative and prophecy of crustacean shell waste sustainable transformation into bioactives and renewable energy. *Renewable and Sustainable Energy Reviews* 184: 113595.
47. Liu X, Liao W, Xia W (2023) Recent advances in chitosan based bioactive materials for food preservation. *Food Hydrocolloids*: 108612.
48. Crini G, Torri G, Lichtfouse E, Kyzas GZ, Wilson LD, Morin-Crini N (2019) Dye removal by biosorption using cross-linked chitosan-based hydrogels. *Environmental Chemistry Letters* 17: 1645–1666.
49. Mohan K, Ganesan AR, Ezhilarasi P, Kondamareddy KK, Rajan DK, Sathishkumar P, Rajarajeswaran J, Conterno L (2022) Green and eco-friendly approaches for the extraction of chitin and chitosan: a review. *Carbohydrate Polymers* 287: 119349.
50. Gupta S, Ghosal A, Goswami A, Bhawana, Nadda AK, Sharma S (2022) The scope of biopolymers in food industry. *Biopolymers: Recent Updates, Challenges and Opportunities*. Springer, pp. 173–198.
51. Recillas-Mota MT, Montiel-Herrera M (2019) Chitosan derivatives: introducing new functionalities with a controlled molecular architecture for innovative materials. *Advances in Chitin/Chitosan Characterization and Applications* 1: 1.
52. Yeh H-Y, Lin J-C (2008) Surface characterization and in vitro platelet compatibility study of surface sulfonated chitosan membrane with amino group protection–deprotection strategy. *Journal of Biomaterials Science, Polymer Edition* 19: 291–310.
53. Liu T, Gou S, He Y, Fang S, Zhou L, Gou G, Liu L (2021) N-methylene phosphonic chitosan aerogels for efficient capture of Cu^{2+} and Pb^{2+} from aqueous environment. *Carbohydrate Polymers* 269: 118355.
54. Malerba M, Cerana R (2019) Recent applications of chitin-and chitosan-based polymers in plants. *Polymers* 11: 839.
55. Liu X, Wu Y, Zhao X, Wang Z (2021) Fabrication and applications of bioactive chitosan-based organic-inorganic hybrid materials: a review. *Carbohydrate Polymers* 267: 118179.

56. Bhowmik S, Agyei D, Ali A (2022) Bioactive chitosan and essential oils in sustainable active food packaging: recent trends, mechanisms, and applications. *Food Packaging and Shelf Life* 34: 100962.

57. Román-Doval R, Torres-Arellanes SP, Tenorio-Barajas AY, Gómez-Sánchez A, Valencia-Lazcano AA (2023) Chitosan: properties and its application in agriculture in context of molecular weight. *Polymers* 15: 2867.

58. El Hadrami A, Adam LR, El Hadrami I, Daayf F (2010) Chitosan in plant protection. *Marine Drugs* 8: 968–987.

59. Badawy ME, Rabea EI (2011) A biopolymer chitosan and its derivatives as promising antimicrobial agents against plant pathogens and their applications in crop protection. *International Journal of Carbohydrate Chemistry* 2011: 1–29.

60. Kong M, Chen XG, Xing K, Park HJ (2010) Antimicrobial properties of chitosan and mode of action: a state of the art review. *International Journal of Food Microbiology* 144: 51–63.

61. Xing K, Zhu X, Peng X, Qin S (2015) Chitosan antimicrobial and eliciting properties for pest control in agriculture: a review. *Agronomy for Sustainable Development* 35: 569–588.

62. Gil G, Del Mónaco S, Cerrutti P, Galvagno M (2004) Selective antimicrobial activity of chitosan on beer spoilage bacteria and brewing yeasts. *Biotechnology Letters* 26: 569–574.

63. Jaber N, Al-Remawi M, Al-Akayleh F, Al-Muhtaseb N, Al-Adham IS, Collier PJ (2022) A review of the antiviral activity of Chitosan, including patented applications and its potential use against COVID-19. *Journal of Applied Microbiology* 132: 41–58.

64. Jäger T, Mokos A, Prasianakis NI, Leyer S (2022) Pore-level multiphase simulations of realistic distillation membranes for water desalination. *Membranes* 12(11): 1112.

65. Yilmaz Atay H (2019) Antibacterial activity of chitosan-based systems. *Functional Chitosan: Drug Delivery and Biomedical Applications*: 457–489.

66. Liu B, Wang D, Yu G, Meng X (2013) Adsorption of heavy metal ions, dyes and proteins by chitosan composites and derivatives—A review. *Journal of Ocean University of China* 12: 500–508.

67. Renault F, Sancey B, Badot P-M, Crini G (2009) Chitosan for coagulation/flocculation processes–an eco-friendly approach. *European Polymer Journal* 45: 1337–1348.

68. Ben Amor I, Hemmami H, Laouini SE, Zeghoud S, Benzina M, Achour S, Naseef A, Alsalme A, Barhoum A (2023) Use of insect-derived chitosan for the removal of methylene blue dye from wastewater: process optimization using a central composite design. *Materials* 16: 5049.

69. May KL, Tangso KJ, Hawley A, Boyd BJ, Clulow AJ (2020) Interaction of chitosan-based dietary supplements with fats during lipid digestion. *Food Hydrocolloids* 108: 105965.

70. Wydro P, Krajewska B, Hac-Wydro K (2007) Chitosan as a lipid binder: A Langmuir monolayer study of chitosan–lipid interactions. *Biomacromolecules* 8: 2611–2617.

Chapter 6

Biological Characteristics of Chitin and Chitosan

Mohamed Belkacem,[a,b] Hadjer Bouosmat,[c] Asmaa Benettayeb,[a] Mustapha Hadj Brahim,[a,b] and Boumediene Haddou[b]

[a]*Laboratory of Chemical Engineering and Heterogeneous Catalysis, Department of Chemical Engineering, Faculty of Chemistry, University of Sciences and Technology-Mohamed Boudiaf, El-M'naouar, Oran, Algeria*
[b]*Laboratory of Physico-Chemical of Materials: Catalysis and Environment, Department of Chemical Engineering, Faculty of Chemistry, University of Sciences and Technology-Mohamed Boudiaf, El-M'naouar, Oran, Algeria*
[c]*Laboratory of Structure, Elaboration, and Application of Molecular Materials, Department of Process Engineering, Faculty of Science and Technology, Abdelhamid Ibn Badis University, Mostaganem, Algeria*

mohamed.belkacem@univ-usto.dz

In the last few decades, a biopolymer comprising cellulose, polysaccharides, chitosan, starch, chitin, and alginate generated a growing demand for the development of natural alternatives to man-made polymers. Chitosan, a biological polymer, has attracted the attention of fundamental and applied scientists since the late 1970s because of its unusual molecular structure and physicochemical characteristics and because its biological functions are distinct from synthetic polymers. Chitin has different

Chitin and Chitosan: Physical and Chemical Properties
Edited by Shakeel Ahmed, Soumeia Zeghoud, Hadia Hemmami, and Ilham Ben Amor

ISBN 978-981-5129-39-7 (Hardcover), 978-1-003-58977-8 (eBook)
www.jennystanford.com

physicochemical characteristics and different biologic properties than other synthetic polymer types. Products made from chitin and its byproducts are used in chemical, agricultural, medical, cosmetics, textile, paper, and other industries. The veterinary, biomedical, beverage, sanitary, health, and personal sectors, as well as the catalysis, chromatography, sewage treatment, and biotechnological industries, all show great interest in chitosan because of its distinct biological properties.

6.1 Introduction

Since French chemist Henri Braconnot first isolated and characterized chitin $(C_8H_{13}O_5N)_n$ from mushrooms, the first known polysaccharide was discovered in 1811 [1] and has since proved to be the world's second most abundant natural biopolymer [2] accounting for about 106–107 tons of marine biomass alone. N-acetyl-D-glucosamine (GlcNAc) is the homo-polymer for a long-chain derivative of 2-acetamido-2-deoxy-D-glucane (GlcNAc) [2–4]. Strong acids can break down the insulating chitin polymer to either chitosan or acetic acid, and chitin is then broken down by different kinds of unspecific immunity agents such as cellulases, fats, proteins, and possibly chitosanases to its two main byproducts: amino glucose and chitosan [5]. In particular, chitin is converted to a non-toxic biopolymer, chitosan, through the process of deacetylation. Both substances are currently attracting a lot of attention, due to their biological activity with practical approach uses. Chitosan and chito oligosaccharides, which are breakdown products of chitin or chitosan generated through enzyme or other acid-based hydrolysis, have been part of a European Union research program [6].

In recent decades, they have been applied in a wide range of healthcare products, inclusive bandages, and illegal substance administration systems, as well as in the food and chemical industries as a source of bioactive material. Biologically, however, both chitin and its byproducts show promise for a wide variety of uses, including membranes, aquaculture, agriculture, dentistry, wastewater, cosmetics, and the food and medical industries [7–10].

In this chapter, the focus is on biological properties, including antibacterial, antioxidant, and others.

6.2 Chitosan and Chitin Structure

N-acetyl-D-glucosamine units in chitin each other with-D (1→4) chemical bonds to form a linear polymer. In terms of chemical reactivity and reduced dissolution, it is comparable to cellulose. It merely differs from cellulose in the acetamido band and substitutes for the hydroxyl in the C-2 group. Although the deacetylation is typically not complete, chitosan contains chitin's N-diacetyl radicals [11].

6.3 Biological Properties

Due to their fascinating intrinsic characteristics, notably biological suitability, biological degradability, and lack of toxicity, there is considerable current enthusiasm for the potential medicinal and therapeutic uses of chitin and chitosan [12]. Other characteristics that have also been verified include hemostatic, anticancer, analgesic, hypocholesterolemic, antioxidant, and antibacterial capacities [13]. The level of deacetylation (DD) has the greatest influence, as most of the chitosan's organic characteristics are linked to its highly cationic nature. However, in certain situations, molecular weight has a significant influence [14, 15].

Chitin and chitosan have intriguing qualities, which have made them excellent for use in the field of bio-medicine, such as matching, non-toxicity, and readily degradable and, as a result, are now attracting a large amount of attention in terms of healthcare products and services for medical and scientific purposes. Other qualities have also been reported: analgesic, anticancer, hemostatic, hypocholesterolemic, antimicrobial, and antioxidant [16].

Chitosan needs to be thoroughly characterized and isolated from auxiliary chemicals to better understand what is behind these characteristics [17]. In more, chitins and chitosans that have been derived in different ways may be deployed to support cellular hypotheses about organic growth. The DD parameter has a greater impact, as most organic characteristics are linked to the chitosan's cationic properties. However, the molecular weight (Mw) parameter sometimes plays a more important role.

Other variables, including the conformation of the chain, the degree of substitution, and solubility, were also taken into account in addition to DD and Mw. In aqueous solutions, chitosans made through inhomogeneous acetylation using block configuration of both acetylated and deacetylated functional groups tend to form aggregates [18]. The number of accessible sites on the chitosan molecule can be reduced by extensive aggregation and intermolecular interactions. In particular, failure by originators to pay particular care and attention to the manufacturing process of these specific chitosan blends or a different method of formulation in these studies may account directly for variations in the effects recorded for this compound [19].

6.3.1 Biodegradability

Mammals lack chitin and chitosan but a number of proteases (papain, lysozyme, pepsin...) can destroy them in vivo. They are biological. Degradation leads to the creation of variable-length, toxic-free trace elements, which can either be removed from or included in glycoproteins and glycosaminoglycans. Also, degradation leads to the generation of variable-length, toxic-free trace elements, which can either be expelled or added to glycosaminoglycans and glycoproteins, among other metabolic pathways [20].

All mammalian tissues include the non-specific protease lysozyme, which appears to participate in the degradation of chitosan to chitin. The level of crystallinity, which is mainly regulated by the extent of acetylation, appears to be inversely related to degradation kinetics. Furthermore, the location of acetyl groups influences biodegradability since very little enzymatic degradation occurs when acetyl groups are absent or evenly distributed (randomly rather than in blocks) [21].

Finally, a number of studies have indicated that the degradation rate is also influenced by chain weight (Mw) [22]. Given that understanding and controlling the rate of chitosan and chitin breakdown-based devices is essential to many applications for the delivery of low and big compounds, as well as for applications in the rejuvenation of structural fabrics, this topic is of considerable interest. Ideally, stage degradation scaffolding must correspond to the level of development of novel fabrics, if not be sufficient to enable the regulated liberation of organic chemicals. It is, therefore, essential to comprehend and regulate the process by which each substance degrades and the rate at which it does so [23].

Suitability is also influenced by degradation rate, as extremely rapid degradation rates lead to an inflammatory reaction and an accumulation of amino sugars. Due to the slow degradation rate, chitosan samples with high DD cause only a modest reaction, while those at poor DD result in an acute inflammatory response [24]. Ibram et al. [25] examined By observing variations in a chitosan solution's viscosity in the presence of lysozyme, one can learn about the enzymatic behaviors of various chitosan species. They discovered with low DD; chitosan tends to degrade more rapidly. Other authors, however, claimed that variations in acetamide position within the chitosan structure were responsible for the differences in degradation [26], due to variations in deacetylation conditions, which modify inter- or intra-molecular repulsion forces and change the viscosity of the chitosan solution [27]. It can, therefore, be deduced that it is impossible to use DD alone to calculate the rate of biodegradation.

6.3.2 Biocompatibility

Both chitosan and chitin show outstanding suitability but sample characteristics (production process, natural source, DD, and Mw) influence this property. Most experiments have used chitosan samples because of their adaptability and superior biological characteristics [28].

Chitin and chitosan are not orally absorbed, although gastrointestinal enzymes can partially degrade both polymers. They are therefore considered non-bioavailable. In experiments on mice, chitosan has an LD_{50} of around 16 g/kg, which is quite comparable to the values for salt and glucose [29]. According to reports, the toxicity of chitosan changes with DD. According to Tsurkan et al. [30], chitosan with DD above 35% shows modest toxicity but those with DD below 35% show dose-dependent toxicity. However, chitosan molar mass has no effect on toxicity.

In vitro, compared to chitin, chitosan is more cytocompatible. The cytocompatibility of chitosan with human myocardium, epithelial and endothelial tissue, fibers, liver cells, chondrocytes, and keratin cells has been demonstrated in vitro [31]. This characteristic seems to be linked to the DD of the samples. Because of the existence within chitosan of certain free amino groups, cell-cell adhesion also increases as the positive charge on chitosan increases. They discovered that keratinocytes and fibroblasts adhered to different chitosan films and grew there. The proportion of cell adhesion in both cells is strongly influenced by DD, which increases with this parameter. Cell type also influenced adhesion, with fibroblasts having a more negatively charged cell wall than keratinocytes. On another front, increasing DD significantly reduced proliferation. This characteristic seems to be linked to the DD of the samples. Because of the existence within chitosan of certain free amino groups, cell-cell adhesion also increases as the positive charge on chitosan increases. Keratinocyte and fibroblast adhesion and proliferation on various chitosan films with various DDs are dependent on the cell type and DD. DD has a significant impact on both cells' proportion of cell adhesion, which increases with this parameter. Cell type also influenced adhesion, with fibroblasts having a more negatively charged cell wall than keratinocytes. On another front, increasing DD significantly reduced proliferation.

Residual chitin and chitosan proteins can cause allergic symptoms, including hypersensitivity. The source of the sample, as well as its preparation technique, has an impact on the protein concentration of the sample [32].

6.3.3 Hemostatic

In vitro experiments have revealed that chitosan and sulfated chitosan oligomers have an anticoagulant effect [15]. As the membranes of red globules have a negative charge and as chitin does not work as well as chitosan, the anticoagulant action of chitosan seems to be linked to its positive charge [33, 34]. Mw chitosan also affects the agglutination or binding of red blood cells [35]. In the new study, chitosan in solid form and chitosan dissolved in a mixture of acetic acid have been compared. A number of chitosan samples with Mw ranging between 2,000 and 400 kDa and DD ranging between 90% and 70% have been examined.

The hemostatic processes used by chitosan in the firm state and the physiological brine solution of chitosan acetic acid were found to differ. Erythrocytes aggregated and changed shape when blood was combined with a saline solution containing chitosan and acetic acid. In contrast to the effect of Mw in the region of 105–106, DD, particularly high DD, in the physiological saline solution of chitosan and acetic acid had a substantial impact on the abnormal red blood cell aggregation and distortion. However, this effect was not visible in the solid-state chitosan liquid. High DD solid-state chitosan was more hemostatic and bound more platelets [36].

6.3.4 Analgesic Effect

Both chitin and chitosan are said to have analgesic properties, according to a number of publications [37]. Both chitin and chitosan have been studied for their ability to reduce the discomfort associated with inflammation caused by administering acetic acid intraperitoneally, and Okamoto et al. [38] proposed a method to account for this event. These researchers discovered that chitosan had a stronger impact than chitin. They discovered that chitosan had a stronger impact than chitin. The two drugs' various modes of action polymers were used to explain this disparity. According to the results, the primary analgesic action of chitosan is the absorption of the proton ions produced by inflammation.

The presence of proton ions allows the protonation of the basic free amino groups of chitosan because of its multifaceted nature, and the decrease in pH is responsible for the analgesic action. The concentration required for chitin to have the identical action as chitosan levels was lower than anticipated, although it is also capable of partially taking up such proton ions. According to the findings of the trial, the absorption of bradykinin, one of the key chemicals associated with pain, was responsible for the analgesic effect [39].

6.3.5 Antitumor Activity

Chitosan has been credited with having an anticancer impact because it inhibits the proliferation of tumor cells mostly as a result of an immune-stimulating effect. But there is a lot of debate about this asset [35].

Chitosan oligomers were shown to exhibit anticancer properties by Jeon and Kim when they examined them in vitro and in vivo [40]. Significant antimetastatic benefits of chitosan against Lewis's lung cancer were found in studies using low molecular weight chitosan in mice. Both partly deacetylated chitin and chitin containing a carboxymethyl group have been shown to inhibit tumor growth [41]. This process involves the activation of cytolytic T cells by the immunostimulatory properties provided by chitin and carboxymethyl compounds. They are thought to possess immunostimulatory activities, which stimulate peritoneal macrophages to become active and enhance unspecific resistance in the target host since this activity increases as the size of the molecules decreases. However, oligomers with a larger molecular weight were able to demonstrate anticancer activity. A similar mechanism involving increased lymphokine synthesis by activated lymphocytes has been proposed to explain their activity [42].

Chitosan's impact on the development and metastasis of tumors was researched. Chitosan's anticancer effects in vivo may be mediated by activating macrophages, whereas the negative effects of chitosan, such as promoting tumor development and invasion, may be caused by its angiogenic stimulating capabilities [43].

6.3.6 Mucoadhesion

Physiological variables and chitosan physicochemical properties are just a few of the factors that determine chitosan mucoadhesion [22]. Mucin, a type of glycoprotein, makes up the majority of mucus. That contains sialic acid residues and has a strong negative charge. Because of the acidic environment in the stomach, through electrostatic contact, the mucin can be attracted by the positively charged chitosan. The Mw and DD of chitosan, as well as the amount of sialic acid in the mucin, determine how much of this union occurs. It has been discovered that as chitosan's Mw rises, more of it penetrates the mucin layer, strengthening the mucoadhesion [44]. On the other hand, when the DD rises, the molecule's charge density rises, and its sticky qualities are more important [45].

Huang et al. [43] have investigated the impacts of several parameters, such as nanoparticles and the cytotoxicity of chitosan molecules in vitro and the influences of DD and Mw on cellular absorption. It was observed that the binding affinity and absorption capacity of chitosan nanoparticles exhibited a reduction when DD and Mw were lowered. The impact of DD was more significant compared to Mw due to its influence on the zeta potential of the nanoparticles. The amount of deacetylation and Mw, however, had less of an influence on the uptake of chitosan molecules.

Mucoadhesive micrometrical chitosan/poly(-caprolactone) films were created by El-Kamel et al. [46] to treat periodontal disorders. These scientists found varying adhesion strengths between films containing different Mw chitosans, although the data was subjected to random statistical testing, which showed that there was no appreciable variation between films in terms of bioadhesion strength. In contrast, demonstrated that medium Mw chitosan had a larger maximum detachment force than either low or high Mw chitosan [47].

6.3.7 Permeation Enhancing Effect

According to studies, chitosan opens epithelial tight junctions to increase permeability [48]. A reorganization of proteins connected to the tight junction results from the chemical reaction of positively charged chitosan with cell membranes., elucidating the underlying mechanism for this phenomenon [49].

Tsurkan et al. [27] employed chitosan hydrochloride salts at pH 5.5 to study the influences of chitosan characteristics (Mw and DD) on their ability to increase in vitro absorption. (Caco-2 cell monolayers). It was discovered that the Mw and the DD affect how well chitosan improves mannitol transport. As a result, highly deacetylated chitosan was effective at both low and high Mws as a booster of permeation but chitosan with low DD proved effective at high Mws values alone. According to subsequent publications in this field, the largest effect on cells in culture is a promoting one when deacetylation exceeds 80% [50].

A study looked at how five distinct chitosan models using differing Mw. The mucociliary transport velocity was modulated by DD. The frog palate-clearing mechanism was demonstrated to be unaffected by any of the five kinds of chitosan that were investigated [51]. Various varieties of chitosan were utilized for nasal delivery applications in guinea pigs for a 28-day study after administering a chitosan solution, the frequency of beating eyelashes; neither chitosan had an impact on eyelash frequency. This demonstrates that various forms of chitosan can be used without risk.

6.3.8 Anticholesterolemic

Chitosan has been proposed to lower cholesterol in a number of different ways. The most recent studies in this field take many possibilities into account. It is thought that a viscous polysaccharide solution's entrapment reduces the absorption of fat and cholesterol from the diet [50]. Conversely, the inclusion of the amino group within the chemical composition of chitosan regulates the electrostatic interaction between chitosan and anionic molecules, such as fatty acids and bile acids, which possess negative charges. According to the proposal [52], bile acids spontaneously produce insoluble chitosan ions, which due to their hydrophobic nature, should be able to attract lipids and cholesterol through hydrophobic contact. To test this notion, an average Mw of 150 kDa, DD 87% commercial food grade chitosan was employed.

The interactions of chitosan with anionic surface-active substances (bile acids, phospholipids) are governed by its three different reactive functional groups, which include the amino group at position C2, the primary and secondary hydroxyl groups at locations C-3 and C-6, and the amino group at position C2. Thongngam et al. [52] have shown that micelle-like clusters form within the chitosan structure when it interacts with a model bile salt. Various mechanisms might potentially explain the formation of a protective layer and the adsorption of chitosan onto the surface of emulsified lipids. The lipase or co-lipase's capacity to bind to the surfaces of droplets and get access to the lipids there may be impeded by this protective layer.

Even though a lot of work has gone into establishing a link between chitosan's physicochemical properties and its binding capacity into fat, several key connections have been demonstrated. Using six commercially available chitosans with different physicochemical properties, No et al. [53] demonstrated that the fat binding capacity had a substantial negative correlation with bulk density while showing a tendency toward a positive correlation with Mw. The same researchers evaluated the capacity of five chitosans with varying Mw (500–800 kDa) to bind fat while keeping a constant DD and observed that the sample with the second-lowest Mw had much higher activity [54]. Irradiation and sonolysis were performed on a chitosan sample in a different investigation, and five decreasing Mw were formed in the range of 25–400 kDa. When applying a biopharmaceutical model of the digestive tract, the samples exhibited a trend of enhancing their ability to bind with fats as Mw decreased [54]. In order to simulate the responses occurring in the stomach and duodenum, a variety of experimental designs have been utilized. The interaction of chitosan with various viscosities, DD, and sunflower oil has been studied using a digestive chemistry model. There were no appreciable variations in chitosan behavior according to its properties, despite a negative association between the percentage of entrapped oil and increasing oil addition [55]. Chitosan adsorption to fat droplets was investigated using an additional in vitro human digestion model. High Mw chitosan was shown to be more strongly adsorbed on the droplet than low Mw chitosan. Several conformations of the chitosan molecule with cationic loops and Mw high tails, as well as its enhanced surface activity, have been supposed as explanations for this phenomenon [57].

In a recent contribution, an exploration was carried out on eleven distinct chitosan formulations. The in vitro fat-binding capacity, bile acid specificity, DD, solution viscosity, and swelling volume of these formulations were all carefully assessed and compared. It was shown that the chitosan sample that bound to a particular bile acid the best did not necessarily have the best binding capacity to other bile acids. Individual bile acid-binding ability did not correspond with any of the other chitosan physicochemical properties studied. These findings imply that it may be hard to predict chitosan's propensity to bind bile acids using Mw, as determined by DD, swelling capacity, or solution viscosity [58].

6.3.9 Antimicrobial Activity

The antibacterial activity of chitosan, chitin, and its derivatives against a range of pathogens, including yeast, bacteria, and fungus, has drawn a lot of interest in recent years. There are two main reasons why chitosan prevents the growth of microbial cells. Being polycationic, it forms an impermeable barrier surrounding the cell when in contact with anionic groups on the cell surface, blocking the transfer of essential solutes. The region of action in Gram (-) bacteria is the outer membrane, according to electron microscopy. Chitosan has been seen to have a permeabilizing effect when protonated. However, this permeabilizing impact is reversible [8].

Permeating into the cell nucleus and blocking the production of RNA and proteins is the second phase. Liu et al. [56] found tagged chitosan oligomers with Mw spanning from 8 to 5 kDa in *E. coli* cells, and they displayed remarkable antibacterial capabilities. In this case, the Mw is the deciding element (Table 6.1).

Table 6.1 Chitosan DD and Mw's effects on antimicrobial activity

Physical and chemical properties	Influence on antimicrobial activity
↑DD	↑Membrane electrostatic bonding
	↑Transmittance effect
↑Mw	↓Permeation into the cell nucleus

Additionally, alternate methods have been suggested. Chitosan may hinder microorganisms from growing at their normal rate by functioning as a chelating agent and blocking metals, trace minerals, or vital nutrients from being accessible to the organism.

Chitosan may also interact with proteins that cause flocculation. However, this interaction is very pH-dependent [57]. According to some studies, a more direct disruption of membrane function might account for the antibacterial effect of chitosan against filamentous fungi [35]. It is unclear, nevertheless, whether cell death or growth inhibition is what leads to chitosan's antibacterial effect.

Chitosan, O-CM-chitosan, and N,O-CM chitosan were shown to have an increase in antibacterial activity in that sequence. Less amino residues are present in the first product, where carboxymethyl groups have replaced the hydroxyl and amino groups. O-CM-chitosan has the same number of amino groups as other forms of the substance. Additionally, its carboxyl group may have had an intra- or intermolecular reaction with the amino groups, charging these groups. The researchers came to the conclusion that chitosan and its carboxymethylated derivatives' antibacterial activity is influenced by the quantity of effective $-NH^{3+}$ groups [61].

Numerous applications demonstrate that chitosan's ability to attach to bacterial cell walls is strengthened by an increase in positive charge. A study by [58] has highlighted the connection between Mw, the amount of charges, and antibacterial action. They demonstrated that O-CM chitosan, which is made from chitin that has been degraded, is more efficient than regular chitin. The COOH group's intra– or intermolecular contact with the NH_2 group to impart a charge was thought to be the cause of the increase in the amount of $-NH_3$ groups. In the context of native chitosan, an elevated concentration of amino groups on O-CM chitosan facilitates the development of a configuration characterized by robust intramolecular hydrogen bonding. This configuration, in turn, diminishes the availability of amino groups for the purpose of binding bacterial surfaces.

However, many researchers have yet to uncover a direct link between deacetylation levels and antibacterial activity. These scientists think that both the chitosan and the microorganism utilized have an impact on the antibacterial activity of chitosan [63]. Three Gram (-) bacteria, five Gram (+) bacteria, and hetero-chitosans with varied DD all showed antibacterial properties, according to the research. Gram (+) bacteria, Mws with 75% deacetylation, demonstrated the most potent antimicrobial activity [59].

6.3.9.1 Chitosan concentration effects on antibacterial effects

Numerous studies have shown that chitosan may stop bacterial growth at various doses [60]. A solution with DD = 7.5% outperformed one with DD = 15% [61]. The amount of chitosan required to suppress bacterial growth is usually linked to the DD of the chitosan.

Chitosan only binds to the negatively charged cell surface of Gram (–) bacteria at low doses. It affects the permeability of the cell membrane in this situation, resulting in the release of internal components and cellular death.

However, at higher quantities, chitosan's positively charged cellular surface may be coated with amino groups., preventing intracellular components from entering the cell. Furthermore, positively charged chitosan and positively charged bacterial cells will reject each other in the case of Gram (+) bacteria, preventing agglutination [62].

The effectiveness of chitosan's Mw and concentration against *E. coli* was examined by Liu et al. [63]. They employed chitosans with a range of molecular weights (5.5×10^4 to 15.5×10^4 Da) and concentrations (20 to 1000 ppm). In the case of important concentrations (>200 ppm), all chitosan samples demonstrated effective antibacterial properties, whereas, at low concentrations (20 ppm), all samples might aid in the development of *E. coli*. This might interpret the possibility that the bactericidal effects of high chitosan concentrations result from bacterial flocculation. Low chitosan concentrations had the opposite effect, favoring bacterial multiplication and enhancing bacterial survival [64].

6.3.9.2 Influence of environment pH on chitosan's antibacterial effect

The ambient pH has a significant impact on the antibacterial efficacy of chitosan and its derivatives. While it primarily affects chitosan solubility, it also has an impact on the electrical charges inside the molecule. This distinguishing property enables chitosan molecules to establish bonds through electrical interactions, emphasizing its significance in promoting molecular connections [70].

Native chitosan is commonly recognized to be insoluble in organic solvents, water, and alkaline environments, while it is soluble in organic acids at pH levels lower than 6 (Fig. 6.1). By neutralizing it with HCl, lactic acid, acetic acid, or formic acid, many water-soluble chitosan salts can be created [71]. The ease with which glucosamine units convert into the soluble form of $R\text{-}NH_3^+$ determines its capacity to dissolve in weak aqueous solutions. Chitosan, which is insoluble in water, has been found to exhibit antibacterial activity in acidic media through experimental data, making it suitable for application as a preservative in acidic food products [65].

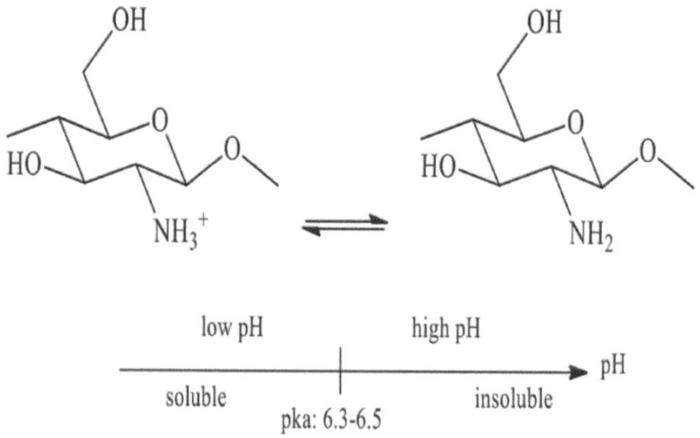

Figure 6.1 Discuss the influence of pH on the solubility of chitosan (reproduced with permission from Nadia et al. [67]).

On higher animals' chitosan has been found to have a number of impacts, and it has been proven to have a variety of effects on higher animals. Once again, the results are significantly influenced using the physicochemical characteristics of chitosan as well as the nature and conditions of the interaction. Chitosan has several present and future applications because of its compatibility with and physiological influences on both humans and animals.

As a result, these chitosan-related elements are well explored, and various reviews on relevant topics are accessible. For a number of species and people, the nutritional effects of chitosan have been studied. Chitosan-supplemented meals for agricultural and laboratory animals, including fish, rats, mice, chickens, and rabbits, have been shown to have nutritional effects. The majority of studies show a pattern of development that is normal or increased and a decrease in the absorption of dietary fat. Due to alterations in the digestive microbiota and lack of appetite, several writers have suggested that chitosan-supplemented diets may have long-term negative consequences [66].

6.3.9.3 Contribution of bacteria type to the antibacterial influence of chitosan

In 1979, a preliminary addition was made to our understanding of the antibacterial effect of chitosan on many species of bacteria [68]. Over time, it was discovered that the amino groups that are directly affected by the DD and DP degrees are responsible for the entire antibacterial effect of chitosan. Because of the differences in the composition of their cell walls, chitosan interacts differently with Gram (-) and Gram (+) bacteria [69]. The increased affinity of amino groups for the anionic radicals present in the cell wall is thought to make Gram (-) bacteria more vulnerable to bactericidal action than Gram (+) bacteria [77]. Previous research has shown that the Gram (-) outer membrane barrier makes Gram (+) bacteria more susceptible to anti-bacterial therapy. When chitosan is absorbed into the surface of bacterial cells, it is assumed to possess antibacterial characteristics. The lipid cell membrane then becomes more porous, allowing essential chemicals to leave the cell and ultimately leading to cell death [67].

Although a number of potential theories for chitosan and its derivatives' antibacterial activity have been put up, the precise mechanism of action is yet unknown. The presence of molecular connections between these chemicals and the cell membrane, however, is amply supported by the available data [70]. Ionic and/or hydrophobic interactions are frequently blamed for cell membrane degradation or rupture [71]. Because of these features, Gram (-) bacteria have far higher hydrophilicity Gram (-) bacteria, which makes them particularly susceptible to the effects of chitosan. Gram (-) bacteria incur more morphological changes in their cell walls as a consequence of antibacterial medication activity than Gram (+) bacteria do. The quantity of chitosan that is adsorbed depends on the density of electrical charges on the surface of the bacterial cell. The alterations in cell membrane permeability and structure become more pronounced the more chitosan is adsorbed. According to electronic microscopy studies [72], chitosan modifies the bacterial outer membrane by generating vesicular structures and altering the cell membrane's barrier function [73]. Chitosan has a polycationic nature and mostly interacts with lipopolysaccharides and proteins found in anionic bacterial walls. The effects of this interaction are shown by the externalization of internal components and the inability of nutrients, which are bound to metals, to enter the bacterial cell [74].

Chelating processes allow amino groups to interact with the metal ions in a reaction. It occurs due to the protonation of amino groups in an acidic medium. They can change the typical function of anionic substances like proteins or anionic residues to induce electrostatic attraction, yet, at the same time, their ability to bind metal cations is diminished [75].

The peptidoglycan layer that forms the hydrophilic surface of Gram (–) bacteria's membrane is a two-dimensional, thin structure (Fig. 6.2). Lipoproteins, phospholipids, and lipopolysaccharides make up the cytoplasmic membrane. Electrostatic interaction is feasible when the protonated amino groups of chitosan come into touch with a certain anionic bacterial surface (phosphate residues, carboxylic residues, etc.), particularly if the chitosan is of low molecular weight. As a result, cell permeability is affected, and the bacterial wall's osmotic stability is reduced. The contract gives rise to complexes that alter the barrier characteristics of the cytoplasmic membrane. These complexes penetrate the cell and impede bacterial physiological functions, leading to the leaking of bacterial enzymes and nucleotides [58]. Finally, Gram (–) bacteria's exterior membrane layer serves as a defense against hydrophobic byproducts and macromolecules, which explains why they are resistant to hydrophobic antibiotics.

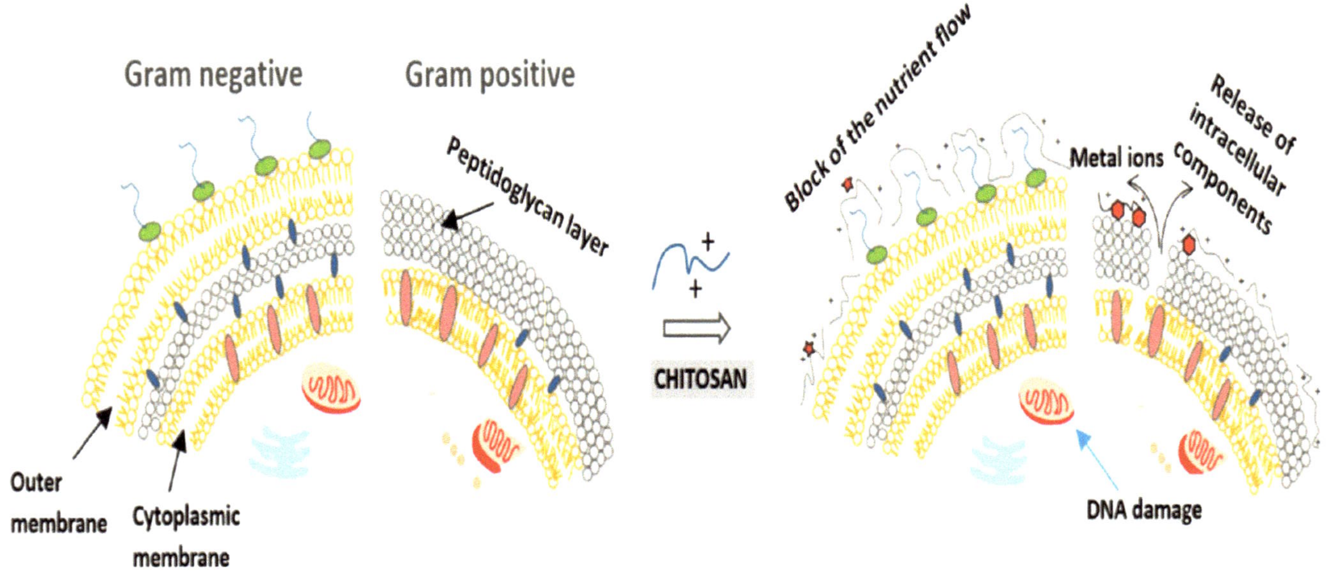

Figure 6.2 Models for how chitosan affects both Gram (-) and Gram (+) bacteria have been proposed (reproduced with permission from Kravanja et al. [77]).

Chitosan has a chelating action that binds necessary metals and prevents the development of microorganisms. It is generally known that chitosan has exceptional metal-binding properties, working with metals via charged amino groups [75]. Toxin generation and bacterial growth are both prevented by this form of interaction between amino functions and divalent ions (Ca^{2+} or Mg^{2+}) in the microorganism cell wall [76].

The molecular size of chitosan and the medium's pH have a significant impact on the permeability of bacterial membranes. These elements were already discussed. In particular, peptidoglycans make up the three-dimensional layer that makes up the cellular wall of Gram (+) bacteria.

The bactericidal effect of chitosan on Gram (+) bacteria may be attributed to its capacity to directly obstruct the passage of nutrients and oxygen into the intracellular area [67]. This method necessitates the use of greater Mw chitosan, which forms a polymer membrane on the bacterial cell's surface.

Low-molecular-weight chitosan's interaction with DNA, which restricts mRNA and protein synthesis after entering microorganism nuclei, is another bactericidal mechanism. The finding of intracellular chito-oligomers in *E. coli* showed this. These are almost certainly preventing DNA transcription [85].

Finally, the variations in chitosan sensitivity between Gram (-) and Gram (+) bacteria are mostly related to changes in cell wall topologies. Gram (-) bacteria have three barrier membranes, as opposed to Gram (+) bacteria, which only have a thin, negatively charged peptidoglycan layer. These are the hydrophobic outer membrane, the peptidoglycan layer, and the cell membrane. Furthermore, variations in the structure, intrinsic (degree of deacetylation, molecular mass), and extrinsic (pH, concentration, contact duration) variables influence the manifestation and effectiveness of chitosan's or its derivatives' bactericidal impact on various bacteria species.

6.3.10 Antifungal Properties

Chitosan exhibits a strong antifungal effect against chitosan-sensitive fungi. It is assumed that chitosan, which carries a positive charge, establishes a dynamic interaction with the anionic phospholipids that are present in the fungal cell membrane. This intricate interplay subsequently instigates the disruption of the membrane integrity, facilitating the orchestrated ingress of the chitosan into the fungal cell. Chitosan is blocked by the cell membrane of the chitosan-resistant fungus, and the polysaccharide stays on the outside surface. Palma et al. discovered the difference in permeability between chitosan-sensitive and chitosan-resistant fungi. Their findings led to the deduction that fungi responsive to chitosan exhibited an augmented presence of polyunsaturated fatty acids within their plasma membranes compared to chitosan-resistant counterparts. This observation strongly suggests a potential correlation between chitosan-induced membrane permeabilization and the fluidity of the lipid bilayer. This hypothesis gained support from an empirical investigation involving a variant of *Neurospora crassa*, which displayed reduced plasma membrane fluidity compared to the wild-type counterpart. Notably, the mutant strain of *N. crassa*, characterized by diminished levels of unsaturated fatty acids, demonstrated heightened resistance to the antifungal effects of chitosan. Chitosan easily penetrates through the cell membrane, causing it to be destroyed and the fungal cell to die.

A study revealed that fungal chitosan, with lower molecular weight and higher degree of deacetylation, showed more significant antifungal activity against *Candida albicans*. In a related study, they showed a correlation between the molecular weight of chitosan and its effects on different strains of *Candida spp.* The inhibitory effect on the development of *Candida tropicalis* and *Candida parapsilosis* increased with the increase in the molecular weight of chitosan. Nevertheless, modifications in the molecular weight did not provide any discernible impact on the growth of *C. albicans*.

6.3.11 Antioxidative Activity

Demonstrating efficacy on par with established commercial antioxidants, chitosan has exhibited a robust capability to neutralize diverse radical species. The potential of extracts derived from chitin sourced from crab shells, and possessing degrees of deacetylation (DD) at 90%, 75%, and 50%, was investigated for their ability to counter alkyl, hydroxyl, superoxide, and 1,1-diphenyl-2-picrylhydrazyl (DPPH) radicals [9]. The anti-superoxide and anti-hydroxyl characteristics of chitosans of various sizes and their sulfate derivatives, on the other hand, were investigated. Table 6.2 shows that chitosan Mw and activity are adversely associated. The chitosan sulfated derivatives were more effective in scavenging peroxide radicals, although the chitosan with the lowest Mw was more effective at chelating ferrous ions than the others [22]. An aspect that positions chitosan as a potential natural antioxidant for enhancing the shelf life of lipid-based foods is its ability to chelate metal ions. Through the process of chelation, chitosan interacts with existing ferrous ions within the system, effectively mitigating their prooxidant effects or transformation into ferric ions. This action, in turn, has the potential to delay the onset of lipid oxidation, thereby extending the product's longevity [89].

Table 6.2 The impact of chitosan DD and Mw on antioxidative activity

Physical and chemical properties	Influence on antioxidative activity
↑ DD	↑ Scavenging influence
↑ Mw	↓ The impact of radical scavenging
	↓ The efficacy of ion chelation

Emphasis must be placed on their provisioning specifically rather than the unique favorable characteristics that chitosan and its derivatives display, like biodegradability, safety, biocompatibility, and intriguing life activities [78]. The biological, biotechnology, food, and pharmaceutical sectors existed before that. The following will cover the antitumor, antimicrobial, and cell-reinforcing effects associated with their enjoyable everyday activities. Chitosan has ideal film-making properties, and these properties must be very important in the field of food packaging. Protection will also make edible antibacterial films rather than needing chemical additives. Chitosan requires additional great obstructive qualities, such as a polymeric ingredient for spectacular cell reinforcement and antibacterial capabilities, as it does not transfer well from the ensuring fresh into a film. It is possible that this has already been confirmed and clarified by more research [79]. For example, Friedman et al. [80] examined the antimicrobial properties of chitosan in solutions, powders, eatable films, and coverings against infections, bacteria that cause food spoilage, and growths for various food categories. Apple and orange juices, eggs, pork, dairy goods, cereal, and fish are a few examples of them.

They assert that at pH levels below 6, low-molecular-weight chitosans present the best circumstances for attaining amicable anti-oxidative and antibacterial preservative actions before fluid, as well as strong meals [81]. The aforementioned study could have described how chitosan and its derivatives are used in the food industry. These biopolymers, they said, offer a broad range of special request-directing characteristics, such as protecting food from microbial decay and forming from claims biodegradable sheets. In a study [82], it was claimed that chitosan films produced in response to demand have a tendency to provide immunity in the event of lipid diffusion and specific per manganese gas corrosion. The gearbox's poor water and water vapor safety, however, is where issues start. This could be a result of chitosan's high hydrophilicity, which encourages water atoms to group together and interact. Therefore, techniques for producing chitosan-based bioactive coatings, such as polymer blending, the use of bio-composites, and multilayer frameworks, require help.

6.4 Relationship Between Biological Activities and Chemical Structure

Instead of the specific positive features that chitosan and its derivatives exhibit, like biodegradability, biocompatibility, safety, and interesting life activities, emphasis must be given to their provisioning, in particular; prior to that, the biomedical, biotechnology, food, and pharmaceutical industries [32]. Antimicrobial, antitumor, and cell-reinforcing actions related to their pleasant daily activities will be reviewed together with the topic of interest below. Due to the perfect film-making characteristics of chitosan, these qualities must be extremely valuable in the field of food packaging. Protection will also refrain from using concoction additives and will produce edible antimicrobial films. Chitosan does not transfer well from the ensuring new into a film and needs extra fantastic obstructive features, much like a polymeric ingredient for superb cell reinforcement and antibacterial capabilities. This may have previously been clarified and established using other investigations [80].

For instance, Friedman et al. [83] were the researchers who looked at the antimicrobial properties of chitosan previously in solutions, eatable films, powders, and coverings against pathogens that are transmitted through food, infections, bacteria that cause food spoilage, and growths for various food categories. These include juices from apples and oranges, dairy products, eggs, pork, fish, and cereal. They contend that low-molecular-weight chitosans exhibit optimal conditions for achieving amiable anti-oxidative and antibacterial preservative effects before fluid and also strong foods [80] at pH levels below 6. The study mentioned may have also included a description of the use of chitosan and its derivatives in the food sector. They explained that these biopolymers suggest a wide range of unique request-directing capabilities, including safeguarding foods from microbial retrogradation and shaping from claiming biodegradable films. According to [84], chitosan films made with a request tend to provide imperviousness in the case of fat propagation and particular gas permanganic corrosive. However, problems begin with its shoddy water and water vapor gearbox safety [85]. This could be due to chitosan's very hydrophilic characteristics, which encourage water atoms to climb together and communicate. For this reason, methods to manufacture chitosan-based bioactive coatings, such as polymer mixing or the utilization of bio-composites and multilayer frameworks, need assistance.

6.5 Conclusion and Prospects

Chitin and chitosan are value-adding items that satisfy customers' need for affordable, natural goods with a favorable influence on health. The unanimous consensus surrounding these polysaccharides arises from their advantageous characteristics, including biocompatibility, biodegradability, non-toxicity, and solubility. These attributes are pivotal for the transformation of polysaccharides into innovative materials with potential applications. Consequently, these materials find utility across diverse industrial and pharmaceutical sectors. Their versatile potential encompasses roles as antibacterial, antifungal, radical scavengers, antioxidants, antitumor agents, antiparasitic agents, cardiovascular aids, anti-hypercholesterolemia agents, anti-inflammatory agents, antiaging agents, antiviral agents, detoxification aids, hepatoprotective agents, hemostatic agents, wound healing promoters, and immunomodulators. The recent strides in exploring polysaccharides integrated with functional nanomaterials (guided by biocompatibility, hydrophilicity, and water swelling ratios via electrostatic interactions) across various applications represent a promising avenue, particularly in the realm of nanotechnology.

References

1. Elieh-Ali-Komi D, Hamblin MR (2016) Chitin and chitosan: production and application of versatile biomedical nanomaterials. *International Journal of Advanced Research* 4: 411.
2. Amor IB, Hemmami H, Laouini SE, Temam HB, Zaoui H, Barhoum A (2023) Biosynthesis MgO and ZnO nanoparticles using chitosan extracted from Pimelia Payraudi Latreille for antibacterial applications. *World Journal of Microbiology and Biotechnology* 39: 19.
3. Amor IB, Hemmami H, Laouini SE, Abdelaziz AG, Barhoum A (2023) Influence of chitosan source and degree of deacetylation on antibacterial activity and adsorption of AZO dye from water. *Biomass Conversion and Biorefinery*: 1–11.
4. Alhamad AA, Zeghoud S, Amor IB, Hemmami H (2023) Chitosan-based hydrogels for wound healing: correspondence. *International Journal of Surgery* 109: 1821–1822.

5. Park BK, Kim M-M (2010) Applications of chitin and its derivatives in biological medicine. *International Journal of Molecular Sciences* 11: 5152–5164.
6. Xia W, Liu P, Zhang J, Chen J (2011) Biological activities of chitosan and chitooligosaccharides. *Food hydrocolloids* 25: 170–179.
7. Dave U, Somanader E, Baharlouei P, Pham L, Rahman MA (2021) Applications of chitin in medical, environmental, and agricultural industries. *Journal of Marine Science and Engineering* 9: 1173.
8. Morin-Crini N, Lichtfouse E, Torri G, Crini G (2019) Applications of chitosan in food, pharmaceuticals, medicine, cosmetics, agriculture, textiles, pulp and paper, biotechnology, and environmental chemistry. *Environmental Chemistry Letters* 17: 1667–1692.
9. Ben Amor I, Hemmami H, Laouini SE, Mahboub MS, Barhoum A (2022) Sol-gel synthesis of ZnO nanoparticles using different chitosan sources: effects on antibacterial activity and photocatalytic degradation of AZO Dye. *Catalysts* 12: 1611.
10. Ben Amor I, Hemmami H, Laouini SE, Zeghoud S, Benzina M, Achour S, Naseef A, Alsalme A, Barhoum A (2023) Use of insect-derived chitosan for the removal of methylene blue dye from wastewater: process optimization using a central composite design. *Materials* 16: 5049.
11. Al-Rooqi MM, Hassan MM, Moussa Z, Obaid RJ, Suman NH, Wagner MH, Natto SS, Ahmed SA (2022) Advancement of chitin and chitosan as promising biomaterials. *Journal of Saudi Chemical Society* 26: 101561.
12. Aranaz I, Mengíbar M, Harris R, Paños I, Miralles B, Acosta N, Galed G, Heras Á (2009) Functional characterization of chitin and chitosan. *Current Chemical Biology* 3: 203–230.
13. Chen Z, Cheng S, Lin H, Wu W, Liang L, Chen X, Zheng X, He Y, Zhang K (2021) Antibacterial, anti-inflammatory, analgesic, and hemostatic activities of Acanthopanax trifoliatus (L.) merr. *Food Science & Nutrition* 9: 2191–2202.
14. Brunner E, Richthammer P, Ehrlich H, Paasch S, Simon P, Ueberlein S, van Pée KH (2009) Chitin-based organic networks: an integral part of cell wall biosilica in the diatom Thalassiosira pseudonana. *Angewandte Chemie International Edition* 48: 9724–9727.
15. Tsurkan D, Wysokowski M, Petrenko I, Voronkina A, Khrunyk Y, Fursov A, Ehrlich H (2020) Modern scaffolding strategies based on naturally pre-fabricated 3D biomaterials of poriferan origin. *Applied Physics A* 126: 1–9.

16. Cheung RCF, Ng TB, Wong JH, Chan WY (2015) Chitosan: an update on potential biomedical and pharmaceutical applications. *Marine Drugs* 13: 5156–5186.
17. Pellis A, Guebitz GM, Nyanhongo GS (2022) Chitosan: sources, processing and modification techniques. *Gels* 8: 393.
18. Sreekumar S, Wattjes J, Niehues A, Mengoni T, Mendes AC, Morris ER, Goycoolea FM, Moerschbacher BM (2022) Biotechnologically produced chitosans with nonrandom acetylation patterns differ from conventional chitosans in properties and activities. *Nature Communications* 13: 7125.
19. Argüelles-Monal WM, Lizardi-Mendoza J, Fernández-Quiroz D, Recillas-Mota MT, Montiel-Herrera M (2018) Chitosan derivatives: introducing new functionalities with a controlled molecular architecture for innovative materials. *Polymers* 10: 342.
20. Varki A (2017) Biological roles of glycans. *Glycobiology* 27: 3–49.
21. Obaid RJ, Mughal EU, Naeem N, Sadiq A, Alsantali RI, Jassas RS, Moussa Z, Ahmed SA (2021) Natural and synthetic flavonoid derivatives as new potential tyrosinase inhibitors: a systematic review. *RSC Advances* 11: 22159–22198.
22. Aranaz I, Alcántara AR, Civera MC, Arias C, Elorza B, Heras Caballero A, Acosta N (2021) Chitosan: an overview of its properties and applications. *Polymers* 13: 3256.
23. Rodríguez-Vázquez M, Vega-Ruiz B, Ramos-Zúñiga R, Saldaña-Koppel DA, Quiñones-Olvera LF (2015) Chitosan and its potential use as a scaffold for tissue engineering in regenerative medicine. *BioMed Research International* 2015: 821279.
24. Abourehab MA, Pramanik S, Abdelgawad MA, Abualsoud BM, Kadi A, Ansari MJ, Deepak A (2022) Recent advances of chitosan formulations in biomedical applications. *International Journal of Molecular Sciences* 23: 10975.
25. Ibram A (2021) Studies on chitosan, chitin and chitooligosaccharides and their biomedical properties. *European Journal of Natural Sciences and Medicine* 4: 99–112.
26. de Queiroz Antonino RSCM, Lia Fook BRP, de Oliveira Lima VA, de Farias Rached RÍ, Lima EPN, da Silva Lima RJ, Peniche Covas CA, Lia Fook MV (2017) Preparation and characterization of chitosan obtained from shells of shrimp (Litopenaeus vannamei Boone). *Marine Drugs* 15: 141.
27. Islam N, Dmour I, Taha MO (2019) Degradability of chitosan micro/nanoparticles for pulmonary drug delivery. *Heliyon* 5: e01684.

28. Piekarska K, Sikora M, Owczarek M, Jóźwik-Pruska J, Wiśniewska-Wrona M (2023) Chitin and chitosan as polymers of the future—obtaining, modification, life cycle assessment and main directions of application. *Polymers* 15: 793.
29. Nowacki K, Stępniak I, Langer E, Tsurkan M, Wysokowski M, Petrenko I, Khrunyk Y, Fursov A, Bo M, Bavestrello G (2020) Electrochemical approach for isolation of chitin from the skeleton of the black coral Cirrhipathes sp. (Antipatharia). *Marine Drugs* 18: 297.
30. Tsurkan MV, Voronkina A, Khrunyk Y, Wysokowski M, Petrenko I, Ehrlich H (2021) Progress in chitin analytics. *Carbohydrate Polymers* 252: 117204.
31. Ehrlich H, Bailey E, Wysokowski M, Jesionowski T (2021) Forced biomineralization: a review. *Biomimetics* 6: 46.
32. Younes I, Rinaudo M (2015) Chitin and chitosan preparation from marine sources. Structure, properties and applications. *Marine Drugs* 13: 1133–1174.
33. Khrunyk Y, Lach S, Petrenko I, Ehrlich H (2020) Progress in modern marine biomaterials research. *Marine Drugs* 18: 589.
34. Kertmen A, Ehrlich H (2022) Patentology of chitinous biomaterials. Part I: Chitin. *Carbohydrate Polymers* 282: 119102.
35. Anitha A, Sowmya S, Kumar PS, Deepthi S, Chennazhi K, Ehrlich H, Tsurkan M, Jayakumar R (2014) Chitin and chitosan in selected biomedical applications. *Progress in Polymer Science* 39: 1644–1667.
36. Yang J, Tian F, Wang Z, Wang Q, Zeng YJ, Chen SQ (2008) Effect of chitosan molecular weight and deacetylation degree on hemostasis. *Journal of Biomedical Materials Research Part B: Applied Biomaterials* 84: 131–137.
37. Casadidio C, Peregrina DV, Gigliobianco MR, Deng S, Censi R, Di Martino P (2019) Chitin and chitosans: characteristics, eco-friendly processes, and applications in cosmetic science. *Marine Drugs* 17: 369.
38. Okamoto Y, Kawakami K, Miyatake K, Morimoto M, Shigemasa Y, Minami S (2002) Analgesic effects of chitin and chitosan. *Carbohydrate Polymers* 49: 249–252.
39. de Sousa Victor R, Marcelo da Cunha Santos A, Viana de Sousa B, de Araújo Neves G, Navarro de Lima Santana L, Rodrigues Menezes R (2020) A review on Chitosan's uses as biomaterial: tissue engineering, drug delivery systems and cancer treatment. *Materials* 13: 4995.

40. Kim S (2018) Competitive biological activities of chitosan and its derivatives: antimicrobial, antioxidant, anticancer, and anti-inflammatory activities. *International Journal of Polymer Science* 2018.
41. Adhikari HS, Yadav PN (2018) Anticancer activity of chitosan, chitosan derivatives, and their mechanism of action. *International Journal of Biomaterials* 2018: 2952085.
42. Azuma K, Osaki T, Minami S, Okamoto Y (2015) Anticancer and anti-inflammatory properties of chitin and chitosan oligosaccharides. *Journal of Functional Biomaterials* 6: 33–49.
43. Ding J, Guo Y (2022) Recent advances in chitosan and its derivatives in cancer treatment. *Frontiers in Pharmacology* 13: 888740.
44. Burhan AM, Klahan B, Cummins W, Andrés-Guerrero V, Byrne ME, O'reilly NJ, Chauhan A, Fitzhenry L, Hughes H (2021) Posterior segment ophthalmic drug delivery: role of muco-adhesion with a special focus on chitosan. *Pharmaceutics* 13: 1685.
45. Zhong Y, Zhuang C, Gu W, Zhao Y (2019) Effect of molecular weight on the properties of chitosan films prepared using electrostatic spraying technique. *Carbohydrate Polymers* 212: 197–205.
46. El-Kamel AH, Ashri LY, Alsarra IA (2007) Micromatricial metronidazole benzoate film as a local mucoadhesive delivery system for treatment of periodontal diseases. *AAPS Pharmscitech* 8: E184–E194.
47. Ways TMM, Lau WM, Khutoryanskiy VV (2018) Chitosan and its derivatives for application in mucoadhesive drug delivery systems. *Polymers* 10: 267.
48. Mercy H, Halim A, Hussein A (2012) Chitosan-derivatives as hemostatic agents: their role in tissue regeneration. *Regenerative Research* 1: 38–46.
49. Mikušová V, Mikuš P (2021) Advances in chitosan-based nanoparticles for drug delivery. *International Journal of Molecular Sciences* 22: 9652.
50. Román-Doval R, Torres-Arellanes SP, Tenorio-Barajas AY, Gómez-Sánchez A, Valencia-Lazcano AA (2023) Chitosan: properties and its application in agriculture in context of molecular weight. *Polymers* 15: 2867.
51. Muzzarelli RA, Orlandini F, Pacetti D, Boselli E, Frega NG, Tosi G, Muzzarelli C (2006) Chitosan taurocholate capacity to bind lipids and to undergo enzymatic hydrolysis: an in vitro model. *Carbohydrate Polymers* 66: 363–371.

52. Wang DQ-H, Tazuma S, Cohen DE, Carey MC (2003) Feeding natural hydrophilic bile acids inhibits intestinal cholesterol absorption: studies in the gallstone-susceptible mouse. *American Journal of Physiology-Gastrointestinal and Liver Physiology* 285: G494–G502.

53. No HK, Lee SH, Park NY, Meyers SP (2003) Comparison of physicochemical, binding, and antibacterial properties of chitosans prepared without and with deproteinization process. *Journal of Agricultural and Food Chemistry* 51: 7659–7663.

54. Jin Q, Yu H, Wang X, Li K, Li P (2017) Effect of the molecular weight of water-soluble chitosan on its fat-/cholesterol-binding capacities and inhibitory activities to pancreatic lipase. *PeerJ* 5: e3279.

55. Coscueta ER, Sousa AS, Reis CA, Pintado M (2021) Chitosan-olive oil microparticles for phenylethyl isothiocyanate delivery: optimal formulation. *PLoS One* 16: e0248257.

56. Tang H, Zhang P, Kieft TL, Ryan SJ, Baker SM, Wiesmann WP, Rogelj S (2010) Antibacterial action of a novel functionalized chitosan-arginine against Gram-negative bacteria. *Acta Biomaterialia* 6: 2562–2571.

57. Muxika A, Etxabide A, Uranga J, Guerrero P, De La Caba K (2017) Chitosan as a bioactive polymer: processing, properties and applications. *International Journal of Biological Macromolecules* 105: 1358–1368.

58. Godoy-Gallardo M, Eckhard U, Delgado LM, de Roo Puente YJ, Hoyos-Nogués M, Gil FJ, Perez RA (2021) Antibacterial approaches in tissue engineering using metal ions and nanoparticles: from mechanisms to applications. *Bioactive Materials* 6: 4470–4490.

59. JE J-Y, KIM S-K, Byun H-G, Moon S-H (2004) Antimicrobial activity of hetero-chitosans and their oligosaccharides with different molecular weights. *Journal of Microbiology and Biotechnology* 14: 317–323.

60. Aliasghari A, Khorasgani MR, Vaezifar S, Rahimi F, Younesi H, Khoroushi M (2016) Evaluation of antibacterial efficiency of chitosan and chitosan nanoparticles on cariogenic streptococci: an in vitro study. *Iranian Journal of Microbiology* 8: 93.

61. Rabea EI, Badawy ME-T, Stevens CV, Smagghe G, Steurbaut W (2003) Chitosan as antimicrobial agent: applications and mode of action. *Biomacromolecules* 4: 1457–1465.

62. Hosseinnejad M, Jafari SM (2016) Evaluation of different factors affecting antimicrobial properties of chitosan. *International Journal of Biological Macromolecules* 85: 467–475.

63. Liu N, Chen X-G, Park H-J, Liu C-G, Liu C-S, Meng X-H, Yu L-J (2006) Effect of MW and concentration of chitosan on antibacterial activity of Escherichia coli. *Carbohydrate Polymers* 64: 60–65.
64. Varlamov V, Il'Ina A, Shagdarova BT, Lunkov A, Mysyakina I (2020) Chitin/chitosan and its derivatives: fundamental problems and practical approaches. *Biochemistry (Moscow)* 85: 154–176.
65. Qin C, Li H, Xiao Q, Liu Y, Zhu J, Du Y (2006) Water-solubility of chitosan and its antimicrobial activity. *Carbohydrate Polymers* 63: 367–374.
66. Lizardi-Mendoza J, Monal WMA, Valencia FMG (2016) Chemical characteristics and functional properties of chitosan. *Chitosan in the Preservation of Agricultural Commodities*. Elsevier, pp. 3–31.
67. Morin-Crini N, Lichtfouse E, Torri G, Crini G (2019) Fundamentals and applications of chitosan. *Sustainable Agriculture Reviews 35: Chitin and Chitosan: History, Fundamentals and Innovations*: 49–123.
68. Ardean C, Davidescu CM, Nemeş NS, Negrea A, Ciopec M, Duteanu N, Negrea P, Duda-Seiman D, Musta V (2021) Factors influencing the antibacterial activity of chitosan and chitosan modified by functionalization. *International Journal of Molecular Sciences* 22: 7449.
69. Chung Y-C, Su YP, Chen C-C, Jia G, Wang HL, Wu JG, Lin JG (2004) Relationship between antibacterial activity of chitosan and surface characteristics of cell wall. *Acta Pharmacologica Sinica* 25: 932–936.
70. Matica MA, Aachmann FL, Tøndervik A, Sletta H, Ostafe V (2019) Chitosan as a wound dressing starting material: antimicrobial properties and mode of action. *International Journal of Molecular Sciences* 20: 5889.
71. Pavinatto A, Souza AL, Delezuk JA, Pavinatto FJ, Campana-Filho SP, Oliveira Jr ON (2014) Interaction of O-acylated chitosans with biomembrane models: probing the effects from hydrophobic interactions and hydrogen bonding. *Colloids and Surfaces B: Biointerfaces* 114: 53–59.
72. Goy RC, Britto Dd, Assis OB (2009) A review of the antimicrobial activity of chitosan. *Polímeros* 19: 241–247.
73. Li Z, Yang F, Yang R (2015) Synthesis and characterization of chitosan derivatives with dual-antibacterial functional groups. *International Journal of Biological Macromolecules* 75: 378–387.
74. Severino R, Ferrari G, Vu KD, Donsì F, Salmieri S, Lacroix M (2015) Antimicrobial effects of modified chitosan based coating containing

nanoemulsion of essential oils, modified atmosphere packaging and gamma irradiation against Escherichia coli O157: H7 and Salmonella Typhimurium on green beans. *Food Control* 50: 215–222.

75. Guibal E (2004) Interactions of metal ions with chitosan-based sorbents: a review. *Separation and Purification Technology* 38: 43–74.
76. Mohd Yusof H, Mohamad R, Zaidan UH, Abdul Rahman NA (2019) Microbial synthesis of zinc oxide nanoparticles and their potential application as an antimicrobial agent and a feed supplement in animal industry: a review. *Journal of Animal Science and Biotechnology* 10: 1–22.
77. Kravanja G, Primožič M, Knez Ž, Leitgeb M (2019) Chitosan-based (Nano) materials for novel biomedical applications. *Molecules* 24: 1960.
78. Frąc M, Hannula SE, Bełka M, Jędryczka M (2018) Fungal biodiversity and their role in soil health. *Frontiers in Microbiology* 9: 707.
79. Muñoz-Tebar N, Pérez-Álvarez JA, Fernández-López J, Viuda-Martos M (2023) Chitosan edible films and coatings with added bioactive compounds: antibacterial and antioxidant properties and their application to food products: a review. *Polymers* 15: 396.
80. Friedman M, Juneja VK (2010) Review of antimicrobial and antioxidative activities of chitosans in food. *Journal of Food Protection* 73: 1737–1761.
81. de Farias BS, Grundmann DDR, Rizzi FZ, Martins NSS, Junior TRSAC, de Almeida Pinto LA (2019) Production of low molecular weight chitosan by acid and oxidative pathways: effect on physicochemical properties. *Food Research International* 123: 88–94.
82. Singh TP, Chatli MK, Sahoo J (2015) Development of chitosan based edible films: process optimization using response surface methodology. *Journal of Food Science and Technology* 52: 2530–2543.
83. Alishahi A, Aïder M (2012) Applications of chitosan in the seafood industry and aquaculture: a review. *Food and Bioprocess Technology* 5: 817–830.
84. Butnaru E, Stoleru E, Brebu MA, Darie-Nita RN, Bargan A, Vasile C (2019) Chitosan-based bionanocomposite films prepared by emulsion technique for food preservation. *Materials* 12: 373.
85. Kardas I, Struszczyk MH, Kucharska M, van den Broek LA, van Dam JE, Ciechańska D (2012) Chitin and chitosan as functional biopolymers for industrial applications. *The European Polysaccharide Network of Excellence (EPNOE) Research Initiatives and Results*: 329–373.

Chapter 7

Physicochemical Properties of Chitosan Products

Thiago Luis Aguayo de Castro, Giovanna Eduarda Machado, Juliana Ketlin Moro da Silva, Jéssica Bassani de Oliveira, Rosangela Maria Ferreira da Costa e Silva, and Claudia Andrea Lima Cardoso

Center of Studies in Natural Resources, State University of Mato Grosso do Sul, Dourados, Mato Grosso do Sul, Brazil

thiagoaguayo@gmail.com

Chitosan, a cationic polyelectrolyte biopolymer, is not only unique in its properties but also versatile in its applications. It is derived from the deacetylation of chitin's acetamide groups, and it exhibits solubility in an acid medium and the ability to form non-toxic, biocompatible complexes. The versatility of chitosan is evident in its wide range of products, spanning from the medical field to materials. The emergence of technologies like nanotechnology has further underscored the significance of this biopolymer. This chapter will delve into the intriguing physicochemical properties of chitosan and its various products, including gel, fiber, membrane, and films. It will also emphasize its role in producing flame retardants and as a drug delivery agent, providing a comprehensive understanding of its versatility.

Chitin and Chitosan: Physical and Chemical Properties
Edited by Shakeel Ahmed, Soumeia Zeghoud, Hadia Hemmami, and Ilham Ben Amor

ISBN 978-981-5129-39-7 (Hardcover), 978-1-003-58977-8 (eBook)
www.jennystanford.com

This chapter delves into the physicochemical characteristics of chitosan, both in its powdered form and in an aqueous medium. It also explores some of its main products, such as gel formulations, chelates, films, and fibers, and their applications in chelating, drug delivery, and flame retardancy. This detailed exploration will provide a comprehensive understanding of chitosan's properties and its potential applications.

7.1 Introduction

Chitosan is a polycationic biopolymer formed by saccharides containing more than 5000 glucosamine units and is widely used because it is biodegradable, non-toxic, and biocompatible with human tissue [1]. It presents between 7% and 8% of nitrogen content in its composition, which is present in the form of amines, thus distinguishing chitosan from cellulose by providing specific properties such as the ability to form polyelectrolyte complexes [1]. In this way, chitosan is the only cationic pseudo-natural polymer.

Synonyms of chitosan: chitopearl; CTFA 04299; deacetylated chitin; Flonac N; Kytex H; polyglusam; Sea Cure F [2]. Chitosan was discovered in the 19th century [3].

Obtaining chitosan occurs through the chitin deacetylation reaction that converts the acetamide groups of the monomers into amines (Fig. 7.1), and this conversion can have different levels of efficiency depending on the process employed; consequently, there is a reduction in molecular mass and an increase water solubility [4]. The degree of acetylation (DD) differentiates between chitosan and chitin. To be considered chitosan, the deacetylation product must have a DA greater than 50% [5].

Chitosan's versatility is a key factor in its widespread use. Several functional groups (amines, hydroxyls, glycosidic bonds, and residual acetamide) make chitosan susceptible to modifications in its chemical structure, which can alter its physicochemical properties [6]. The presence of primary amines in its structure enhances chitosan's reactivity, leading to more significant biological activity [7]. This variety of properties makes chitosan an attractive material for developing products, including metal chelating, biocompatibility, non-toxicity, antimicrobial, antitumoral, bone regenerating, and drug delivery.

Another relevant aspect regarding chitosan is that it is not a chemical entity, with its composition directly affected by the production method and raw material [2]. Despite its long history of use, the research on chitosan and its diverse applications has seen a significant surge in recent times (Fig. 7.2), highlighting its growing importance in various fields.

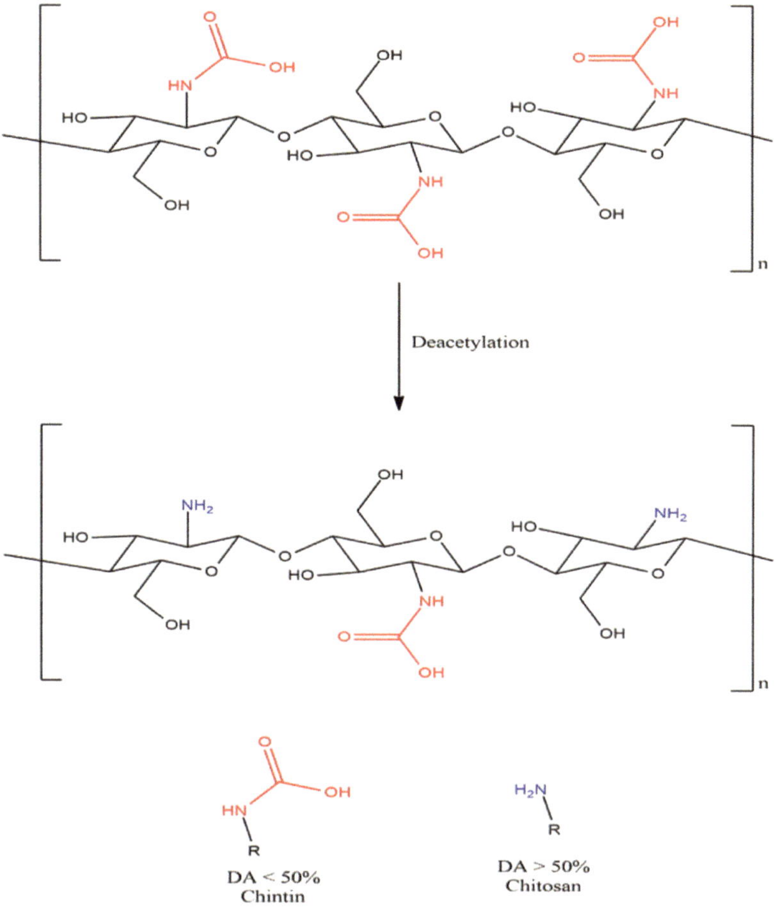

Figure 7.1 Chisotine deacetylation reaction. Source: The authors (2023).

This chapter will present the physicochemical characteristics of chitosan in powder and aqueous medium, as well as some of its main products (gel formulations, chelates, films, and fibers) and their application as chelating, drug delivery, and flame retardants.

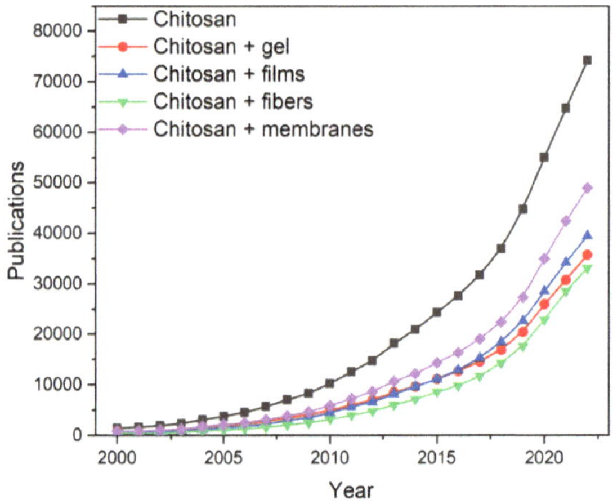

Figure 7.2 Number of research articles with chitosan. Prepared based on data obtained from Dimensions AI (https://www.dimensions.ai/). Source: The authors (2023).

7.2 Chitosan Powder and Liquid

Chitosan has a semicrystalline polymeric structure. A chitosan crystal where complete deacetylation has occurred shows orthorhombic conformation ($P2_12_12_1$) [8]. This same conformation can also be observed in anhydrous [9] or hydrated [10] crystals. The stability of the crystalline structure occurs mainly through hydrogen bonds between hydroxyls 3 and 5 and the bond between the amine group 2 and hydroxyl 6 (Fig. 7. 3) [9, 10].

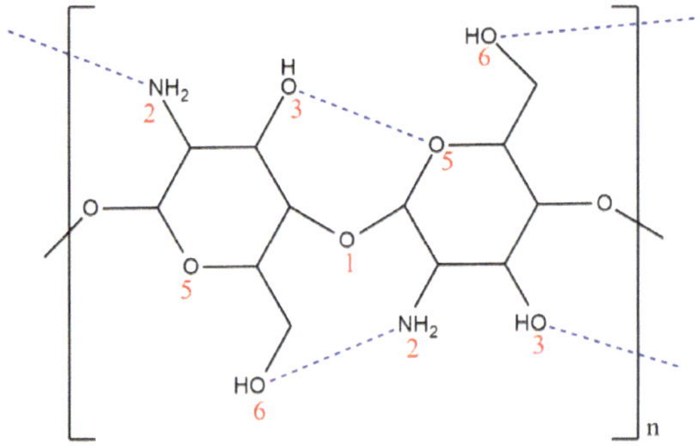

Figure 7.3 Intermolecular interactions of crystalline chitosan. Hydrogen bonds are represented in blue dotted lines. Adapted from Yui et al. [9] and Okuyama et al. [10].

The chitin deacetylation process reduces differences in the crystalline arrangement of samples from different sources and the crystallinity index of the chitosan structure [11], which is one factor in the increase in the solubility of chitosan [12].

There is a polymorphy difference in commercial chitosan [13]. Jang et al. [14] studied the structure of chitins and classified the structures present in α, β, and γ, with the produced chitosan presenting identical conformations. Chitosan α is formed by two arranged polysaccharide chains followed by two in reverse positions joined by solid intermolecular forces and with high crystallinity, β is composed of polysaccharide chains arranged in parallel joined by weak intermolecular forces, and γ is composed of three parallel polysaccharide chains with two being in the same direction and one in the opposite direction [15] (Fig. 7.4).

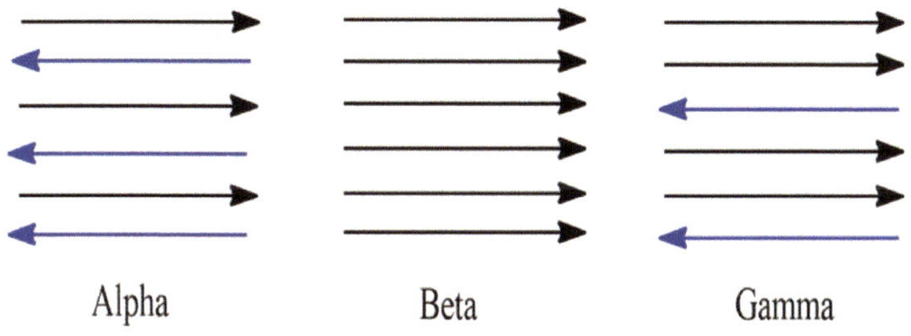

Figure 7.4 Polymorph types of chitin and chitosan. Source: The authors (2023).

The residual crystallinity of chitosan directly affects the accessibility of the polymer's internal sites for hydration, a parameter that varies according to the product's origin [16].

The ionization of chitosan in solution is dependent on pH and degree of deacetylation (DD), as the number of amines present in the polymer structure, their positions, and the number of acetamide groups performing internal hydrogen bonds in the structures directly influence the solubility [12].

The pH directly affects the properties of chitosan when it is solubilized due to the presence of amines that have their protonation state affected, becoming positively charged and giving cationic polyelectrolyte properties to the polymer in an acid medium; however, becoming insoluble at pH greater than 6, due to deprotonation of the nitrogen of the polymeric structure (Fig. 7.5) [17]. The polycationic characteristic of chitosan is associated with its antimicrobial activity, as the positive charge accumulated in the nitrogen interacts with the plasma membrane of gram-negative bacteria [18].

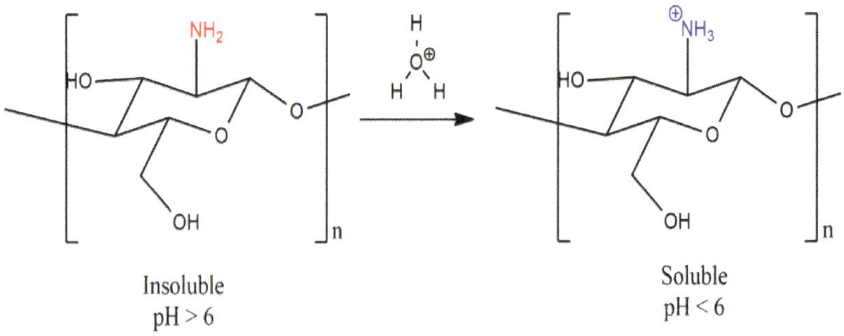

Figure 7.5 Effect of deprotonation on chitosan solubility. Source: The authors (2023).

Chitosan is soluble in solvents and organic acids and partially soluble in inorganic acids, except for phosphoric and sulfuric acids. Chitosan products have a high viscosity, resembling natural gums [2].

Reducing the molecular mass of chitosan chains results in increased biological activities and water solubility [19]; however, biodegradability increases with molecular weight reduction, thus being a crucial factor to consider according to the desired application [20]. According to Dash et al. [20], chitosan is marketed as having low molecular mass (20 to 190 kDa) and high molecular mass (190 to 375 kDa). Table 7.1 summarizes the effect of molecular mass and DD on chitosan properties.

Chitosan stability is influenced by different factors: molecular weight, DD, purity, and moisture content [21]. This process begins with depolymerizing one of the chitosan monomers, reducing molecular weight, destroying functional groups through secondary reactions, and forming oxidizing radicals (Fig. 7.6) [22]. The degradation of chitosan occurs by acid catalysis in the stomach and by enzymatic action [20]. Chitinases and chitosanase enzymes act in the depolymerization of chitosan [23].

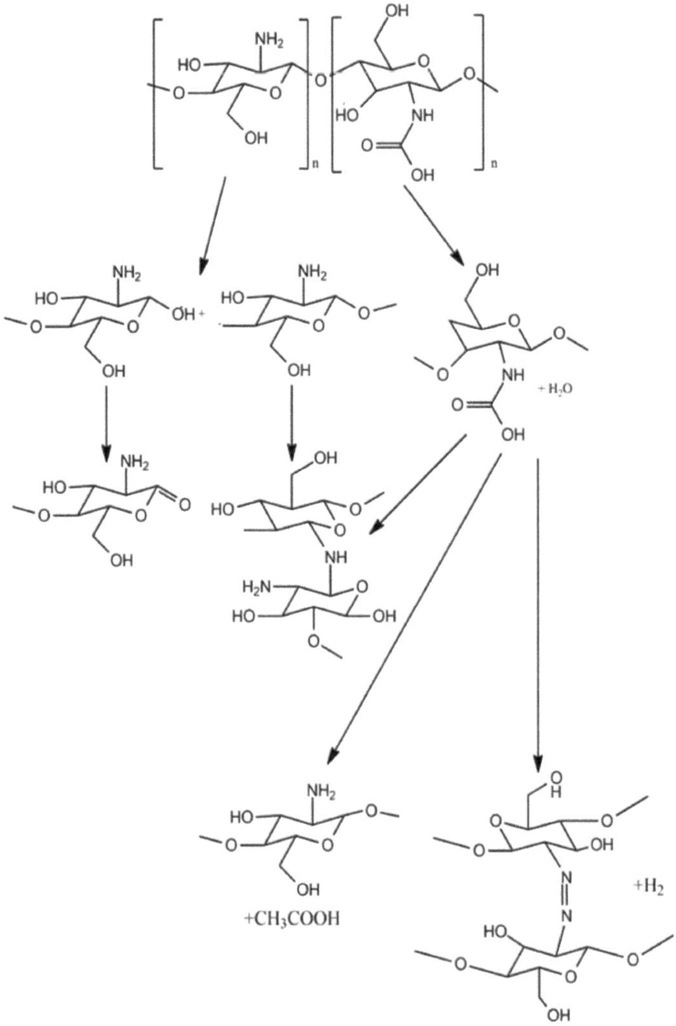

Figure 7.6 Chitosan degradation process. Adapted from Mucha et al. [22].

The degradation of chitosan occurs by acid catalysis in the stomach and by enzymatic action [20]. Chitinases and chitosanase enzymes act in the depolymerization of chitosan [23].

Table 7.1 Effect of molecular mass and DD on chitosan properties

Property	Molecular mass effect	Effect of degree of acetylation
Solubility	Proportional	Not reported
Crystallinity	Inversely proportional	Not reported
Biodegradability	Inversely proportional	Inversely proportional
Viscosity	Proportional	Not reported
Biocompatibility	Proportional	Not reported
Mucoadhesion	Proportional	Proportional
Analgesic	Proportional	Not reported
Antimicrobial	Proportional	Proportional
Permeation enhancing effect	Proportional	Not reported
Antioxidant	Proportional	Inversely proportional
Hemostatic	Proportional	Not reported

Fonte: Dash et al. [20]

Chitosan is considered a food additive [24] and a functional ingredient [25] by Brazilian legislation. The use of claims that products with chitosan help reduce the absorption of fat and cholesterol is allowed by the Brazilian regulatory agency ANVISA (National Health Surveillance Agency) and by the Commission Regulation in the European Union, if the product contains at least 3 g of chitosan, it must be present the content in the nutritional table [26, 27]. The fact that chitosan is classified as a supplement in the United States means that there is no regulation by the Food and Drug Administration [2]. However, a quality parameter can be checked in the United States Pharmacopeia 34-NF 29 [28] to commercialize chitosan (Table 7.2).

Chitosan is also used in the production of nanoparticles. In nanotechnology, materials have dimensions less than 100 nm [16]. Nanomaterials have advantages over conventional ones due to their high surface area and cellular absorption capacity [29].

Chitosan is highly versatile, being prepared in different ways and possibly copolymerized with other substances and chemical modifications. In this way, numerous products are produced with liquid or powdered chitosan.

Table 7.2 Parameters determined by the United States Pharmacopeia [28]

Parameter	The United States Pharmacopeia
Degree of deacetylation	70–90%
Distribution of molecular weight	0.85–1.15
Loss on drying	≤5%
Insolubles/Impurities	≤1.0%
Heavy metal	≤10 ppm
Iron	≤10 ppm
Protein	≤0.2%
Microbiological contamination	Absence of *Pseudomonas aeruginosa* and *Staphylococcus aureus*
Aerobic microbial	10^3 cfu
Molds and yeasts	10^2 cfu

7.3 Chitosan Gel

Gels are polymeric materials capable of incorporating large amounts of water, air, and oils into their three-dimensional structure [30]. Unlike other marine polysaccharides, chitin and its derivatives do not naturally occur in gels, so researchers have developed different strategies for gelling chitosan.

The formation of chitosan is based on cross-linking by "chemical" or "physical" means. The chemical form mainly explores intermolecular covalent interactions for forming permanent networks through interactions with chemical reagents [31]. In this way, gel with chitosan can be formed in different ways: acetylation with acetic anhydride, the addition of β-glycerol phosphate under heating, the addition of bases under controlled conditions, the addition of complexing agents or ionic crosslinkers [32]. The physical electrostatic and hydrophobic interaction is associated with hydrogen bonds through controlled altering factors such as temperature, pH, and ionic strength [31].

The deacetylation of chitosan with acetic anhydride generates additional hydrophobic intermolecular interactions, leading to the formation of a cloudy chitin gel [32], while the addition of bases under controlled temperature results in different types of gels depending on the solvent used and amount of chitosan [33].

It is possible to form gels through complexation with polyanions such as DNA, with the process being controlled by the reagents' molecular weight, the solvent used, ionic strength, pH, temperature, and the density of the molecules' charges [32].

The classification of the gels takes place through stretching under oscillation at different frequencies using rheometry, considering that viscous liquids have a storage modulus (G'') lower than the loss modulus (G'') when the angular frequency (ω) is equal to 0, while strong gels present $G_0 >> G''$ with almost constant values at different frequencies, acting as an elastic solid [31].

Chitosan hydrogels produced by physical means are safe, flexible, and soft but do not have stability, mechanical resistance, and pore sizes that are difficult to control. In contrast, chitosan hydrogels produced by cross-linking are soft, flexible, and with sizes of stable pores but may show toxicity [34].

Marudova et al. [35] produced a pectin gel with the addition of chitosan, observing the pectin load's dependence on the process's efficiency. In this study, chitosan acts as a crosslinker for pectin, with the interaction occurring efficiently at pH 5.6 and depending on the degree of pectin esterification.

Another method was explored by Chen et al. [36], who used enzymes to catalyze the formation of gels. Transglutaminase formed stronger and faster gels in the presence of chitosan, but such gels could not perform reversible sol-gel transitions. The other enzyme explored by the authors was mushroom tyrosinase, which formed a gel only in the presence of chitosan, which is weaker than that formed with transglutaminase, which can undergo a collagen-like coil-to-helix transition.

7.4 Films

Supercritical fluid (SCF) drying, starch/chitosan blend under the action of irradiation, and chitosan-potato starch using microwave treatment. More details about the methodologies can be consulted in the review by Dutta et al. [37].

A study by Moura et al. [38] investigated the formation of chitosan films. The study revealed that chitosan with higher molecular weights and degrees of deacetylation exhibited the best physicochemical properties. In this study, chitosan showed tensile strength of 22.3–39.8 MPa, elongation-at-break of 9.5–15.7%, and water vapor permeability of 0.62–1.44 g Pa^{-1} s^{-1} m^{-1} 9×10^{-11}.

The increase in DD does not compromise the biocompatibility of the chitosan film. However, it does influence cell growth and cell adhesion, with low-DD chitosan films proving advantageous. These films adhere to fibroblasts and promote the proliferation of keratinocytes, thereby facilitating epidermal regeneration [39].

Balau et al. [40] produced a chitosan film by dry phase inversion, obtaining a transparent, flexible, and smooth material with a thickness of less than 20 μm. Thermal decomposition occurred at two different temperatures: 180 °C and 540 °C.

Sionkowska et al. [41] studied the effect of UV radiation (λ = 254 and 248 nm) on the surface of chitosan films, observing an increase in polarity due to photooxidation and observing that KrF laser caused significant damage by forming more foam about the UV radiation emitted, with a slight reduction in AD in both cases.

Vilela et al. [42] prepared chitosan films with ellagic acid, obtaining a material with high mechanical properties, thermal stability between 215 °C and 220 °C, moderate water vapor permeability, potential antioxidant, antimicrobial and protective action against UVA and UVB, which can be applied for food packaging. Yadav et al. [43] also produced chitosan film incorporating cellulose nanocrystals, increasing the photoprotective potential, increasing tensile strength and Young's modulus, and reducing moisture absorption, water solubility, and water vapor permeation. Salama et al. [44] optimized the obtainment of carboxymethyl cellulose films with guanidinylated-modified chitosan and titanium oxide nanoparticles, significantly increasing the preservation of green pigmentations.

Isa et al. [45] produced a film of chitosan, low linear density polyethylene, and antimicrobial agent (AM), obtaining greater stiffness with 10% chitosan with AM and lower tensile strength and more excellent elongation at break with 5% chitosan and 5% from AM.

7.5 Fibers

Chitosan fibers have the advantage of being a suitable coating material with the possibility of building multiple layers. However, such layers are difficult to manufacture, requiring many steps [34].

Fibers can be produced by electrospinning to obtain nanofibers, produced with uniaxial stretching in a capillary tube with a needle under high tension with a metal collector [46]. Table 7.3 summarizes the mechanical properties of some chitosan fibers.

7.6 Membranes

The membranes have high porosity and imitate the skin and extracellular matrix but are difficult to manufacture with pure chitosan [34].

A promising application of chitosan membranes is in fuel cells as an eco-friendly substitute for classical membranes. Chitosan membranes are low-cost, have low methanol permeability, and exhibit high water retention at elevated temperatures [71]. The low conductivity of chitosan is a major disadvantage that makes its use unfeasible for this application, which is why numerous studies have been exploring chemical modifications to circumvent this problem (Table 7.4).

Wan et al. [103] prepared a three-layer chitosan membrane (two reticulated solids with a porous interior) incorporated with potassium hydroxide; such a membrane has a conductivity of 10^{-2} S cm^{-1} after 1 h of hydration at room temperature, with possible application in a cell of hydrogen.

Table 7.3 Mechanical properties of some chitosan fibers. Adapted from Ilyas et al. [34]

Composition	Tensile strength (Mpa)	Tensile modulus (Mpa)	Ref.
*Chitosan/Cellulose-modified	22–80	236–3316	[47]
*Chitosan/Bamboo charcoal	25–75	4600–5400	[48]
*Chitosan/Modified bamboo charcoal	75–110	5400–7000	[48]
*Chitosan/Thyme	5.59–12.2	–	[49]
*Chitosan/Clove	6.54–12.2	–	[49]
*Chitosan/Cinnamon	12.2–21.35	–	[49]
*Chitosan/PLA/CS	30.95	4.10	[50]
*Chitosan/PLA/CS/ENR	10.0	4.70	[50]
**Chitosan/Polyhydroxybutyrate	7.5–11	1044–2499	[51]
**Chitosan/Deacetylated chitosan	59.8–117.1	2.1–4.1 GPa	[51]
**Chitosan/CMC-CH-OL	7.0 ± 0.8	–	[52]
**Chitosan/CMC-CH-OL-CEO	4.8 ± 0.9	–	[52]
**Chitosan/Cellulose nanowhiskers	21.6–31.25	399.5–535.76	[53]
**Chitosan/Cellulose nanowhiskers	21.6–38.25	399.5–644	[53]
**Chitosan/Glycerol-free	28–44.5	1.05–1.15 GPa	[54]
**Chitosan/Glycerol-plasticized	22.5–33	0.6–1.0 GPa	[54]
**Chitosan/Nano diamond (4.5–1%)	100 ± 2.5	3314 ± 416	[55]
**Chitosan/Biogenic silver nanoparticles	65.04 ± 1.46	–	[56]
**Chitosan/Polyvinyl alcohol	24–43	–	[57]

(Continued)

Table 7.3 (*Continued*)

Composition	Tensile strength (Mpa)	Tensile modulus (Mpa)	Ref.
**Chitosan/Glycerol	23.87–61.82	–	[58]
***Chitosan/Sisal fiber reinforced with hybrid polymer sandwich composite	110–146	5800–6646	[59]
***Chitosan/Calcium phosphate-flexible chitosan	45.7	–	[60]
***Chitosan/Clay–chitosan hybrid	2.25–2.70	0.2–1.5	[61]
***Bioactive calcium phosphate-flexible chitosan	1.6–45.7	10.2–77.3	[60]
***Chitosan/Hydroxyapatite	3.12	73.67	[62]
***Chitosan/CS fiber porous scaffold	0.68	3.40	[62]
***Chitosan/Sodium montmorillonite and zinc oxide nanoparicles	22.34	1.750	[63]
***Chitosan/Nano-ZnO nanocomposite	30.49	2.190	[63]
***Chitosan/Nano-ZnO and organoclay anocomposite-C4	38.86	2.410	[63]
***Chitosan/Grape pomace extract	9.89–13.58	0.13–0.20	[64]
***Chitosan/Potato starch	9.27–12.5	–	[65]
***Chitosan/Cellulose nanocrystal	79.3–104.7	1607–2068	[43]
***Chitosan/Galangal rhizome extract	46.1–67.5	–	[66]
***Chitosan (coating material)/Soy protein isolated and human hair fibers	11.67–24.54	–	[67]
***Chitosan/Viscose rayon filaments	105–151	1.94–2.43 GPa	[68]
***Chitosan/Corn starch and flax fabric	17.64–24.03	0.63–0.66 GPa	[69]
***Chitosan (coating material)/Soy protein and sisal fiber	11.67–23.70	–	[70]

*Chitosan-based green composite, **Chitosan-blend composites, ***Chitosan hybrid composites.

Table 7.4 Information on the performance of different modified chitosan membranes. Adapted from Muhmed et al. [72]

Material	Σ (10^{-3})	Ts	WU	SW	IEC	Ref.
Polydopamine-functionalized carbon nanotubes into chitosan	28	30.5	76.1	68.5	–	[73]
Chitosan/exfoliated molybdenum disulfide	2.9	74	25.7	11.4	0.51	[74]
Sulfonated poly(vinylidene fluoride)/chitosan	285*	31.7–16	–	–	–	[75]
Methanesulfonic acid and sodium salt of dodecylbenzene sulfonic acid-doped cross-linked chitosan	0.467	–	98	–	0.24	[76]
Sulfonated chitosan/sulfonated graphene oxide	7.2	140	65.2	48	1.20	[77]
CHI/PVS-Nafion	78.9	–	25.8	–	–	[78]
Chitosan-/silica-coated carbon nanotubes	25	–	110	–	–	[79]
Chitosan/sulfonated graphene oxide	66	–	56	26	0.3	[80]
Chitosan-/zwitterion-functionalized titania–silica	40.8	–	70	30	0.59	[81]
Chitosan/(SHNTs)	18.6	–	48	32	0.27	[82]
Chitosan/(SHNTs)	119.7	–	69.5	45.7	1.23	[83]
Chitosan-polyvinyl alcohol-sulfonated polyethersulfone	12	8.3	–	–	0.69	[84]
Organophosphorylated titania submicron spheres/chitosan	11.4	–	48	37.5	0.2	[85]
H_2SO_4-modified chitosan	–	–	65.1	–	–	[86]
Chitosan and PSSA-MA copolymer	22	–	67	–	0.55	[87]
Chitosan sulfate	30	–	49.3	–	–	[88]

(Continued)

Table 7.4 (*Continued*)

Material	Σ (10^{-3})	Ts	WU	SW	IEC	Ref.
Chitosan/zeolite beta	15.5	–	60	–	0.11	[89]
Chitosan-crown ether membrane	52.18	13.7	–	–	1.31	[90]
Glycine betaine-intercalated layered double hydroxide/QCS//PVA	35.7	23.6	–	–	–	[91]
Quaternized chitosan/poly(diallyldimethylammonium chloride	38	25.3	161.3	–	2.33	[92]
Chitosan/quaternary phosphonium-functionalized graphene oxide	5	44.7	-	84.5	0.44	[93]
Chitosan/1-ethenyl-3-methyl-1h-imidazolium chloride-1-ethenyl-2-pyrrolidone	9	6.95	76	–	1.23	[94]
Chitosan/poly(3-methyl-1-vinyl imidazolium chloride)-co-(1-vinylpyrrolidone)	8.6	10.45	80	–	–	[95]
Quaternized chitosan/bi-imidazolium	41.9	26.4	~55	–	1.25	[96]
N-2[(2-hydroxy-3-trimethylammonium)propyl] chitosan hydroxide	10	-	33	–	0.86	[97]
Quaternized poly(vinyl alcohol)/chitosan/MoS2	30	23.8	–	–	0.90	[98]
Guanidinium chitosan	15	25.2	62.4	–	–	[99]
Quaternized chitosan/tetrabutyl titanate	~13	41.2	~130	–	1.17	[100]
Polyvinyl alcohol/chitosan	20	44	–	–	–	[101]
N-[(2-hydroxy-3-trimethylammonium)propyl] chitosan chloride (HTCC)/quaternized PVA	16	–	198	–	1.7	[102]

σ, ionic conductivity (mS cm^{-1}); Ts, tensile strength (MPa); WU, water uptake (%); SW, swelling (%); IEC, ion exchange capacity (mmol g^{-1}).

* 30 °C.

7.7 Chelates

The main mechanism of chelation is based on the theory of hard and soft acids and bases (HSB) [104] since the ability of chitosan to form chelates is associated with the presence of amine groups in its structure, as this function group presents reactivity with metal ions. When the amine group is deprotonated, it presents unpaired electrons capable of performing complexation bonds with cationic metal ions. In contrast, protonated amines in an acid medium can attract metal ions and anionic dyes due to the difference in charge (Fig. 7.7) [105]. Thus, higher DD results in more significant amounts of free amine groups interacting in the chelation process [105].

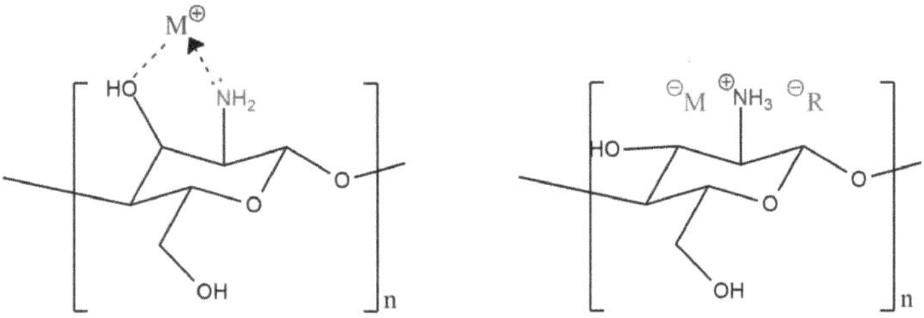

Figure 7.7 Anionic and cationic chelation mechanism with chitosan. Source: The authors (2023).

Another relevant factor for the chemical interaction between metallic ions and chitosan is the crystallinity index, as this factor affects the number of free amines for chemical interactions, as demonstrated by Jaworska et al. [106]. The chitosan ion chelation process makes it widely used in water purification as it results from the coagulation and flocculation of contaminants [107]. Chitosan tends to have the following order of preference for metal ions: Pd > Au > Hg > Pt > Cu > Ni > Zn > Mn > Pb > Co > Cr > Cd > Ag [108]. Chitosan is also associated with alkaline and alkaline earth metals because it does not have unsaturated *d* and *f* orbitals [105].

According to Guibal et al. [105], two main explanations exist for forming chitosan chelates with metals. The first model mentioned is the "bridge model," where the metal coordinates with several amine groups simultaneously in an inter and intramolecular way, while the "pendant model" considers that chelation occurs only with one amine group. The hydroxyl can act by performing hydrogen bonds to aid stability.

The adsorbent surface's absorption rate and the active site diffusion control the adsorption kinetics [109]. According to Salehi et al. [109], the pseudo-second-order model is the most recurrent absorption, due to a vacant orbital in the valence and electrostatic interactions.

Despite chitosan's great potential, industrial application for effluent treatment can be complex due to its cost, the natural variability of product characteristics, and availability [105].

Negm et al. [110] grouped the types of cross-linking for chitosan modification for use as adsorbents: (1) Substituted chitosan derivatives, phosphorylated and N-phthalate derivatives, (2) cross-linked chitosan derivatives, (3) carboxylic acid derivatives of chitosan obtained of carboxy alkylation, acrylation, methacrylation, and benzoylation of chitosan, (4) ionic derivatives of chitosan, and (5) chitosan linked to specific molecules. Several authors have also explored the formulation of resin with chitosan to remove various ions. Table 7.5 lists examples of different chitosan products used for ion removal.

Azzam et al. [129] explored another way to use chitosan's properties, the use in water treatment, producing gold or silver nanoparticles in clay and demonstrating the potential for Cu^{2+} removal. Another method of using nanoparticles was studied by Kong et al. [130], who studied the addition of chitosan for the chelation of metals followed by the formation of nanoparticles with sodium phytate removable with centrifugation.

Emik et al. [131] produced a resin with poly(2-Dimethylaminoethyl) methacrylate and carboxymethyl chitosan networks, presenting an isothermal adsorption mechanism for acidic and basic dyes.

Another strategy that has been used is the ion-imprinted one, which is produced through the complexation of the metallic ion to a polymerizable ligand, followed by the polymerization of the complex and ending the removal of the template ion [132], increasing the selectivity and stability [133].

Table 7.5 Use of chitosan resins for ion removal

Cross-linking	Target ions	Ref.
Thiourea	Ag^{+}	[111]
Carboxymethyl	Cu^{2+}	[112]
Tetraethylenepentamine	UO_2^{2+}	[113]
Carboxymethyl-chitosan	Pb^{2+}	[114]
Ethylenediamine	Cr^{6+}	[115]
Glycine	Au^{3+}, Pt^{4+} and Pd^{2+}	[116]
Carbon disulfide	Pd^{2+}	[117]
Onto composite dextran	Cu^{2+}, Co^{2+}, Ni^{2+}, Pb^{2+} and Cd^{2+}	[118]
Tripolyphosphate	Cu^{2+}	[119]
Thiourea	Au^{2+}	[120]
N-methylene phosphonic	Al^{3+}	[121]
Glycine	Cd^{2+}	[122]
Molybdate oxoanions	As^{5+}	[123]
Epichlorohydrin	Cr^{4+}	[124]
Glycine	Mn^{2+}	[125]
Glycine	Fe^{3+}	[126]
Glutaraldehyde	Pd^{2+}	[127]
Phenylthiourea	Hg^{2+}, Cd^{2+} and Zn^{2+}	[128]

Li et al. [134] prepared an ion-imprinted chitosan with TiO_2 producing a material capable of retaining methyl orange and adsorption for Ni^{2+}. Already Liu et al. [135] prepared an ion-imprinting hydrogel of chitosan, polyvinyl alcohol, and ethylene glycol diglycidyl ether for UO^{2+} ion removal. Another selective material was prepared by Birlik et al. [136] using phthalate, which presents selectivity for Th^{4+}. The composite synthesized by Ren et al. [137] with mycelium and chitosan fungal residues and with a Fe_3O_4 nanoparticle template obtained selectivity for ions in order for $Cu^{2+}>Zn^{2+}>Co^{2+}>Ni^{2+}$.

Xi et al. [138] have already taken advantage of chitosan's chelating characteristic for use in immobilized metal affinity chromatography (IMAC) columns. The IMAC technique is used to purify proteins through affinity with metal ions. Xi et al. [138] bound copper ions on a silica base, demonstrating the efficiency in the purification of bovine serum albumin.

Other authors have already used chitosan to produce other columns to purify other proteins. Wu et al. [139] used chitosan bound with Cu^{2+}, Zn^{2+}, and Ni^{2+} ions to purify Trypsin. Chitosan has also been used to bind with Ti^{4+} to enrich phosphopeptide solutions [140]. In the study by Inanan [141], the material was prepared with chitosan modified with 2-hydroxyethyl methacrylate linked with nanostructured Cu^{2+} to purify enzymes.

Le Dévédec et al. [142] used chitosan affinity with metal ions to purify chitosan oligomers (a chitosan derivative) by IMAC. Marques Jr. et al. [143] used chitosan biofilms to remove Al^{3+} and Fe^{3+} in industrial effluent, such as use in drug delivery systems.

Chitosan presents pH sensitivity, biocompatibility with the human body, low toxicity [144], and is metabolizable by human enzymes such as lysozyme [145]. In this way, this biopolymer can be applied to the estimate for drug delivery. The release rate in systems with chitosan is greatly influenced by the pH of the medium [144]. In addition, chitosan has other specific properties that make it attractive for drug delivery systems: mucoadhesive, in situ gelling, transfection enhancing, efflux pump inhibitory, and permeation enhancing properties [146]. Changing the medium's pH induces the protonation/deprotonation of amino groups; therefore, this is a pH-regulated drug carrier soluble in acidic pH and insoluble in basic pH [147, 148].

The incorporation of drugs in chitosan polymers adds numerous benefits to the formulation, such as increased circulation time and bioavailability of the drug in the body, targeting and efficiency of drug delivery at the site to be treated, reduction of resistance of microorganisms to drugs, reduction of the dosage to be applied, and consequently the reduction of all side effects about the pure drug [149, 150].

The amino and hydroxyl groups present in chitosan can also be modified to increase the formulation's specificity to the tissue to be treated, regulate the polymer's solubility, interact with the drug to be transported, and affect other properties of the formulation [150]. The ease of modifying the structure of chitosan has enabled a wide variety of controlled drug release systems, where chitosan is the main vehicle or one of the components of nanocomposites or micro composites, with preparation methodologies ranging from simple processes to more elaborate processes, using new technologies [7, 149].

The interaction and effects of each drug with the body are unique. Therefore, solutions for forming composites or modifying the chitosan polymer aim to introduce specific properties to meet the demands of each drug and its applications and limitations. Drugs used in cancer treatments often carry significant toxicity. The biocompatibility of chitosan polymers acts as a shield, minimizing the drug's impact outside the diseased tissues. This targeted delivery reduces the necessary doses, enhancing the drug's efficiency in the diseased tissue. The most prevalent controlled release systems for chitosan, as documented in the literature, include tablets, microspheres, nanoparticles, hydrogels, nanofibers, and films [149, 151, 152].

Chitosan tablets can be prepared using the polymer alone or associated with other polymers such as xanthan gum or sodium alginate. The tableting techniques used are wet granulation or direct compression techniques. The three-dimensional gel formed in the body controls the release of the drug into the body, increases the circulation time and the bioavailability of the drug, and consequently reduces the dosages necessary to maintain the therapeutic effect [149].

Chitosan microspheres can control drug release and specific target sites and increase the lifetime of the pharmaceutical active in the body. The main methods of preparing microspheres are spray drying, solvent evaporation, ionic gelation, and emulsion [7, 148]. Ji et al. [153] developed an injectable delivery system composed of porous chitosan microspheres carrying cariogenic and dimethylallyl glycine encapsulated in a thermosensitive hydrogel of hydroxypropyl chitin. The composite system promoted cartilage and bone regeneration in subchondral experiments using Sprague-Dawley rats, a potential candidate in the osteochondral regeneration system.

Nanofibers with chitosan-polycaprolactone core-shell structures were prepared by coaxial electrospinning for tetracycline loading. The resulting structures were characterized by transmission electron microscopy (TEM) and scanning electron microscopy

(SEM). The resulting nanofibers had a smooth structure without granules and an average diameter of 285 ± 75 nm. The antimicrobial activity of tetracycline-carrying nanofibers was confirmed, with potential for applications in controlled delivery systems of antibiotics for use in dressings [154].

Chitosan's ability to form films has many applications as a carrier of drugs such as antibiotics, nucleic acids, and proteins, especially in healing diseased tissues. These films resist water, traction, humidity, color retention, and thermal stability. The bacteriostatic properties increase the percentage of healing, while the hemostatic properties increase with application, mainly in healing processes in the dermal area, tissue engineering, and dentistry. Solution casting methods are the main way of preparing chitosan films [155, 156].

The effectiveness of thymol loaded onto alginate nanoparticles in chitosan/gelatin films was demonstrated through in vitro and in vivo evaluations. These films showed efficient antibacterial action against gram-positive and negative bacteria and were non-cytotoxic to L929 fibroblastic cells. The application of thymol films on rat wounds accelerated recovery, tissue regeneration, and healing, underscoring the practical applications of this research [157].

The use of small-sized chitosan nanoparticles offers, in addition to the general benefits described for chitosan, the possibility of overcoming various biological membranes and reaching the target more efficiently. The cationic character helps direct the drug to the target of pharmaceutical formulations. Several methodologies for preparing nanoparticles containing chitosan may vary according to the drug and the composition of the nanoparticle formers. The main methods described in the literature are sieving, ionic gelation, emulsion, precipitation, and reverse micellar [148, 158].

Chitosan nanoparticles are mainly used in drug delivery for cancer treatment by reducing drug toxicity, reducing liver uptake, and directing controlled release to tumor tissue, which is more acidic than healthy tissues and drug release. In acid pH, be favored [7]. The coating of magnetic nanoparticles of iron oxide with chitosan enables the protection of magnetite and maguemite phases sensitive to the presence of oxygen in the body, reducing the agglomeration of these magnetic nanoparticles, which results in loss of activity when subjected to magnetic fields, for applications in treatments of localized cancer through a synergistic association of magnetic hyperthermia and chemotherapy techniques [7, 148].

Chitosan nanoparticles loaded with organic curcumin and antimicrobial peptide were evaluated for antibacterial action against *E. coli*, inflammatory responses, intestinal morphology, and intentional microflora in rats. This resulted in biocompatible formulations with the intestinal microflora, which are therefore an alternative to prevent intestinal microflora disorders, intestinal barrier, bacterial resistance, and intestinal microflora imbalance [152].

The pH-sensitive nanocomposites containing an association of chitosan hydrogels, gelatin, and carbon quantum dots were used as curcumin carriers for cancer treatment. The formulation was capable of increasing the half-life of turmeric circulation, reaching release times of up to 22 hours, being biocompatible, increasing the stability of turmeric in biological media, and presenting zeta potential values and other physicochemical properties compatible with the system, stable, and monodisperse [159].

7.8 Flame Retardant

The flame-retardant process can occur in different ways: formation of char inhibiting the release of flammable gases, formation of a porous foam acting as a barrier to heat, interfering with flame propagation reactions, decomposing endothermally by cooling the surface and/or releasing non-flammable compounds that dilute compounds susceptible to pyrolysis [160].

The treatment of materials with chitosan for preparing flame retardant products is already classic [161], acting in char formation during burning [162]. Struszczyk and Kivekäs [163] already reported in their work that textile companies use chitosan impregnation to prepare fireproof fabrics. Teli et al. [164] used chitosan from bio-waste to finish cotton, obtaining a multi-functional material that, in addition to having flame-retardant properties, was also wrinkle-free and antimicrobial.

Other authors use cross-linking and copolymerization to improve flame-retardant properties. Charuchinda et al. [165] used chitosan in cotton fabric to increase the washing resistance of the flame-retardant sodium polyphosphate. Chitosan forms a protective film that prevents the removal of sodium polyphosphate without the need for a cross-linking reaction.

The cationic property of chitosan was taken advantage of by Srikulkit et al. [166] when producing a polymer with anionic compound polyphosphoric acid through electrostatic interactions through the formation by layer-by-layer (LbL) deposition with flame-retardant properties; however, this method several immersion cycles is not applicable at an industrial level. This type of interaction was also explored by Laufer et al. [167] using chitosan and poly(vinyl sulfonic acid sodium salt) deposition on polyurethane by LbL, creating a system that, in addition to retarding flames, also stops the dripping of melted foam.

Abou-Okeil et al. [168] studied the improvement of cotton fabric's thermal properties by adding chitosan phosphate. A decrease in the decomposition temperature was observed, improving strength properties when used with a low formaldehyde N-methylol finishing agent. El-Tahlawy [169] also studied the addition of chitosan phosphate in cotton fabric and observed that the treatment increases the residue by 32% after burning at 500 °C, indicating protection against burning. El-Tahlawy et al. [170] used a mixture of stannate/phosphate with chitosan, considering that stannate acts as a catalyst in the cellulose dehydration process, forming a protective film, with an increase of 39.24% of the residue at 500 °C.

The study by Hu et al. [171] altered the amine group of chitosan by adding phosphorous acid associated with nickel ions, obtaining a material that dehydrates cellulose and reduces the amount of flammable material volatilized during combustion. Hu et al. [172] produced a chitosan phosphate acrylate material, successfully incorporating it into epoxy acrylate and improving its thermal stability. This material inhibits the release of flammable gases and has a flame-retardant effect.

The degradation product of chitosan phosphate reduces the rate of CO/CO_2 formation, catalyzes the dehydration of cellulose, and forms a protective matrix that prevents contact of oxygen with the flammable material [170].

Chitosan can be polymerized with phosphazene because it has a lower temperature of the maximum exothermic effect and the maximum mass loss rate [161].

Carosio et al. [173] prepared a chitosan coating based on ammonium polyphosphate through LbL on cotton and polyester fabrics, resulting in better preservation of the material than the material elaborated using silica nanoparticles.

Another application is as a flame retardant in paper, with precipitation with sodium hydroxide and sodium silicate being the most efficient methods [174]. Laufer et al. [175] prepared a clay nanoplatelet material with chitosan using the LbL method, and such a product reduces oxygen permeability and interrupts the melting of flexible polyurethane foam, indicating a possible use for food packages. Hassan et al. [176] also studied using clay, incorporating salt of chitosan phosphate into organically modified montmorillonite, verifying the protective effect against flames in low-density polyethylene by forming a protective compact char layer.

Laufer et al. [177] produced thin chitosan and phytic acid films through LbL on cotton fabric, reducing the peak heat release rate (pkHRR) between 50% and 76% relative to the uncoated control. Carosio et al. [173] prepared used deoxyribonucleic acid (DNA) and chitosan to retard flames in cotton by the LbL method. Viscose fabric was studied for chitosan incorporation, where a flame-retardant effect was also observed [178]. In the work by Apaydin et al. [179], Polyallylamine-montmorillonite with chitosan was incorporated into polyamide-6 films by LbL.

This property of chitosan has also been exploited for incorporation into polyvinyl alcohol. Hu et al. [180] used urea salt of chitosan phosphate to incorporate polyvinyl alcohol, delaying the flame by forming charcoal at a lower temperature and increasing the dehydration of the material. Hu et al. [180] also obtained positive results when incorporating melamine chitosan phosphate with chitosan in polyvinyl alcohol.

Liu et al. [181] performed derivatizations of chitosan with benzaldehyde (BCS), salicylaldehyde (SCS), and hydroxybenzaldehyde (HCS), associating these compounds with ammonium polyphosphate (APP) to determine the best product for use as flame retardants. Their results indicate that there may be chemical interactions between SCS and APP that result in fine char microstructures resulting in an efficient flame-retardant effect.

Wang et al. [182] nanostructured montmorillonite polymerized with chitosan for application as a flame retardant and sensors, controlling the morphology of nanoparticles through changes in the functional group of chitosan.

7.9 Physical and Chemical Characterization Methods

The principal quality parameter of chitosan is DA, which is influenced by the analytical method. These methods can be hydrogen bromide titrimetric analysis, Fourier transform infrared spectroscopy (FT-IR), first derivative UV-spectrophotometric analysis, nuclear magnetic resonance spectroscopy (NMR), or X-ray. Currently, the main methods used are NMR and FT-IR.

The hydrogen bromide titrimetric analysis method is based on the precipitation and purification of chitosan hydrobromide salt through the reaction between chitosan and hydrobromic acid, with subsequent titration with sodium hydroxide using phenolphthalein as an acid-base indicator [183].

An alternative titrimetric method was proposed by Dutta and Priyanka [184] through acid-base titration, protonating the amine group with hydrochloric acid to subsequently perform the titration with sodium hydroxide, with the turning point being indicated by the pH indicator methyl orange. The use of linear potentiometric titration was also proposed by Jiang et al. [185], based on acid-base titration with a pH meter.

FTIR is a technique that analyzes the absorption of radiation in the infrared region by molecules, considering that this energy is absorbed in specific lengths for each molecule, being converted into stretching and bending, which depends on the functional groups present [186]. Table 7.6 shows the main bands observed in chitosan.

Table 7.6 Main bands observed in chitosan and the vibrations of associated functional groups

Wavelength (cm^{-1})	Assignment	Ref.
3000–3700	Stretching of N–H and O–H	[187–192]
2918–2960	Symmetric stretching CH	[187, 193, 194]
2855–2877	Asymmetric stretching CH	[187, 193, 194]
2864	N–H amide II blend	[188]
1635–1664	Stretching C=O amide I	[187, 188, 191, 195–198]
1490–1551	Deformation N–H amide II	[190, 196]
1312–1383	Stretching C–N amide III	[187, 188, 194, 196, 199]
1580–1597	Primary Amine N–H blend	[196]
1381–1423	Flexion CH_2	[187, 194]
1375–1425	Symmetric deformation CH_3	[187, 194]
1150–1167	Asymmetric stretching C–O–C	[191, 192, 196, 199–201]
1020–1095	Stretching C–O	[190–192, 197, 199, 201, 202]

A method using UV spectroscopy has also been proposed by Tan et al. [203], which uses D-glucosamine as a standard curve with monitoring at a wavelength of 203 nm. Liu et al. [204] also used UV spectroscopy using two standards (N-acetylglucosamine and glucosamine) using the absorbance ratio at 201 nm and the molar concentration of the standards.

Yan and Evenocheck [205] used liquid chromatography with a UV detector to analyze chitosan through chitosan hydrolysis, followed by derivatization with N-(9-fluorenylmethoxycarbonyloxy) succinimide. The authors could separate and quantify glucosamine, allowing the determination of chitosan content in blends with other compounds.

X-ray diffraction has also been proposed as an alternative, considering that DD presents linearity with a crystallinity index. However, this method is not feasible for sample mixing [206].

Abou-Shoer et al. [207] proposed another colorimetric method based on the reaction between the amino group and the acid dye bromocresol purple to determine the chitosan content in commercial products. This method uses a chitosan standard for the standard curve and monitors the change in absorbance at the length wavelength of 589 nm.

7.10 Conclusion

Chitosan is a versatile biopolymer derived from chitin. It has gained significant attention due to its unique physical and chemical properties and potential applications in various fields, including pharmaceuticals, food science, agriculture, and others.

Acknowledgment

Coordenação de Aperfeiçoamento de Pessoal de Nível Superior (CAPES) (TLAC - Finance Code 001) and Conselho Nacional de Desenvolvimento Científico e Tecnológico (CNPq) (CALC – process 312671/2021-0).

References

1. Ahmed, S., Ikram, S. (2015). Chitosan & its derivatives: A review in recent innovations. *Int. J. Pharm. Scienc. Res.*, 6(1), pp. 14–30.
2. National Institute of Health. (2023). Nomination Background: Chitosan. https://ntp.niehs.nih.gov/ntp/htdocs/chem_background/exsumpdf/chitosan_508.pdf.
3. Stasińska-Jakubas, M., Hawrylak-Nowak, B. (2022). Protective, biostimulating, and eliciting effects of chitosan and its derivatives on crop plants. *Molecules*, 27(9), e2801.
4. Hahn, T., Tafi, E., Paul, A., Salvia, R., Falabella, P., Zibek, S. (2020). Current state of chitin purification and chitosan production from insects. *J. Chem. Technol. Biotechnol.*, 95(11), pp. 2775–2795.
5. Chen, Y., Liu, Y., Dong, Q., Xu, C., Deng, S., Kang, Y., Fan, M. (2023). Application of functionalized chitosan in food: A review. *Int. J. Biol. Macromol.*, 235, e123716.
6. Jha, R., Mayanovic, R. A. (2023). A Review of the preparation, characterization, and applications of chitosan nanoparticles in nanomedicine. *Nanomater*, 13, e1302.
7. Rostami, E. (2020). Progresses in targeted drug delivery systems using chitosan nanoparticles in cancer therapy: A mini-review. *J. Drug Deliv. Sci. Technol.*, 58, e101813.
8. Cartier, N., Domard, A., Chanzy, H. (1990). Single crystals of chitosan. *Int. J. Biol. Macromol.*, 12, pp. 289–294.

9. Yui, T., Imada, K., Okuyama, K., Obata, Y., Suzuki, K., Ogawa, K. (1994). Molecular and crystal structure of the anhydrous form of chitosan. *Macromol*, 27, pp. 7601–7605.
10. Okuyama, K., Noguchi, K., Miyazawa, T., Yui, T., Ogawa, K. (1997). Molecular and crystal structure of hydrated chitosan. *Macromol.*, 30, pp. 5849–5855.
11. Novikov, V. Y., Derkach, S. R., Konovalova, I. N., Dolgopyatova, N. V. (2023). Mechanism of heterogeneous alkaline deacetylation of chitin: A review. *Polymer*, 15(7), e1729.
12. Roy, J. C., Salaün, F., Giraud, S., Ferri, A. (2017). Solubility of chitin: solvents, solution behaviors and their related mechanisms. In: *Solubility of Polysaccharides*. IntechOpen.
13. Oyatagun, G. M., Esan, T. A., Akpan, E. I., Adeosun, S. O., Popoola, A. P. I., Imasogie, B. I., Soboyejo, W. O., Afonja, A. A., Ibitoye, S. A., Abere, V. D., Oyatogun, A. O., Oluwasegun, K. M., Akinwole, I. E., Akinluwade, K. J. (2020). Chitin, chitosan, marine to market. In: Gopi, S., Thomas, S., Pius, S. *Handbook of Chitin and Chitosan. Volume 3: Chitin and Chitosan based Polymer Materials for Various Applications*. USA, Elsevier.
14. Jang, M.-K., Kong, B.-G., Jeong, Y.-I., Lee, C. H., Nah, J.-W. (2004). Physicochemical characterization of α-chitin, β-chitin, and γ-chitin separated from natural resources. *J. Polym. Sci. Part A: Polym. Chem.*, 42(14), pp. 3423–3432.
15. Wang, W., Xue, C., Mao, X. (2020). Chitosan: Structural modification, biological activity and application. *Int. J. Biol. Macromol.*, 164, pp. 4532–4546.
16. Shukla, S. K., Mishra, A. K., Arotiba, O. A., Mamba, B. B. (2013). Chitosan-based nanomaterials: A state-of-the-art review. *Int. J. Biol. Macromol.*, 59, pp. 46–58.
17. Pillai, C. K. S., Paul, W., Sharma, C. P. (2009). Chitin and chitosan polymers: Chemistry, solubility and fiber formation. *Prog. Polym. Sci.*, 34, pp. 641–678.
18. Hu, Z., Gänzle, M. (2018). Challenges and opportunities related to the use of chitosan as a food preservative. *J. Appl. Microbiol.*, 126, pp. 1318–1331.
19. Kou, S. G., Peters, L. M., Mucalo, M. R. (2021). Chitosan: A review of sources and preparation methods. *Int. J. Biol. Macromol.*, 169, pp. 85–94.

20. Dash, M., Chiellini, F., Ottenbrite, R. M., Chellini, E. (2011). Chitosan—A versatile semi-synthetic polymer in biomedical applications. *Prog. Polym. Sci.*, 36, pp. 981–1014.
21. Szymańska, E., Winnicka, K. (2015). Stability of chitosan — A challenge for pharmaceutical and biomedical applications. *Mar. Drugs*, 13, pp. 1819–1846.
22. Mucha, M., Pawlak, A. (2002). Complex study on chitosan degradability. *Polimery*, 47(7–8), pp. 509–516.
23. Zhang, H., Neau, S. H. (2001). In vitro degradation of chitosan by a commercial enzyme preparation: Effect of molecular weight and degree of deacetylation. *Biomater*, 22, pp. 1653–1658.
24. ANVISA. (2004). Resolução RE nº. 1, January 8, 2004 a. Concede o registro de alimentos e bebidas, registro de alimentos e bebidas importado, registro único de alimentos e bebidas, retificação de publicação de registro, recurso interposto deferido. Diário Oficial da União; Poder Executivo, January 9, 2004.
25. ANVISA. (2006). Alimentos com Alegações de Propriedades Funcionais e ou de Saúde, Novos Alimentos/Ingredientes, Substâncias Bioativas e Probióticos VIII - Lista das Alegações Aprovadas. 2006. Available on: http://www.anvisa.gov.br/alimentos/comissoes/tecno_lista_alega.htm.
26. ANVISA. (2019). Alegacoes de propriedade funcional aprovadas. https://www.gov.br/agricultura/pt-br/assuntos/inspecao/produtos-vegetal/legislacao-de-produtos-origem-vegetal/biblioteca-de-normas-vinhos-e-bebidas/alegacoes-de-propriedade-funcional-aprovadas_anvisa.pdf.
27. Official Journal of the European Union. (2012). Commission Regulation (EU) No 432/2012 of 16 May 2012. https://eur-lex.europa.eu/legal-content/EN/ALL/?uri=CELEX%3A32012R0432.
28. The United States Pharmacopeia; USP 34–NF 29; The United States Pharmacopeial Convention: Rockville, MD, USA (2011). pp. 5361–5365, Second Supplement.
29. Amor, I. B., Emran, T. B., Hemmami, H., Zeghoud, S., Laouini, S. E. (2023). Nanomaterials based on chitosan for skin regeneration: An update. *Int. J. Surg.*, 109(3), pp. 594–596.
30. Cao, Y., Mezzenga, R. (2020). Design principles of food gels. *Nature Foods*, 1(2), pp. 106–118.
31. Sacco, P., Furlani, F., de Marzo, G., Marsich, E., Paoletti, S., Donati, I. (2018). Concepts for developing physical gels of chitosan and of chitosan derivatives. *Gels*, 4, e67.

32. Nilsen-Nygaard, J., Strand, S. P., Varum, K. M., Draget, K. I., Nordgård, C. T. (2015). Chitosan: Gels and interfacial properties. *Polymer*, 7, pp. 552–579.

33. Ladet, S., David, L., Domard, A. (2008). Multi-membrane hydrogels. *Nature*, 452, pp. 76–79.

34. Ilyas, R. A., Aisyah, H. A., Nordin, A. H., Ngadi, N., Zuhri, M. Y. M., Asyraf, M. R. M., Sapuan, S. M., Zainudin, E. S., Sharma, S., Abral, H., Asrofi, M., Syafri, E., Sari, N. H., Rafidah, M., Zakaria, S. Z. S., Razman, M. R., Majid, N. A., Ramli, Z., Azmi, A., Bangar, S. P., Ibrahim, R. (2022). Natural-fiber-reinforced chitosan, chitosan blends and their nanocomposites for various advanced applications. *Polymer*, 14, e874.

35. Marudova, M., MacDougall, J., Ring, S. G. (2004). Pectin–chitosan interactions and gel formation. *Carbohydr. Res.*, 339, pp. 1933–1939.

36. Chen, T., Embree, H. D., Brown, E. M., Taylor, M. M., Payne, G. F. (2003). Enzyme-catalyzed gel formation of gelatin and chitosan: Potential for in situapplications. *Biomater*, 24, pp. 2831–2841.

37. Dutta, P. K., Tripathi, S., Mehrotro, G. K., Dutta, J. (2009). Perspectives for chitosan based antimicrobial films in food applications. *Food Chem.*, 114, pp. 1173–1182.

38. Moura, J. M., Farias, B. S., Rodrigues, D. A. S., Moura, C. M., Dotto, G. L., Pinto, L. A. A. (2015). Preparation of chitosan with different characteristics and its application for biofilms production. *J. Polym. Environ.*, 23, pp. 470–477.

39. Chatelet, C., Damour, O., Domard, A. (2001). Influence of the degree of acetylation on some biological properties of chitosan. *Biomater.* 22, pp. 261–268.

40. Balau, L., Lisa, G., Popa, M. I., Tura, V., Melnig, V. (2004). Physico-chemical properties of Chitosan films. *Cent. Eur. J. Chem.*, 2(4), pp. 638–647.

41. Sionkowska, A., Kaczmarek, H., Wisniewski, M., Skopinska, J., Lazare, S., Tokarev, V. (2006). The influence of UV irradiation on the surface of chitosan films. *Surf. Sci.*, 600, pp. 3775–3779.

42. Vilela, C., Pinto, R. J. B., Coelho, J., Domingues, M. R. M., Daina, S., Sadocco, P., Santos, S. A. O., Freire, C. S. R. (2017). Bioactive chitosan/ellagic acid films with UV-light protection for active food packaging. *Food Hydrocoll.*, 73, pp. 120–128.

43. Yadav, M., Behera, K., Chang, Y.-H., Chiu, F.-C. (2020). Cellulose nanocrystal reinforced chitosan based UV barrier composite films for sustainable packaging. *Polym.*, 12, e202.

44. Salama, H. E., Aziz, M. S. A. (2020). Optimized carboxymethyl cellulose and guanidinylated chitosan enriched with titanium oxide nanoparticles of improved UV-barrier properties for the active packaging of green bell pepper. *Int. J. Biol. Macromol.*, 165, pp. 1187–1197.
45. Isa, S. A. B., Mohamed, R. (2015). Physical and mechanical properties of chitosan and polyethylene blend for food packaging film. *Int. J. Mech. Prod. Eng.*, 3(10), pp. 51–55.
46. Elsabee, M. Z., Naguib, H. F., Morsi, R. E. (2012). Chitosan based nanofibers, review. *Mater. Sci. Eng. C*, 32, pp. 1711–1726.
47. Grząbka-Zasadzińska, A., Amietszajew, T., Borysiak, S. (2017). Thermal and mechanical properties of chitosan nanocomposites with cellulose modified in ionic liquids. *J. Therm. Anal. Calorim.*, 130, pp. 143–154.
48. Nitayaphat, W., Jiratumnukul, N., Charuchinda, S., Kittinaovarat, S. (2009). Mechanical properties of chitosan/bamboo charcoal composite films made with normal and surface oxidized charcoal. *Carbohydr. Polym.*, 78, pp. 444–448.
49. Hosseini, S. M. H., Razavi, S. H., Mousavi, M. (2009). Antimicrobial, physical and mechanical properties of chitosan-based films incorporated with thyme, clove and cinnamon essential oils. *J. Food Process. Preserv.*, 33, pp. 727–743.
50. Zakaria, Z., Islam, S., Hassan, A., Haafiz, M. K. M., Arjmandi, R., Inuwa, I. M., Hasan, M. (2013). Mechanical properties and morphological characterization of PLA/chitosan/epoxidized natural rubber composites. *Adv. Mater. Sci. Eng.*, e629092. https://doi.org/10.1155/2013/629092.
51. Rajan, R., Sreekumar, P. A., Joseph, K., Skrifvars, M. (2012). Thermal and mechanical properties of chitosan reinforced polyhydroxybutyrate composites. *J. Appl. Polym. Sci.*, 124, pp. 3357–3362.
52. Noshirvani, N., Ghanbarzadeh, B., Gardrat, C., Rezaei, M. R., Hashemi, M., Le Coz, C., Coma, V. (2017). Cinnamon and ginger essential oils to improve antifungal, physical and mechanical properties of chitosan-carboxymethyl cellulose films. *Food Hydrocoll.*, 70, pp. 36–45.
53. Thou, C. Z., Khan, F. S. A., Mubarak, N., Ahmad, A., Khalid, M., Jagadish, P., Walvekar, R., Abdullah, E., Khan, S., Khan, M., Hussain, S., Ahmad, I. (2021). Surface charge on chitosan/celulose nanowhiskers composite via functionalized and untreated carbon nanotube. *Arab. J. Chem.*, 14, e103022.

54. Cobos, M., González, B., Fernández, M. J., Fernández, M. D. (2018). Study on the effect of graphene and glycerol plasticizer on the properties of chitosan-graphene nanocomposites via in situ green chemical reduction of graphene oxide. *Int. J. Biol. Macromol.*, 114, pp. 599–613.
55. Delavar, Z., Shojaei, A. (2017). Enhanced mechanical properties of chitosan/nanodiamond composites by improving interphase using thermal oxidation of nanodiamond. *Carbohydr. Polym.*, 167, pp. 219–228.
56. Kadam, A., Momin, B., Palamthodi, S., Lele, S. (2019). Physicochemical and functional properties of chitosan-based nanocomposite films incorporated with biogenic silver nanoparticles. *Carbohydr. Polym.*, 211, pp. 124–132.
57. Bonilla, J., Fortunati, E., Atarés, L., Chiralt, A., Kenny, J. (2014). Physical, structural and antimicrobial properties of poly vinyl alcohol–chitosan biodegradable films. *Food Hydrocoll.*, 35, pp. 463–470.
58. Leceta, I., Molinaro, S., Guerrero, P., Kerry, J., de la Caba, K. (2015). Quality attributes of map packaged ready-to-eat baby carrots by using chitosan-based coatings. *Postharvest Biol. Technol.*, 100, pp. 142–150.
59. Arumugam, S., Kandasamy, J., Shah, A. U. M., Sultan, M. T. H., Safri, S. N. A., Majid, M. S. A., Basri, A. A., Mustapha, F. (2020). Investigations on the mechanical properties of glass fiber/sisal fiber/chitosan reinforced hybrid polymer sandwich composite scaffolds for bone fracture fixation applications. *Polymer*, 12, e1501.
60. Ding, S.-J. (2007). Biodegradation behavior of chitosan/calcium phosphate composites. *J. Non-Cryst. Solids*, 353, pp. 2367–2373.
61. Liu, K.-H., Liu, T.-Y., Chen, S.-Y., Liu, D.-M. (2007). Effect of clay content on electrostimulus deformation and volume recovery behavior of a clay–chitosan hybrid composite. *Acta Biomater.*, 3, pp. 919–926.
62. Guo, Y.-P., Guan, J.-J., Yang, J., Wang, Y., Zhang, C.-Q., Ke, Q.-F. (2015). Hybrid nanostructured hydroxyapatite–chitosan composite scaffold: Bioinspired fabrication, mechanical properties and biological properties. *J. Mater. Chem. B*, 3, pp. 4679–4689.
63. Rodrigues, C., de Mello, J. M. M., Dalcanton, F., Macuvele, D. L. P., Padoin, N., Fiori, M. A., Soares, C., Riella, H. G. (2020). Mechanical, thermal and antimicrobial properties of chitosan-based-nanocomposite with potential applications for food packaging. *J. Polym. Environ.*, 28, pp. 1216–1236.

64. Ferreira, A., Nunes, C., Castro, A., Ferreira, P., Coimbra, M. A. (2014). Influence of grape pomace extract incorporation on chitosan films properties. *Carbohydr. Polym.* 113, pp. 490–499.

65. Wu, H., Lei, Y., Lu, J., Zhu, R., Xiao, D., Jiao, C., Xia, R., Zhang, Z., Shen, G., Liu, Y., Li, S., Li, M. (2019). Effect of citric acid induced cross-linking on the structure and properties of potato starch/chitosan composite films. *Food Hydrocoll.*, 97, e105208.

66. Thakhiew, W., Devahastin, S., Soponronnarit, S. (2013). Physical and mechanical properties of chitosan films as affected by drying methods and addition of antimicrobial agent. *J. Food Eng.*, 119, pp. 140–149.

67. Deepmala, K., Jain, N., Singh, V. K., Chauhan, S. (2018). Fabrication and characterization of chitosan coated human hair reinforced phytagel modified soy protein-based green composite. *J. Mech. Behav. Mater.*, 27, pp. 1–8.

68. Gorade, V., Chaudhary, B., Parmaj, O., Kale, R. (2020). Preparation and characterization of chitosan/viscose rayon filament biocomposite. *J. Nat. Fibers*, 19, pp. 1–12.

69. Prabhakar, M. N., Song, J.-I. (2018). Fabrication and characterisation of starch/chitosan/flax fabric green flame retardant composites. *Int. J. Biol. Macromol.*, 119, pp. 1335–1343.

70. Verma, A., Singh, C., Singh, V., Jain, N. (2019). Fabrication and characterization of chitosan coated sisal fiber–Phytagel modified soy protein-based green composite. *J. Compos. Mater.*, 53, pp. 2481–2504.

71. Ma, J., Sahai, Y. (2013). Chitosan biopolymer for fuel cell applications. *Carbohydr. Polym.*, 92, pp. 955–975.

72. Muhmed, S. A., Nor, N. A. M., Jaafar, J., Ismail, A. F., Othman, M. H. D., Rahman, M. A., Aziz, F., Yusof, N. (2020). Emerging chitosan and cellulose green materials for ion exchange membrane fuel cell: A review. *Energy Ecol. Environ.*, 5(2), pp. 85–107.

73. Wang, X., Zhang, L., Ma, C., Song, R., Hou, H., Li, D. (2009). Enrichment and separation of silver from waste solutions by metal ion imprinted membrane. *Hydrometallurgy*, 100, pp. 82–86.

74. Divya, K., Rana, D., Alwarappan, S., Saraswathi, S. S. A., Nagendran, A. (2019). Investigating the usefulness of chitosan based proton exchange membrane tailored with exfoliated molybdenum disulfide nanosheets for clean energy applications. *Carbohydr. Polym.*, 208, pp. 504–512.

75. Vijayakumar, V., Khatsgir, D. (2018). Chitosan/partially sulfonated poly(vinylidene fluoride) blends as polymer electrolyte membranes for direct methanol fuel cell applications. *Cellulose*, 25, pp. 661–681.

76. Vijayalekshmi, V., Khastgir, D. (2017). Eco-friendly methanesulfonic acid and sodium salt of dodecylbenzene sulfonic acid doped cross-linked chitosan based green polymer electrolyte membranes for fuel cell applications. *J. Membr. Sci.*, 523, pp. 45–59.

77. Shirdast, A., Sharif, A., Abdollahi, M. (2016) Effect of the incorporation of sulfonated chitosan/sulfonated graphene oxide on the proton conductivity of chitosan membranes. *J. Power Sources*, 306, pp. 541–551.

78. Xue, Y., Chan, S. (2015). Layer-by-layer self-assembly of CHI/PVS–Nafion composite membrane for reduced methanol crossover and enhanced DMFC performance. *Int. J. Hydrog. Energy*, 40, pp. 1877–1885.

79. Liu, H., Gong, C., Wang, J., Liu, X., Liu, H., Cheng, F., Wang, G., Zheng, G., Qin, C., Wen, S. (2016). Chitosan/silica coated carbon nanotubes composite proton exchange membranes for fuel cell applications. *Carbohydr. Polym.*, 136, pp. 1379–1385.

80. Liu, Y., Wang, J., Zhang, H., Ma, C., Liu, J., Cao, S., Zhang, X. (2014). Enhancement of proton conductivity of chitosan membrane enabled by sulfonated graphene oxide under both hydrated and anhydrous conditions. *J. Power Sources*, 269, pp. 898–911.

81. Yin, Y., Xu, T., Shen, X., Wu, H., Jiang, Z. (2014). Fabrication of chitosan/zwitterion functionalized titania–silica hybrid membranes with improved proton conductivity. *J. Membr. Sci.*, 469, pp. 355–363.

82. Bai, H., Li, Y., Zhang, H., Chen, H., Wu, W., Wang, J., Liu, J. (2015). Anhydrous proton exchange membranes comprising of chitosan and phosphorylated graphene oxide for elevated temperature fuel cells. *J. Membr. Sci.*, 495, pp. 48–60.

83. Palani, P. B., Abidin, K. S., Kannan, R., Sivakumar, M., Wang, F.-M., Rajashabala, S., Velraj, G. (2014). Improvement of proton conductivity in nanocomposite polyvinyl alcohol (PVA)/chitosan. *RSC Adv.*, 4, pp. 61781–61789.

84. Meenakshi, S., Bhat, S. D., Sahu, A. K., Sridhar, P., Pitchumani, S., Shukla, A. K. (2011). Chitosan-polyvinyl alcoholsulfonated polyethersulfone mixed-matrix membranes as methanolbarrier electrolytes for DMFCs. *J. Appl. Polym. Sci.*, 124, pp. 73–82.

85. Wu, H., Hou, W., Wang, J., Xiao, L., Jiang, Z. (2010). Preparation and properties of hybrid direct methanol fuel cell membranes by

embedding organophosphorylated titania submicrospheres into a chitosan polymer matrix. *J. Power Sources*, 195, pp. 4104–4113.

86. Osifo, P. O., Masala, A. (2010). Characterization of direct methanol fuel cell (DMFC) applications with H_2SO_4 modified chitosan membrane. *J. Power Sources*, 195, pp. 4915–4922.
87. Seo, J. A., Koh, J. H., Roh, D. K., Kim, J. H. (2009). Preparation and characterization of cross-linked proton conducting membranes based on chitosan and PSSA-MA copolymer. *Solid State Ion*, 180, pp. 998–1002.
88. Xiang, Y., Yang, M., Guo, Z., Cui, Z. (2009). Alternatively chitosan sulfate blending membrane as methanol-blocking polymer electrolyte membrane for direct methanol fuel cell. *J. Membr. Sci.*, 337, pp. 318–323.
89. Wang, J., Zheng, X., Wu, H., Zheng, B., Jiang, Z., Hao, X., Wang, B. (2008) Effect of zeolites on chitosan/zeolite hybrid membranes for direct methanol fuel cell. *J. Power Sources*, 178, pp. 9–19.
90. Zheng, X., Shang, C., Yang, J., Wang, J., Wang, L. (2019). Preparation and characterization of chitosan-crown ether membranes for alkaline fuel cells. *Synth Met.*, 247, pp. 109–115.
91. Hu, Y., Tsen, W.-C., Chuang, F.-S., Jang, S.-C., Zhang, B., Zheng, G., Wen, S., Liu, H., Qin, C., Gong, C. (2019) Glycine betaine intercalated layered double hydroxide modified quaternized chitosan/polyvinyl alcohol composite membranes for alkaline direct methanol fuel cell. *Carbohydr. Polym.*, 213, pp. 320–328.
92. Yuan, Y., Shen, C., Chen, J., Ren, X. (2017). Synthesis and characterization of cross-linked quaternized chitosan/poly(diallyldimethylammonium chloride) blend anion-exchange membranes. *Ionics*, 24(4), pp. 1–8.
93. Liu, L., Sun, G. (2017) Promoting the OH-ion conductivity of chitosan membrane using quaternary phosphonium polymer brush functionalized graphene oxide. *Int. J. Electrochem. Sci.*, 12, pp. 9262–9278.
94. Song, F., Fu, Y., Gao, Y., Li, J., Qiao, J., Zhou, X.-D., Liu, Y. (2015) Novel alkaline anion-exchange membranes based on chitosan/ethenylmethylimidazoliumchloride polymer with ethenylpyrrolidone composites for low. *Electrochim. Acta*, 177, pp. 137–144.
95. Song, F., Chen, S., Gao, Y., Liu, Y., Qiao, J. (2015a). Cross-linked anion exchange membranes composed of imidazolium salt for alkaline fuel cell. *ECS Trans.*, 66, pp. 99–104.
96. Wang, J.-L., Wang, L.-L., Feng, R.-J., Zhang, Y. (2015). Synthesis and characterization of novel anion exchange membranes containing

bi-imidazolium-based ionic liquid for alkaline fuel cells. *Solid State Ion*, 278, pp. 144–151.

97. Wan, Y., Peppley, B., Creber, K. A. M., Bui, V. T. (2010). Anion-exchange membranes composed of quaternized-chitosan derivatives for alkaline fuel cells. *J. Power Sources*, 195, pp. 3785–3793.
98. Jiang, X., Sun, Y., Zhang, H., Hou, L. (2018). Preparation and characterization of quaternized poly(vinyl alcohol)/chitosan/MoS_2 composite anion exchange membranes with high selectivity. *Carbohydr. Polym.*, 180, pp. 96–103.
99. Sajjad, S. D., Liu, D., Wei, Z., Sakri, S., Shen, Y., Hong, Y., Liu, F. (2015). Guanidinium based blend anion exchange membranes for direct methanol alkaline fuel cells (DMAFCs). *J. Power Sources*, 300, pp. 95–103.
100. Wang, L. L., Wang, J. L., Zhang, Y., Feng, R. J. (2015). Alkaline hybrid composite membrane for direct methanol fuel cells application. *J. Electroanal. Chem.*, 759, pp. 174–183.
101. Ming Yang, J., Chih Chiu, H. (2012). Preparation and characterization of polyvinyl alcohol/chitosan blended membrane for alkaline direct methanol fuel cells. *J. Membr. Sci.*, 419–420, pp. 65–71.
102. Feketefoldi, B., Cermenek, B. (2016). Chitosan-based anion exchange membranes for direct ethanol fuel cells. *Int. J. Membr. Sci. Technol.*, 6, pp. 1–9.
103. Wan, Y., Creber, K. A. M., Bui, V. T., Halliop, E. (2006). Preliminary evaluation of an alkaline chitosan-based membrane fuel cell. *J. Power Sources*, 162, pp. 105–113.
104. Pearson, R. G. (1963). Hard and soft acids and bases. *J. Am. Chem. Soc.*, 85(22), pp. 3533–3539.
105. Guibal, E. (2004). Interactions of metal ions with chitosan-based sorbents: A review. *Sep. Purif. Technol.*, 38, pp. 43–74.
106. Jaworska, M., Sakurai, K., Gaudon, P., Guibal, E. (2003). Influence of chitosan characteristics on polymer properties. I: Crystallographic properties. *Polym. Int.*, 52, pp. 198–205.
107. De-Paz-Arroyo, G., Picos-Corrales, L., Pérez-Sicairos, S., Licea-Claverie, A. (2023). Flocculants based on responsive polymers and chitosan for removal of metallic nanoparticles as contaminants of emerging concern present in water. *Colloids Surf. A: Physicochem. Eng.*, 675, e132045.
108. Krajewska, B. (2005). Membrane-based processes performed with use of chitin/chitosan materials. *Sep. Purif. Technol.*, 41, pp. 305–312.

109. Salehi, E., Daraei, P., Shamsabadi, A. A. (2016). A review on chitosan-based adsorptive membranes. *Carbohydr. Polym.*, 152, pp. 419–432.

110. Negm, N. A., Hefni, H. H. H., Abd-Elaal, A. A. A., Badr, E. A., Abou Kana, M. T. H. (2020). Advancement on modification of chitosan biopolymer and its potential applications. *Int. J. Biol. Macromol.*, 152, pp. 681–702.

111. Wang, L., Xing, R., Liu, S., Yu, H., Qin, Y., Li, K., Feng, J., Li, R., Li, P. (2010). Recovery of silver (I) using a thiourea-modified chitosan resin. *J. Hazard. Mater.*, 180, pp. 577–582.

112. Sun, S., Wang, A. (2006). Adsorption properties of carboxymethyl-chitosan and cross-linked carboxymethyl-chitosan resin with Cu(II) as template. *Sep. Purif. Technol.*, 49(3), pp. 197–204.

113. Elwakeel, K. Z., Atia, A. A., Guibal, E. (2014). Fast removal of uranium from aqueous solutions using tetraethylenepentamine modified magnetic chitosan resin. *Bioresour. Technol.*, 1606, pp. 107–114.

114. Sun, S., Wang, L., Wang, A. (2006). Adsorption properties of cross-linked carboxymethyl-chitosan resin with Pb(II) as template ions. *J. Hazard. Mater.*, 136(3), pp. 930–937.

115. Hu, X., Wang, J.-S., Liu, Y.-G., Li, X., Zeng, G.-M., Bao, Z.-L., Zeng, X.-X., Chen, A.-W., Long, F. (2011). Adsorption of chromium (VI) by ethylenediamine-modified cross-linked magnetic chitosan resin: Isotherms, kinetics and thermodynamics. *J. Hazard. Mater.*, 185(1), pp. 306–314.

116. Ramesh, A., Hasegawa, H., Sugimoto, W., Maki, T., Ueda, K. (2008). Adsorption of gold(III), platinum(IV) and palladium(II) onto glycine modified cross-linked chitosan resin. *Bioresour. Technol.*, 99(9), pp. 3801–3809.

117. Chen, L., Wu, K., Zhang, M., Liu, N., Li, C., Qin, J., Zhao, Q., Ye, Z. (2023a). Synthesis of carbon disulfide modified chitosan resin and its adsorption properties for palladium(II) in wastewater. *Chem. Eng. J.*, 466, e143082.

118. Liu, Y., Hu, L., Tam, B., Li, J., Gao, X., He, Y., Du, X., Zhang, W., Wang, W. (2019). Adsorption behavior of heavy metal ions from aqueous solution onto composite dextran-chitosan macromolecule resin adsorbent. *Int. J. Biol. Macromol.*, 141, pp. 738–746.

119. Lee, S.-T., Mi, F.-L., Shen, Y.-J., Shyu, S.-S. (2001). Equilibrium and kinetic studies of copper(II) ion uptake by chitosan-tripolyphosphate chelating resin. *Polym.*, 42, pp. 1879–1892.

120. Guo, J., Fan, X., Li, Y., Yu, S., Zhang, Y., Wang, L., Ren, X. (2021). Mechanism of selective gold adsorption on ion-imprinted chitosan resin modified by thiourea. *J. Hazard. Mater.*, 415, e125617.

121. Fan, J., Yu, L., Zhou, X., Liu, J. (2022). Synthesis and characterization of cross linked N-methylene phosphonic chitosan resin chelated with Al(III) for use as adsorbent for fluoride removal from aqueous solutions. *Korean J. Chem. Eng.*, 39, pp. 377–388.

122. Mohamed, N. M., Hamouda, A. S., Kader, S. A., Khalil, M. M. H. (2021). Removal of cadmium from aqueous solution using modified magnetic glycine modified cross-linked chitosan resin. *Egypt. J. Aquat. Biol. Fish.*, 25(1), pp. 555–570.

123. Elwakeel, K. Z. (2014). Removal of arsenate from aqueous media by magnetic chitosan resin immobilized with molybdate oxoanions. *Int. J. Environ. Sci. Technol.*, 11, pp. 1051–1062.

124. Wu, Z., Li, S., Wan, J., Wang, Y. (2012). Cr(VI) adsorption on an improved synthesised cross-linked chitosan resin. *J. Mol. Liq.*, 170, pp. 25–29.

125. Al-Wakeel, K. Z., El Monem, H. A., Khalil, M. M. H. (2015). Removal of divalent manganese from aqueous solution using glycine-modified chitosan resin. *J. Environ. Chem. Eng.*, 3, pp. 179–186.

126. Khalil, M. M. H., Al-Wakeel, K. Z., El Rehim, S. S. A., El Monem, H. A. (2014). Adsorption of Fe(III) from aqueous medium onto glycine-modified chitosan resin: Equilibrium and kinetic studies. *J. Dispers. Sci. Technol.*, 35, pp. 1691–1698.

127. Nagireddi, S., Katiyar, V., Uppaluri, R. (2017). Pd(II) adsorption characteristics of glutaraldehyde cross-linked chitosan copolymer resin. *Int. J. Biol. Macromol.*, 94, pp. 72–84.

128. Monier, M., Abdel-Latif, D. A. (2012). Preparation of cross-linked magnetic chitosan-phenylthiourea resin for adsorption of Hg(II), Cd(II) and Zn(II) ions from aqueous solutions. *J. Hazard. Mater.*, 209–210, pp. 240–249.

129. Azzam, E. M. S., Eshaq, G., Rabie, A. M., Bakr, A. A., Ebd-Elaal, A. A., El Metwally, A. E., Tawfik, S. M. (2016). Preparation and characterization of chitosan-clay nanocomposites for the removal of Cu(II) from aqueous solution. *Int. J. Biol. Macromol.*, 89, pp. 507–517.

130. Kong, A., Ji, Y., Ma, H., Song, Y., He, B., Li, J. (2018). A novel route for the removal of Cu(II) and Ni(II) ions via homogeneous adsorption by chitosan solution. *J. Clean. Prod.*, 192, pp. 801–808.

131. Emik, S., Işık, S., Yıldırım, E. (2021). Simultaneous removal of cationic and anionic dyes from binary solutions using carboxymethyl chitosan based IPN type resin. *J. Polym. Environ.*, 29, pp. 1963–1977.

132. Saatçılar, Ö., Şatıroğlu, N., Say, R., Bektaş, S., Denizli, A. (2006). Binding behavior of Fe^{3+} ions on ion-imprinted polymeric beads for analytical applications. *J. Appl. Polym. Sci.*, 101(5), pp. 3520–3528.

133. Branger, C., Meouche, W., Margaillan, A. (2013). Recent advances on ion-imprinted polymers. *React. Funct. Polym.*, 73, pp. 859–875.

134. Li, Q., Su, H., Tan, T. (2008). Synthesis of ion-imprinted chitosan-TiO_2 adsorbent and its multi-functional performances. *Biochem. Eng. J.*, 38, pp. 212–218.

135. Liu, Y., Cao, X., Hua, R., Wang, Y., Liu, Y., Pang, C., Wang, Y. (2010). Selective adsorption of uranyl ion on ion-imprinted chitosan/PVA cross-linked hydrogel. *Hydrometallurgy*, 104, pp. 150–155.

136. Birlik, E., Büyüktiryaki, S., Ersöz, A., Denizli, A., Say, R. (2006). Selective separation of thorium using ion imprinted chitosan-phthalate particles via solid phase extraction. *Sep. Sci. Technol.*, 41(14), pp. 3109–3121.

137. Ren, Y., Zhang, M., Zhao, D. (2008). Synthesis and properties of magnetic Cu(II) ion imprinted composite adsorbent for selective removal of copper. *Desalination*, 228, pp. 135–149.

138. Xi, F., Wu, J. (2004). Macroporous chitosan layer coated on non-porous silica gel as a support for metal chelate affinity chromatographic adsorbent. *J. Chromatogr. A*, 1057, pp. 41–47.

139. Wu, J., Luan, M., Zhao, J. (2006). Trypsin immobilization by direct adsorption on metal ion chelated macroporous chitosan-silica gel beads. *Int. J. Biol. Macromol.*, 39, pp. 185–191.

140. Wu, T., Shi, J., Zhang, C., Zhang, L., Du, Y. (2016). Highly specific phosphopeptide enrichment by titanium(IV) cross-linked chitosan composite. *J. Chromatogr. B*, 1008, pp. 234–239.

141. Inanan, T. (2019). Chitosan Co-polymeric nanostructures for catalase immobilization. *React. Funct. Polym.*, 135, pp. 94–102.

142. Le Dévédec, F., Bazinet, L., Furtos, A., Venne, K., Brunet, S., Mateescu, M. A. (2008). Separation of chitosan oligomers by immobilized metal affinity chromatography. *J. Chromatogr. A*, 1194, pp. 165–171.

143. Marques Jr., J. L., Lütke, S. F., Frantz, T. S., Spinelli Jr., J. B. S., Carapelli, R., Pinto, L. A. A., Cadaval Jr., T. R. S. (2018). Removal of Al(III) and Fe(III) from binary system and industrial effluent using chitosan films. *Int. J. Biol. Macromol.*, 120, pp. 1667–1673.

144. Laranjeira, M. C. M., Fávere, V. T. (2009). Quitosana: Biopolímero funcional com potencial industrial biomédico. *Quím. Nova*, 32(3), pp. 672–678.

145. Muzzarelli, R. A. (1997). Human enzymatic activities related to the therapeutic administration of chitin derivatives. *Cell. Mol. Sci.*, 53, pp. 131–140.

146. Bernkop-Schnürch, A., Dünnhaupt, S. (2012). Chitosan-based drug delivery systems. *Eur. J. Pharm. Biopharm.*, 81, pp. 463–469.

147. Wang, R., Shou, D., Lv, O., Kong, Y., Deng, L., Shen, J. (2017). pH-controlled drug delivery with hybrid aerogel of chitosan, carboxymethyl cellulose and graphene oxide as the carrier. *Int. J. Biol. Macromol.*, 103, pp. 248–253.

148. Gerami, S. E., Pourmadadi, M., Fatoorechi, H., Yazdian, F., Rashedi, H., Nigjeh, M. N. (2021). Preparation of pH-sensitive chitosan/polyvinylpyrrolidone/α-Fe_2O_3 nanocomposite for drug delivery application: Emphasis on ameliorating restrictions. *Int. J. Biol. Macromol.*, 173, pp. 409–420.

149. Raafat, D., Sahl, H.-G. (2009). Chitosan and its antimicrobial potential—a critical literature survey. *Microb. Biotechnol.*, 2(2), pp. 186–201.

150. Liu, Z., Wang, K., Peng, X., Zhang, L. (2022). Chitosan-based drug delivery systems: Current strategic design and potential application in human hard tissue repair. *Eur. Polym. J.*, 166, e110979.

151. Ali, A., Ahmed, S. (2018). A review on chitosan and its nano-composites in drug delivery. *Int. J. Biol. Macromol.*, 109, pp. 273–286.

152. Wan, X., Liu, L., Zhu, Z. (2022). Fabrication of bio-engineered chitosan nanoformulations to inhibition of bacterial infection and to improve therapeutic potential of intestinal microflora, intestinal morphology, and immune response in infection induced rat model. *Drug Deliv.*, 29(1), pp. 2002–2016.

153. Ji, X., Shao, H., Li, X., Ullah, M. W., Luo, G., Xu, Z., Ma, L., He, X., Lei, Z., Li, Q., Jiang, X., Yang, G., Zhang, Y. (2022). Injectable immunomodulation-based porous chitosan microspheres/HPCH hydrogel composites as a controlled drug delivery system for osteochondral regeneration. *Biomater*, 285, e121530.

154. Ghazalian, M., Afshar, S., Rostami, A., Rashedi, S., Bahrami, S. H. (2022). Fabrication and characterization of chitosan-polycaprolactone core-shell nanofibers containing tetracycline hydrochloride. *Colloids Surf. A: Physicochem. Eng. Asp.*, 636, e128163.

155. Kavianinia, I., Plieger, P. G., Kandile, N. G., Harding, D. R. K. (2014). Preparation and characterization of chitosan films, cross-linked with symmetric aromatic dianhydrides to achieve enhanced thermal properties. *Polym. Int.*, 64(4), pp. 556–562.

156. Flórez, M., Guerra-Rodríguez, E., Cazón, P., Vázquez, M. (2022). Chitosan for food packaging: Recent advances in active and intelligent films. *Food Hydrocoll.*, 124(part B), e107328.

157. Ahmady, A. R., Razmjooee, K., Saber-Samandari, S., Toghraie, D. (2022). Fabrication of chitosan-gelatin films incorporated with thymol-loaded alginate microparticles for controlled drug delivery, antibacterial activity and wound healing: In vitro and in vivo studies. *Int. J. Biol. Macromol.*, 223, pp. 567–582.

158. Voon, S. H., Tiew, S. X., Kue, C. S., Lee, H. B., Kiew, L. V., Misran, M., Kamkaew, A., Burgess, K., Chung, L. Y. (2016). Chitosan-coated poly(lactic-co-glycolic acid)-diiodinated boron-dipyrromethene nanoparticles improve tumor selectivity and stealth properties in photodynamic cancer therapy. *J. Biomed. Nanotechnol.*, 12(7), pp. 1431–1452.

159. Ostovar, S., Pourmadadi, M., Shamsabadipour, A., Mashayekh, P. (2023). Nanocomposite of chitosan/gelatin/carbon quantum dots as a biocompatible and efficient nanocarrier for improving the Curcumin delivery restrictions to treat brain cancer. *Int. J. Biol. Macromol.*, 242(Part 3), e124986.

160. Malucelli, G. (2020). Flame retardant systems based on chitosan and its derivatives: State of the art and perspectives. *Molecules*, 25, e4046.

161. Allan, G. G., Struszczyk, H. (1981). Thermal properties of chitosan-phosphazene polymers. *J. Therm. Anal.*, 21, pp. 113–117.

162. Feng, X., Wang, X., Wing, W., Yu, B., Song, L., Hu, Y. (2013). Simultaneous reduction and surface functionalization of graphene oxide by chitosan and their synergistic reinforcing effects in PVA films. *Ind. Eng. Chem. Res.*, 52, pp. 12906–12914.

163. Struszczyk, H., Kivekäs, H. (1990). Microcrystalline chitosan-some areas of application. *Br. Polym. J.*, 23, pp. 261–265.

164. Teli, M. D., Sheikh, J., Bhavsar, P. (2013). Multi-functional finishing of cotton using chitosan extracted from bio-waste. *Int. J. Biol. Macromol.*, 54, pp. 125–130.

165. Charuchinda, S., Srikulkit, K., Mowattana, T. (2005). Co-application of sodium polyphosphate and chitosan to improve flame retardancy of cotton fabric. *J. Sci. Res. Chula. Univ.*, 30(1), pp. 97–110.

166. Srikulkit, K., Lamsamai, C., Dubas, S. T. (2006). Development of flame retardant polyphosphoric acid coating based on the polyelectrolyte multilayers technique. *J. Met. Mater. Miner.*, 16(2), pp. 41–45.

167. Laufer, G., Kirkland, C., Morgan, A. B., Grunlan, J. C. (2013). Exceptionally flame retardant sulfur-based multilayer nanocoating for polyurethane prepared from aqueous polyelectrolyte solutions. *ACS Macro Lett.*, 2, 361–365.

168. Abou-Okeil, A., El-Shafie, A., Hebeish, A. (2007). Chitosan phosphate induced better thermal characteristics to cotton fabric. *J. Appl. Polym. Sci.*, 103(3), pp. 2021–2026.

169. El-Tahlawy, K. (2008). Chitosan phosphate: A new way for production of eco-friendly flame retardant cotton textiles. *J. Text. Inst.*, 99(3), pp. 185–191.

170. El-Tahled, K., Eid, R., Sherif, F., Hudson, S. (2008). Chitosan: A new route for increasing the efficiency of stannate/phosphate flame retardants on cotton. *J. Text. Inst.*, 99(2), pp. 157–164.

171. Hu, S., Song, L., Pan, H., Hu, Y. (2012). Thermal properties and combustion behaviors of chitosan based flame retardant combining phosphorus and nickel. *Ind. Eng. Chem. Res.*, 51, pp. 3663–3669.

172. Hu, S., Song, L., Pan, Y., Hu, Y., Gong, X. (2012). Thermal properties and combustion behaviors of flame retarded epoxy acrylate with a chitosan based flame retardant containing phosphorus and acrylate structure. *J. Anal. Appl. Pyrolysis*, 97, pp. 109–115.

173. Carosio, F., Di Blasio, A., Alongi, J., Malucelli, G. (2013). Green DNA-based flame retardant coatings assembled through layer by layer. *Polymer*, 54, pp. 5148–5153.

174. Basta, A. H., Zhan, H., Chen, J. (2004). Enhancement of the strength and flame retardantproperties of handsheets made from deinked pulp. *Appita J.*, 57(6), pp. 465–471.

175. Laufer, G., Kirkland, C., Cain, A. A., Grunlan, J. C. (2012a). Clay–chitosan nanobrick walls: Completely renewable gas barrier and flame retardant nanocoatings. *ACS Appl. Mater. Interfaces.*, 4, pp. 1643–1649.

176. Hassan, M., Nour, M., Abdelmonem, Y., Makhlouf, G., Abdelkhalik, A. (2016). Synergistic effect of chitosan-based flame retardant and modified clay on the flammability properties of LLDPE. *Polym. Degrad. Stab.*, 133, pp. 8–15.

177. Laufer, G., Kirkland, C., Morgan, A. B., Grunlan, J. C. (2012b). Intumescent multilayer nanocoating, made with renewable polyelectrolytes, for flame retardant cotton. *BioMacromol.*, 13, pp. 2843–2848.

178. Zhao, Z., Li, Q., Do, Y., Wang, X., Hong, W. (2012). Effect of chitosan as antimicrobial agent on flame retardant protein viscose fiber. *Adv. Mater. Res.*, 427, pp. 32–37.

179. Apaydin, K., Laachachi, A., Ball, V., Jimenez, M., Bourbigot, S., Toniazzo, V., Ruch, D. (2013). Polyallylamine-montmorillonite as super flame retardant coating assemblies by layer-by layer deposition on polyamide. *Polym. Degrad. Stab.*, 98, pp. 627–634.

180. Hu, S., Song, L., Hu, Y. (2013). Preparation and characterization of chitosan-based flame retardant and its thermal and combustible behavior on polyvinyl alcohol. *Polym. Plast. Technol. Eng.*, 52, pp. 393–399.

181. Liu, X., Gu, X., Sun, J., Zhang, S. (2017). Preparation and characterization of chitosan derivatives and their application as flame retardants in thermoplastic polyurethane. *Carbohydr. Polym.*, 167, pp. 356–363.

182. Wang, S., Chen, L., Tong, Y. (2006). Structure–property relationship in chitosan-based biopolymer/montmorillonite nanocomposites. *J. Polym. Sci.*, 44(1), pp. 686–696.

183. Taghizadeh, S. M., Davari, G. (2006). Preparation, characterization, and swelling behavior of N-acetylated and deacetylated chitosans. *Carbohydr. Polym.*, 64, pp. 9–15.

184. Dutta, J., Priyanka (2022). A facile approach for the determination of degree of deacetylation of chitosan using acid-base titration. *Heliyon*, 8, e09924.

185. Jiang, X., Chen, L., Zhong, W. (2003). A new linear potentiometric titration method for the determination of deacetylation degree of chitosan. *Carbohydr. Polym.*, 54, pp. 457–463.

186. Pavia, D. L., Lampman, G. M., Kriz, G. S., Vyvyan, J. R. (2010). *Introdução à espectrocopia*. 4th ed., Cengage Learning: USA.

187. Petrovici, A. R., Anghel, N., Dinu, M. V., Spiridon, I. (2023). Dextran-chitosan composites: Antioxidant and anti-inflammatory properties. *Polymer*, 15(9), e1980.

188. Aghris, S., Chhaibi, B., Ouatmane, F. Z., Alaoui, O. Tahiri, Laghrib, F., Farahi, A., Bakasse, M., Loukili, A., Lahrich, S., El Mhammedi, M. A. (2004). An electrochemical sensor based on clay/graphene oxide decorated on chitosan gel for the determination of flubendiamide insecticide. *Mater. Chem. Phys.*, 296, e127243.

189. Liu, H., Adhikari, R., Guo, Q., Adhikari, B. (2013). Preparation and characterization of glycerol plasticized (high-amylose) starch–chitosan films. *J. Food Eng.*, 116(2), pp. 588–597.

190. Rahmi, R., Lelifajri, L., Fathurrahmi, F., Fathana, H., Iqhrammullah, M. (2023) Preparation and characterization of PEGDE-EDTA-modified magnetic chitosan microsphere as an eco-friendly adsorbent for methylene blue removal. *South Afr. J. Food Eng.*, 43, pp. 296–302.

191. Espinosa-Andrews, H., Sandoval-Castilla, O., Vázquez-Torres, H., Vernon-Carter, E. J., Lobato-Calleros, C. (2010). Determination of the gum Arabic–chitosan interactions by Fourier transform infrared spectroscopy and characterization of the microstructure and rheological features of their coacervates. *Carbohydr. Polym.*, 79(3), pp. 541–546.

192. Antonino, R. Q., Fook, B. L., Lima, V. O., Rached, R. F., Lima, E., Lima, R. S., Covas, C. P., Fook, M. L. (2017). Preparation and characterization of chitosan obtained from shells of shrimp (*Litopenaeus vannamei* Boone). *Mar. Drugs*, 15(5), p. 141.

193. Khan, A., Khan, R. A., Salmieri, S., Tien, C. L., Riedl, B., Bouchard, J., Chauve, G., Tan, V., Kamal, M. R., Lacroix, M. (2012). Mechanical and barrier properties of nanocrystalline cellulose reinforced chitosan based nanocomposite films. *Carbohydr. Polym.*, 90(4), pp. 601–1608.

194. Queiroz, M. F., Melo, K., Sabry, D., Sassaki, G., Rocha, H. (2014). Does the use of chitosan contribute to oxalate kidney stone formation. *Mar. Drugs*, 13(1), pp. 141–158.

195. Jithendra, P., Rajam, A. M., Kalaivani, T., Mandal, A. B., Rose, C. (2013) Preparation and characterization of aloe vera blended collagen-chitosan composite scaffold for tissue engineering applications. *ACS Appl. Mater. Interfaces*, 5(15), pp. 7291–7298.

196. Huang, P., Cao, M., Liu, Q. (2012). Adsorption of chitosan on chalcopyrite and galena from aqueous suspensions. *Colloids Surf. A: Physicochem. Eng. Asp.*, 409, pp. 167–175.

197. Huang, G., Sun, Y., Xiao, J., Yang, J. (2012) Complex coacervation of soybean protein isolate and chitosan. *Food Chem.*, 135(2), pp. 534–539.

198. Sarmento, B., Ribeiro, A., Veiga, F., Ferreira, D. (2006). Development and characterization of new insulin containing polysaccharide nanoparticles. *Colloids Surf B: Biointerfaces*, 53(2), pp. 193–202.

199. Kumari, S., Rath, P., Kumar, A. S. H., Tiwari, T. N. (2015) Extraction and characterization of chitin and chitosan from fishery waste by chemical method. *Environ. Technol. Innov.*, 3, pp. 77–85.

200. Kriegel, C., Kit, K. M., Mcclements, D. J., Weiss, J. (2009). Electrospinning of chitosan–poly(ethylene oxide) blend nanofibers in the presence of micellar surfactant solutions. *Polymer*, 50(1), pp. 189–200.

201. Mohamed, A., E., Elgammal, W., E., Eid, A. M., Dawaba, A. M., Ibrahim, A., G., Fouda, A., Hassan, S. M. (2022). Synthesis and characterization of new functionalized chitosan and its antimicrobial and in vitro release behavior from topical gel. *Int. J. Biol. Macromol.*, 207, pp. 242–253.

202. Cheng, S., Li, F., Mei, X. (2022). Structure, mechanical and physical properties of hordein/chitosan composite films. *LWT*, 163, e113596.

203. Tan, S. C., Khor, E., Tan, T. K., Wong, S. M. (1998). The degree of deacetylation of chitosan: Advocating the first derivative UV-spectrophotometry method of determination. *Talanta*, 45, pp. 713–719.

204. Liu, D., Wei, Y., Yao, P., Jiang, L. (2006). Determination of the degree of acetylation of chitosan by UV spectrophotometry using dual standards. *Carbohydr. Res.*, 341, pp. 782–785.

205. Yan, X., Evenocheck, H. M. (2012). Chitosan analysis using acid hydrolysis and HPLC/UV. *Carbohydr. Polym.*, 87, pp. 1774–1778.

206. Zhang, Y., Xue, C., Xue, Y., Gao, R., Zhang, X. (2005). Determination of the degree of deacetylation of chitin and chitosan by X-ray powder diffraction. *Carbohydr. Res.*, 340, pp. 1914–1917.

207. Abou-Shoer, M. (2010). A simple colorimetric method for the evaluation of chitosan. *Am. J. Anal. Chem.*, 2, pp. 91–94.

Chapter 8

Chitin and Chitosan Characterization: Spectroscopic Methods

Hadia Hemmami,[a,b] Soumeia Zeghoud,[a,b,c] and Ilham Ben Amor[a,b] Shakeel Ahmed[d,e]

[a]*Department of Process Engineering and Petrochemical, Faculty of Technology, University of El Oued, El Oued, Algeria*
[b]*Renewable Energy Development unit in Arid Zones, University of El Oued, El Oued, Algeria*
[c]*Laboratory of Applied Chemistry and Environment, Faculty of Exact Sciences, University of El Oued, El Oued, Algeria*
[d]*Postgraduate Department of Chemistry, Government Postgraduate College Rajouri, Jammu and Kashmir 185133, India*
[e]*University Centre for Research and Development (UCRD), Chandigarh University, Mohali, Punjab, India*

ilhambenamor97@gmail.com

Chitin is the second most common natural polymer that scientists have recognized as an adaptable biopolymer for a variety of uses in the medical field, agricultural sector, and food business. The outstanding physical, chemical, and biological characteristics of chitosan and chitin, as well as the almost unlimited availability of raw materials for their manufacture, are the main reasons for this. Depending on the source of the chitin and terms for the formation of the chitosan, these polymers have a wide range of varied physical and chemical properties. These biopolymers are

Chitin and Chitosan: Physical and Chemical Properties
Edited by Shakeel Ahmed, Soumeia Zeghoud, Hadia Hemmami, and Ilham Ben Amor

ISBN 978-981-5129-39-7 (Hardcover), 978-1-003-58977-8 (eBook)
www.jennystanford.com

able to undergo a variety of chemical changes because there are functional interactive groups and the biopolymers' polysaccharide nature. Without the utilization of spectroscopic methods, chitosan, chitin, and derivatives thereof cannot be fully characterized chemically and physically. This chapter discusses the different methods for characterizing chitin and chitosan.

8.1 Introduction

Chitin serves as a natural structural polysaccharide, and it is the second most widely distributed natural polymer by scientists [1]. It is a key constituent in the shells, crusts, and carapaces of crustaceans, including shrimp, crabs, and lobsters, as well as in the cell walls of fungus and yeast [2–4]. Its annual output is predicted to be 1010–112 tons. Linear polymer chitin is mostly made up of 2-acetamido-2-deoxy-d-glucopyranose units and is also partially made up of 2-amino-2-deoxy-d-glucopyranose units. Chitin, in this form, is only soluble in specialized solvents like hexafluoro-2-propanol and hexafluoroacetone and is insoluble in water. When less than 50% of chitin is N-acetylated, when it dissolves in aqueous acidic solutions, it is known as chitosan [5]. Chitosan is a term used to refer to a collection of chitins that have been fully or partially deacetylated, although a strict nomenclature regarding the level of N-deacetylation between chitosan and chitin has not been established [6].

Chitosan and chitin should be classified according to their insolubility and solubility in a solution of acetic acid with a concentration of 0.1M; the soluble substance is called chitosan, and chitin is an insoluble substance [9]. The weight of the chitosan and chitin molecules can reach several million g/mol. Chitosan, which is readily accessible for purchase, has an amended molecular weight that ranges from 3,800 to 500,000 $g \cdot mol^{-1}$, and the degree of deacetylation (DD) ranges from 2% to 40% [7].

Because of its loud nitrogen content (6.89%) and exceptional qualities like biocompatibility, adsorptive capacities, non-toxicity, and biodegradability, chitin, and chitosan are of economic importance [8]. Chitin has a narrow range of applications since it is a substance that is mostly insoluble and possesses little chemical reactivity. Because of its various and wide-ranging uses in biomedicine, cosmetics, wastewater treatment, fiber, and food industries, chitosan has recently gained new attention [1]. The majority of chitosan's distinguishing qualities are a result of its high concentration of initial amino groups, which has a pKa = 6.3. Chitosan, due to its positive charge, acts as a cationic polyelectrolyte and is water-soluble at lower pH levels because of the NH_3^+ groups; however, the amino groups' positive charge is eliminated as the pH goes over 6.0, rendering chitosan insoluble [9]. Chitosan's soluble-insoluble transition takes place at the pKa of its main amino groups, which is between pH 6.0 and 6.5. Chitosan solubility is reliant on both DD and the technique used for N-deacetylation since the pKa is extremely reliant on DD [10]. Additionally, chitosan undergoes a number of simple chemical alterations due to the amino groups, which predisposes its continued development for several uses. Numerous scientists [11, 12] have shown that chemically modified chitosan and chitin have better solubility in organic solvents and water.

Parameters including polydispersity, molecular weight (Mw), pattern of acetylation (PA), DD, and content of impurities (heavy metallics and protein) are often unknown for the majority of commercially available chitosans [13]. Understanding the structure-property-activity correlations in chitosan samples thus requires knowledge of their microstructure and specific focus is given in this regard to the chitosans employed in biomedical applications [10]. Another critical step is the comprehensive structural characterization of synthetically produced chitin and chitosan that are physiologically active and water-soluble or insoluble.

The current chapter discusses physicochemical characterization and the structural perspective of chitosan, chitin, and its derivatives using spectroscopic techniques. The most crucial methods include UV-Vis spectroscopy, nuclear magnetic resonance spectroscopy (NMR), X-ray spectroscopy, mass spectrometry, and Fourier-transform infrared (FTIR). It describes how effective these techniques are at establishing and validating molecular structures, tracking reactions, and regulating the purity of these molecules.

8.2 Spectroscopic Methods for Structural Analysis

8.2.1 X-Ray Spectrometry

Unquestionably, the most flexible and popular technique for analyzing materials of all kinds is X-ray spectroscopy [14]. The electronic structure focuses on the valence and primary electrons, which regulate the physical and chemical properties, among other things, and engineering structure, which provides information around the locations of all or a specific group of atomic resolution of the atoms in a molecule, are the two main categories of structural data that can be studied using X-ray spectrometry. This approach combines a number of spectroscopic methods, such as X-ray photoelectron spectroscopy (XPS), X-ray emission spectroscopy (XES), X-ray Auger spectroscopy, and X-ray absorption spectroscopy (XAS), to ascertain the geometric and electronic structures of materials by X-ray excitation. The determination of the target information is contingent upon its geometric properties, electronic, or related to oxidation states, a different type of it is used for X-ray spectroscopy. For instance, the XAS technique, invented by de Broglie, is used to look into the geometries of molecules or local structures as well as empty states, while the XPS technique, invented by Siegbahn, is used to consider occupied electronic states [15]. Thus, X-ray spectroscopy is an effective and versatile instrument that works well in conjunction with several structural investigation methods like IR, UV-Vis, Raman, or NMR.

8.2.1.1 Typical X-ray measurement circumstances

In order to measure X-ray diffraction, a crystal is put on a goniometer and slowly rotated while being irradiated with X-rays [41]. This results in a diffraction pattern of reflections, which are a pattern of spots with regular spacing. The electron density within the crystal is represented in three dimensions by converting the two-dimensional pictures acquired at different rotational angles. This conversion is achieved by the application of Fourier transforms, a mathematical technique, in conjunction with the chemical data collected from the sample. If large enough single crystals cannot be found, additional X-ray techniques, such as powder and fiber diffraction, can be used to capture less precise data.

The earliest X-ray diffraction (XRD) crystal investigations of chitin and chitosan were performed by Clark and Smith in 1937 [7]. They conducted those tests using a commercial copper-target diffraction tube that produced primarily Cu-K lines when run at 30 kV and 25 mA. The sample was 5.0 cm away from the flat film, which was perpendicular to the beam and used to record the diffraction patterns. Later studies mainly modified Clark and Smith's settings [7] while taking X-ray measurements of chitosan and chitin. For instance, to prevent X-ray scattering that would have resulted in a "dirty" background on X-ray film, X-ray diffraction analyses were performed at a relative humidity in a 100% helium environment. Modern X-ray diffractometers are used to conduct X-ray investigations of chitosan, chitin, and its derivatives.

XPS is a widely employed method in X-ray spectroscopy. It has been extensively utilized for the purpose of ascertaining the bonding energies associated with oxygen (O), nitrogen (N), and carbon (C) atoms present on the surface of chitosan. Additionally, XPS has been employed in several other studies pertaining to chitosan and chitin [16]. XES, which is ideal for analyzing chemical bonds, has also been used.

8.2.1.2 Chitosan and chitin X-ray spectrum

The diffractive patterns of chitosan and chitin have been discovered by several X-ray spectroscopy investigations [17]. These patterns have been described by several authors as having variously indexed crystalline peaks, which may be identified using either *d*, the crystallites' center-to-center spacing, or Miller indices, such (020). The use of angle values to describe the diffraction pattern is also rather prevalent.

X-ray diffraction, for instance, of a sample of chitin and -chitosan purified from *Doryteuthis spp.* Three crystalline peaks were found for the chitin sample at angles of about 8°, 20°, and 27°, which were assigned to the (020), (110), and (120) planes, respectively [18].

Miller indices of the chitin-specific diffraction peaks were (020), (110), (120), (101), and (130) [7], notwithstanding the claim that chitin lacked a robust (020) peak. The usual chitin diffraction pattern, when expressed in angle form, revealed significant reflections at 2 about 9–10° and 2–21° as well as weak reflections at higher 2 values, such as at 26.4° and above [19].

Two significant crystalline reflection peaks are visible along the 2 angles in the α-, β-, and γ-chitin [20]. According to these XRD spectra, all of the biopolymers are semi-crystalline materials. The principal crystalline reflection peaks for the α-chitin are shown in Fig. 8.1 to be at 9.6° (020 plane) and 19.6° (110 planes), 9.1° (020 plane) and 20.3° (110 plane), and 9.6° (020 plane) and 19.80° (110 plane). The second reflection peak is seen in β-chitin and γ-chitin at 12.74°, whereas it is also visible in β-chitin but with somewhat less intensity at B12.29°. For α-chitin, the crystallinity index (CrI) value is more than 90%; whereas, for β-chitin, it is lower (greater than 75%–89%). The γ-chitin has a lower CrI value than the α- and β-chitin at B68.6%. Shrimp, lobster, prawn, and king crab chitin have crystalline peaks at 9.28°, 19.36°, and 26.20°, respectively. Squid chitin has crystalline peaks at 9.8° and 19° (Fig. 8.2) [20].

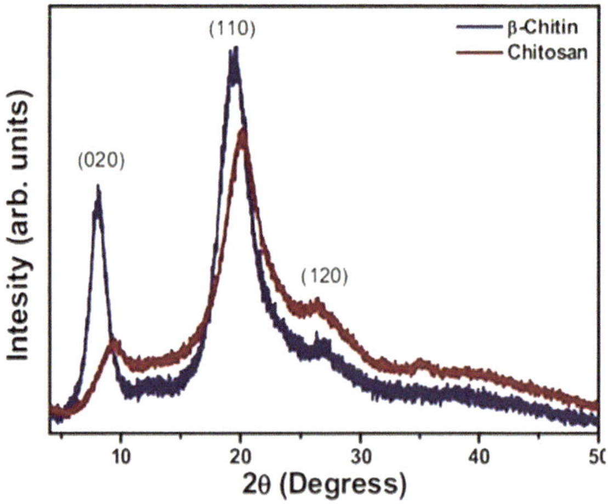

Figure 8.1 Chitosan and α-chitin samples' XRD patterns (reproduced with permission from El Knidriet al. [20]).

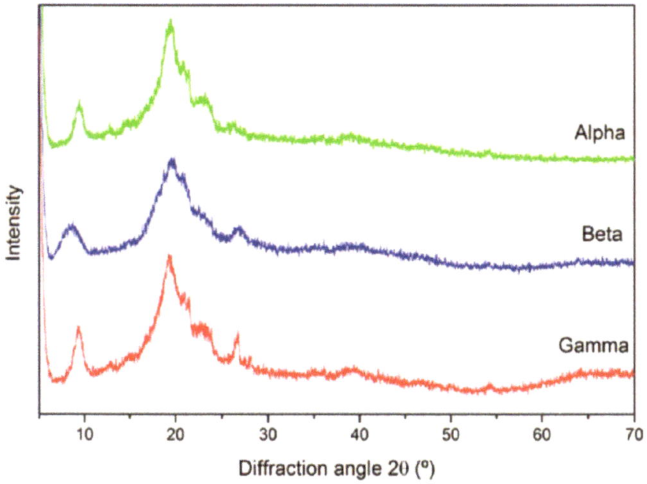

Figure 8.2 XRD spectra for β, α, and γ-chitin (reproduced with permission rom Elieh et al. [21]).

8.2.2 Infrared Spectroscopy

One of the most significant and popular analytical methods accessible to researchers studying chitin and chitosan is infrared (IR) spectroscopy. It is based on how a molecule's atoms vibrate. The process of measuring the amount of incident infrared radiation that is absorbed in a given energy while passing through a material with a permanent or induced dipole moment is known as the infrared spectrum [22]. A sample's specific bond types may be identified qualitatively because each peak in a spectrum of absorption has a certain energy that corresponds to the frequency of a component molecule's vibration. Electromagnetic radiation's energy that passes through a material and is measured by an IR spectrometer as a function of frequency or wavenumber. Today, the entire spectrum is subjected to an interference analysis before being transformed mathematically through the Fourier transform into the wavenumber or frequency range. The quality of infrared spectra has significantly increased thanks to FTIR spectroscopy, which has also reduced the amount of time needed to collect data [22]. The literature reviewed the advancements in contemporary infrared spectroscopy [22].

8.2.2.1 Conditions typical for FTIR spectroscopy of chitosan and chitin

The major focus of this chapter is on the practical ramifications of using the most well-liked spectroscopic technique for the structural study and physicochemical analysis of chitosan and chitin. Additionally, the various spectroscopic methods are discussed. It should also be noted that the simultaneous use of at least two different spectroscopic methods is necessary for chitosan and chitin structure analysis [23]. The main argument in favor of this is that it is challenging to collect varied but comparable data on the composition and physical and chemical properties of these molecules when employing several spectroscopic approaches. Spectrophotometric approaches have revolutionized the physicochemical characterization and chemical of chitosan and chitin materials and are now widely employed by researchers working in the area as a result of continual advances in sampling practices, analytic software, and instrumentation gear [24].

8.2.2.2 Use of infrared spectroscopy to characterize chitin and chitosan

In addition to characterizing isolated chitin, FTIR has also been used to identify the source of chitin. Small differences were noticed, such as A. Caribbeana's stronger IR absorption bands. Even though the FTIR absorption spectra of the unprocessed natural materials revealed comparable distribution patterns for each coral type and verified the existence of chitin in both kinds, the results nonetheless demonstrated that chitin was absent from both species, by using FTIR analysis to identify the lack of a free hydroxyl in *A. Caribbean*, it was discovered that the chitin chains were arranged in sheets and hydrogen-bonded to neighboring chains, which promotes a denser chitin fiber packing. As a result, it was possible to explain why *A. Caribbean* coral in its natural state was more difficult to grind, and based on the FTIR data, it takes a long time to remove proteins from *A. panacea*. IR also supported the discovery of chitin in multicellular sclerites.

Additionally, using two different extraction techniques, extraction in an open reactor after 1 hour after 24 hours of a basic reaction, an acidic extraction, and a reaction in a closed reactor. FTIR has been utilized to assess the purity and yield of chitin extracted from silkworm pupae [25]. In this work, IR also looked at the effectiveness of chitosan synthesis by chitin undergoing N-deacetylation. The band at 1655 cm^{-1} gradually shrank while the band at 1590 cm^{-1} grew, showing the preponderance of NH_2 groups during the N-deacetylation of chitin. Figure 8.3 displays the spectra of chitosan and chitin that were produced after chitin was N-deacetylated in the presence of $NaBH_4$ in a solution of NaOH (40 wt%) for 5 hours.

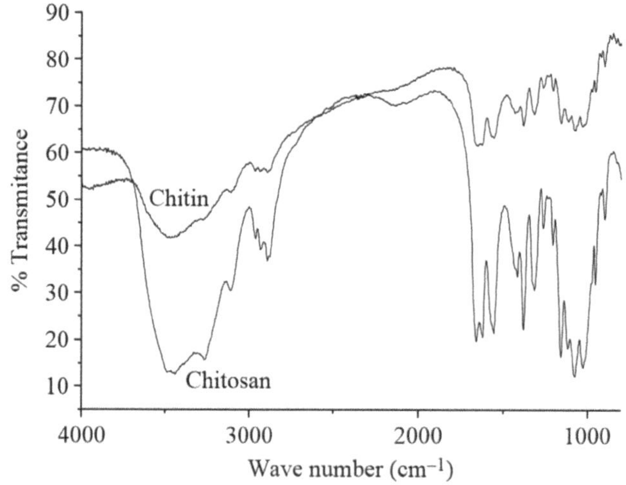

Figure 8.3 FTIR of silkworm pupae-produced chitin and chitosan; range: 1400–1700 cm^{-1} (reproduced with permission from Gharsallaoui et al. [25]).

Ilham et al. [26] compared chitosan extracted from several sources in their investigation (Fig. 8.4). The analysis of these spectra reveals a wider band between 3100 and 3500 cm^{-1} that is linked to the vibrations caused by the stretching of water molecules and free NH groups, which comprise, respectively, NH and atoms. The absorption peak, which occurred between 2850 and 2950 cm^{-1}, was caused by the C-H stretch. The band at 1623 cm^{-1} (deacetylated chitin) was an amide generated as a result of the interactions between hydroxyl and hydrogen groups as a result of the removal of the acetyl group. The relative DD% for *Blaps Lethifera*, *Musca Domestica*, and *Pimelia Fernandezlopezi* was 87.1%, 84.2%, and 88.1%, respectively; the DD value depends on several factors, including sample source, sample preparation method, and the classification of devices utilized in the analysis method.

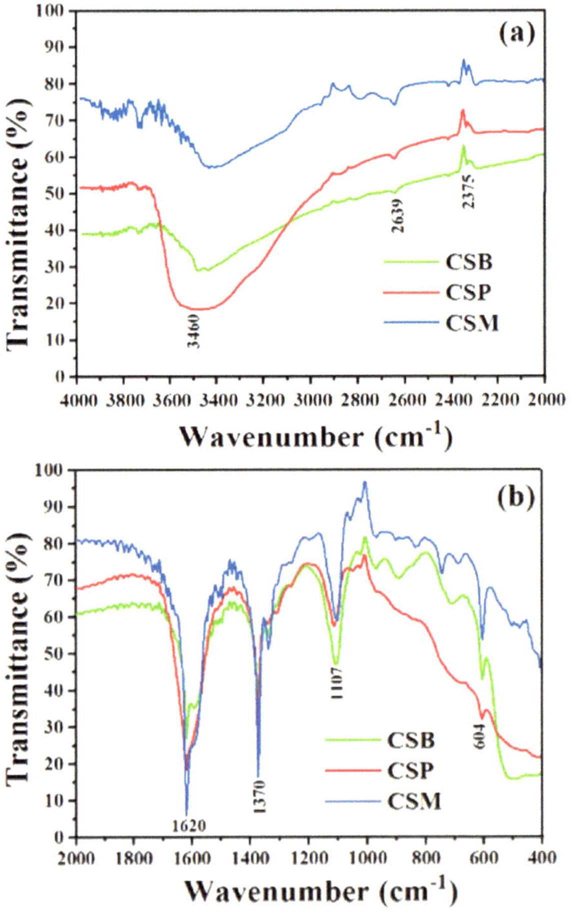

Figure 8.4 Spectres of chitosan derived from insects in two wavenumber ranges in the FTIR spectrum: (a) 4000–2000 cm^{-1} and (b) 2000–400 cm^{-1} (reproduced with permission from ilham et al. [26]).

8.2.3 UV-Visual Spectroscopy

Two factors make ultraviolet/visible (UV-Vis) spectroscopy a strong analytical tool. It may be used for assaying as well as, first, identifying certain functional groups in molecules. UV-Vis spectroscopy, in contrast to IR spectroscopy, involves the up-excitation of electrons to higher energy states after absorbing electromagnetic radiation in the 200–800 nm region [27]. The only functional groups (chromophores) in organic compounds that can absorb ultraviolet and visible light are those with low excitation energy valence electrons. As a result of vibrational and rotational transitions being overlaid on electronic transitions, the UV-Vis spectra are extremely complicated and give the impression of being a continuous absorption band. Today, it is also possible to record the individual observation of electron transfers without adjacent vibrational bands superimposing [28]. Investigating electron transfers between bands or orbitals of molecules, atoms, and ions that are present in the gaseous, solid, and liquid phases is achievable with UV-Vis spectroscopy. Analyses of liquids and crystals are often conducted in transmission, but diffuse reflection spectroscopy (DRS) is frequently used to analyze materials that are in powder form. Due to the wide bandwidths, dispersive spectrometers are virtually solely utilized in UV-Vis spectroscopy, as opposed to IR spectroscopy, which mostly uses Fourier transform methods. Numerous publications and books give further information about UV-Vis and its use [29].

8.2.3.1 Typical UV-Vis measurement circumstances

A commonly used method for obtaining UV-Vis spectra of chitosan and chitin in aqueous acid solutions (such as phosphoric acid, hydrochloric acid, acetic acid, and perchloric acid) involves using a quartz cell with a length of 1.0 cm at ambient temperature [7]. Sometimes solutions of DMSO, an aqueous base, or water are used. Samples that are in the form of powder or films have their diffuse reflectance UV-visible (DRUV) spectra measured. It has also been used to analyze data in the vacuum ultraviolet to near-infrared region.

8.2.3.2 Chitosan and chitin UV-Vis spectroscopy

Because N-acetyl-glucosamine (GlcNAc) and glucosamine (GlcN), two far-UV chromophoric groups, are present in different ratios in chitin and chitosan, they have non-zero extinction coefficients at wavelengths less than around 225 nm. The chitin/chitosan chain's GlcNAc and GlcN residues do not appear to interact. Therefore, the monomer units just add up to the overall absorbance of these polymers at a given wavelength. As shown in Fig. 8.5, the UV spectra of combinations of GlcNAc and GlcN hydrochloride are fairly comparable to those of chitosan. The maximum wavelength is 201 nm in hydrochloric acid solution, with a concentration of 0.1 M [30].

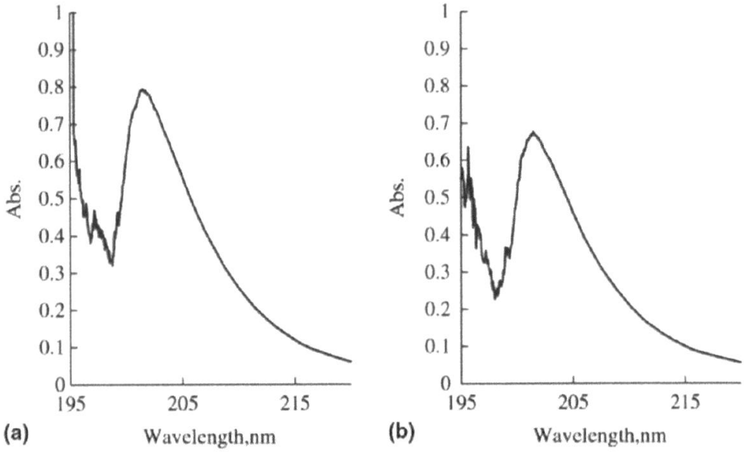

Figure 8.5 UV spectra of 0.1 M hydrochloric acid solution with chitosan (a) and a combination of GlcNAc and GlcN hydrochloride (b) (reproduced with permission from [30]).

For the optical analysis of the chitin source, UV-Vis spectroscopy can be performed [31]. *Antipathes pennacea* and *Antipathes caribbeana* both have optical spectra that display a broad range from 300 nm to 500 nm with minor species-specific variations. The spectra also demonstrated the presence of Mn^{+3} in both species and Fe^{+3} in *A. Caribbean.*

8.2.4 Mass Spectrometry

A destructive analytical method for determining the properties of specific molecules is mass spectrometry (MS). The molecular mass of a chemical, which is discovered by calculating the mass-to-charge ratio (m/z) of its ion, is the fundamental piece of knowledge discovered by mass spectrometric investigation

[32]. The whole details of a molecule's chemical structure can be discovered via the ionization approach. Small molecules to complex macromolecules like nucleic acids, peptides, proteins, or carbohydrates may all be analyzed using the MS technique. Three basic components make up a mass spectrometer: an ion source, an analyzer, and a reagent. The sample must first be introduced to the ion source, which is the center of each mass spectrometer. Through a direct intake (probe, syringe, septum inlet, etc.) or an indirect inlet (liquid chromatograph, capillary electrophoresis column, or gas chromatograph), there are two ways to do this. The sample molecules are ionized in the ion source using a number of different methods, such as chemical ionization (CI), electrospray ionization (ESI), electron impact (EI), and matrix-aided laser desorption ionization (MALDI). The ions are then sent from the stream to the analyzer, where they undergo m/z sorting and separation. Quadrupoles, ion traps, and time-of-flight analyzers are the most often utilized types. When performing tandem mass analyses (MSn, MS/MS), two or more analyzers are combined. Three different operating modes for the MS are available: full scan (the process of doing a scan within a certain range of masse), more sensitivity SIM (the selected ion monitor is SIR minus one ion recording), and the MRM (multi-reaction tracking) that is most sensitive. When the energy of the ions is transformed into electric signals that the computer can interpret, the ions eventually arrive at the detector. Finally, mass spectra are produced, making chemical identification feasible [33, 34].

8.2.4.1 Standard settings for chitin/chitosan mass spectrometric analysis

Depending on the data needed, many mass spectrometer types might be used. Investigations into chitin, chitosan, and its derivatives employ spectrometers with a variety of intake systems, ion sources, and analyzers. The positive ion mode is recommended in most circumstances.

Pyrolysis-mass spectrometry was the first technique employed [35]. Although that was in 1982, the technique is still used often when paired with gas chromatography [36]. The initial sample in GC/MS is pyrolyzed (450–615 °C), and Following separation on a capillary column at a temperature program of 35–50 °C and changing speeds up to 220–300 °C, the volatile pyrolysis products are then separated. The electron impact mode at an energy level of 70 eV and the complete scan mode (from 15–35 to 300–650 $m \cdot z^{-1}$) are frequently employed in GC/MS analysis. Products are recognized based on their mass spectra and GC retention characteristics [36].

Chitosan or any of its impurities can be examined using GLC-MS. First, though, all of the target compounds must be transformed into volatile derivatives (reduction, chitin hydrolysis, impurity extraction, trimethylsililation, or acetylation) [37]. A GLC capillary column that is nonpolar or moderately polar with a gradient temperature program typically from 60 °C to 280 °C with various heating rates is used to separate the products. MS with EI ionization is used for detection [38].

The majority of studies discussing the mass spectrometric analysis of chitosan and chitin state that the ionization technique chosen was MALDI. The time of flight analyzer (MALDI-TOF) is most frequently used in conjunction with this ion source [39]. A nitrogen laser that emits at 337 nm is installed in the MS ion source. Positive ion mass spectra are obtained for these studies using an accelerating potential of 20–30 kV. For chitin/chitosan research, a modified variant of MALDI-TOF called MALDI-TOF PSD MS (mass spectrometry with post-source decay) was introduced. Tandem mass spectrometry is another MS method that uses a MALDI source, and it is a great use for an ion trap analyzer. Using an experimentally verified collision energy, one analyzer enables the acquisition of MS2 and MS3 spectra. Tandem mass spectrometry analysis of chitin and chitosan can also use MALDI-TOF/TOF equipment. The acceleration of precursor ions is at 8 kV, whereas the acceleration of their fragment ions is at 19 kV [40].

Another technique used to study chitosan is secondary ion time-of-flight mass spectrometry. Gallium serves as the principal ion beam in this sort of spectrometer, which operates at 15 kV. The secondary ions that are present are extracted (10 kV) and go to the detector through the analyzer (TOF type) [41].

The second-most significant ionization method for chitosan and chitin analysis, after MALDI, is electrospray ionization (ESI) [42]. Though a quadrupole kind of analyzer may also be utilized, the ESI ion source is typically connected to an ion trap analyzer [42]. Several factors, including the source voltage, ranges from 3.5 to 4 kV, taking into account the flow of nitrogen gas and temperature of a capillary, should be tuned for sample readings. For direct sample injection, a syringe pump (3–4 L/min) is employed. To get additional structural data, tandem mass spectrometry with ESI ionization is desirable. The first quadrupole in a triple, quad system (QqQ) chooses the precursor ions, the second quadrupole serves as the collision chamber, and while argon acts as the neutral collision gas, the third quadrupole analyses the ion products. Quadrupole and time-of-flight (Q-TOF) are sometimes found to work well together. Another benefit is the ability to connect liquid chromatography to the MS system. The analysis is simplified by separating chitosan-containing solutions on a chromatographic column. The same modalities of mass spectrometry (negative and positive ion detection, scanning, SIM mode, electrospray ionization, etc.) are used for LC-MS analysis. The ion source settings are typically the same as those used in traditional MS analysis, which produces sodium adducts $[M+Na]^+$ and protonated products $[M+H]^+$. A different kind of column is utilized for HPLC separation, and the mobile phase frequently consists of water or acetonitrile with formic acid added [43, 44]. Investigation of chitosan derivatives is well suited to LC-MS.

Another soft ionization technique helpful in research on chitin and chitosan is fast atom bombardment (FAB). Thioglycerol, m-nitrobenzyl alcohol, glycerol, dithiothreitol, dithiothreitol-thioglycerol, or 2,4-dinitrobenzyl alcohol can all serve as the matrix for this purpose [45]. The mass spectrometer typically runs in positive-ion mode, generating $[M+H]^+$ ions most frequently, although it may also produce $[M+Na]^+$ or $[M+K]^+$ ions. With xenon serving as the blasting gas, the atom cannon is operated at 6–8 kV. Tandem mass spectrometry using collision-induced dissociation (FAB-CID) observations is compatible with this kind of ion source. Helium collisions and a 10 kV accelerating voltage are employed to produce FAB-CID-MS/MS spectra [46].

For chitin/chitosan analysis, less common ionization techniques are occasionally utilized [47]. One of them uses ammonia or acetic acid as the reactive gas and is known as chemical ionization (CI). In both situations, MS produces $[M+NH_4]^+$ and $[M+H]^+$ in the positive ion mode, respectively. Ionization can also be accomplished by the process of plasma desorption, which is performed using californium fission products in a time-flow mass spectrometer for biochemistry.

Chitin/chitosan research hardly ever uses TC-EA/IRMS (mass spectrometer with high-temperature conversion-elemental analysis). The quantitative conversion of hydrogen to H_2 and oxygen to CO at a high temperature constitutes the initial stage in this process. The GC separates the gaseous products, which are then passed to the IRMS for analysis of the stable hydrogen and oxygen isotope ratios in chitin [48].

The material is heated quickly from 25 °C to 600 °C before the gaseous byproducts are introduced into the mass spectrometer to begin the analysis of chitosan using TG-MS (thermogravimetry-mass spectrometry). Using a channeltron detector and quadrupole analyzer, MS is used to study ions having m/z values between 40 and 130 [49].

8.2.4.2 Measurement of chitosan and chitin polymerization levels by mass spectrometry

Chitosan, a naturally occurring copolymer of glucosamine and N-acetylglucosamine, has antibacterial and anticancer effects and promotes plant development [50]. The molecular weight is intimately connected to this action. If one intends to employ chitosan in biological or pharmacological applications, one must take this aspect into account. Depolymerization can be used to produce chitooligosaccharides. This can be done enzymatically, chemically, physically, or both. In each scenario, the degree of chitosan depolymerization and the kinetics may be tracked by MS [51].

Sonolysis is one of the physical processes used to depolymerize chitosan. It was discovered that chitosan underwent random depolymerization following ultrasonic treatment, resulting in the production of several distinct chitooligosaccharides (COS). The products were found using the ESI-MS and MALDI-TOF MS (Fig. 8.6) techniques. Sonolysis was proven to be an effective alternative or enzymatic chemical process for producing oligosaccharides for commercial application, with the core chemical structure having remained unaltered [52].

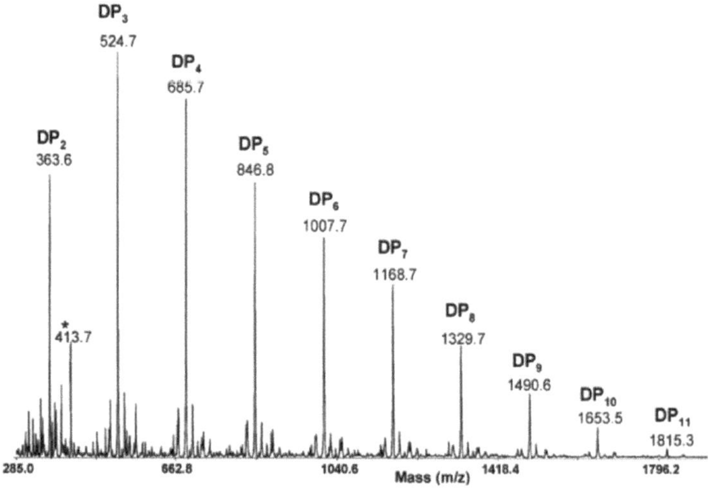

Figure 8.6 Deacetylated chitooligosaccharides from chitosan that have undergone ultrasound treatment, as detected by MALDI-TOF MS (reproduced with permission from [52]).

8.2.5 Nuclear Magnetic Resonance Spectroscopy

One of the more effective methods for studying physicochemical properties and the structure of organic substances, including tiny molecules and polymers, is NMR spectroscopy. It appears to be ideally suited for research on chitosan and chitin. However, there is a significant issue with these compounds' solubility. In contrast to chitosan, which is soluble in an acidic aqueous solution, higher DD chitin is essentially insoluble in all solvents and mixes of various solvents. The amount of N-acetylation, crystallinity, polymerization, distribution of GlcN and GlcNAc throughout the polymer chain, ionization potential of the solvent, pH, and chitosan's solubility is directly related to the concentration of the substance in the solution [53].

Chitosan, chitin, and its derivatives have been studied using various NMR methods, including ^{1}H, ^{31}P, and ^{13}C liquid-state NMR, as well as solid-state NMR using ^{13}C and ^{15}N. However, because this method does not necessitate solubilizing the polymer, it is only possible to analyze chitins with a high DD using ^{13}C and ^{15}N solid-state NMR.

8.2.5.1 Explanation of characterization NMR methods

8.2.5.1.1 *^{1}H NMR spectroscopy*

Finding a suitable solvent is crucial; it should have appropriate solubility characteristics toward the target substance [54]. No quantitative analysis is precise and repeatable when the material is only partly soluble. Additionally, the solvent's residual signal should not cross over with the signals from the sample. For liquid-state NMR spectroscopy, the most used solvents are D2O/DCl, D2O/CD3COOD, and D2O/DCOOD.

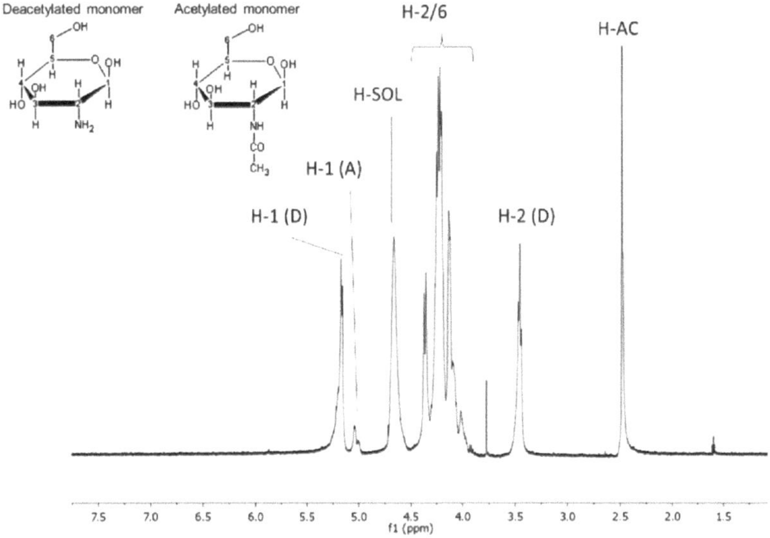

Figure 8.7 Chitosan 1H NMR spectrum (reproduced with permission from [55]).

Figure 8.7 displays a typical 1H NMR spectrum of chitosan. Chitosan's distinctive peaks [29, 30] have been found. The aldohexoses' H-2, H-4, H-5, H-3, and H-6 signals are found between 3.8 and 4.2 ppm. A peak was defined as glucosamine anomeric H (H-1 (D)), and as well as the signal at 5 ppm was determined to be anomeric H-1-N acetyl glucosamine (H-1 (A)). The hydrogens in N-acetylglucosamine's (H-AC) methyl group and H-2 glucosamine (H-2 (D)) were respectively found as the peaks at 2.5–3.5 ppm. The solvent's H was represented by the peak at 4.75 ppm [55].

8.2.5.1.2 ^{13}C *NMR spectroscopy*

Both liquid-state and solid-state ^{13}C NMR spectroscopy are available. The first has the same constraints as ^{1}H NMR spectroscopy, namely that most materials under investigation are not sufficiently soluble [56]. Moreover, due to the characteristics of the carbon nucleus and the fact that the ^{13}C isotope is only 1% abundant in nature, ^{13}C NMR spectroscopy is significantly less sensitive than ^{1}H NMR spectroscopy. The same types of solutions that are used to produce ^{1}H NMR spectra are often utilized to produce liquid-state ^{13}C NMR spectra, but the required number of scans must be significantly higher. It can range from a few dozen to hundreds of thousands [57]. The ^{13}C NMR spectroscopy performed on solids is a far more potent method. Without special solubilization of the polymer, chitin can be analyzed. This implies that a sizeable sample may be utilized, which fixes the issue with the ^{13}C NMR spectroscopy's poor sensitivity. Solid sample ^{13}C NMR spectra are often captured using cross-polarization (CP) and magic-angle spinning (MAS). MAS generates highly resolved spectra by averaging out chemical shift anisotropy and dipolar interactions. By decreasing the relaxation time brought on by the transfer of magnetization from the ^{1}H to the ^{13}C spins, CP significantly boosts the sensitivity of the method. The kinetics of the CP process have an impact on the ^{13}C NMR signal intensities because variable contact periods have an impact on ^{13}C NMR resonance intensities. Therefore, using the appropriate contact and relaxation delay durations is crucial [58]. During repeated assignments of DD of chitin and deacetylated chitin materials, the contact time (1 ms) and the relaxation delay (5 s) were changed. For each carbon atom used in the mathematical formula, care must be taken to make sure that CP builds up in the same way for all of them. In addition, many chitin/chitosan composites frequently include magnetic centers, which may reduce the strength of some ^{13}C NMR signals [59].

8.2.5.1.3 ^{15}N NMR spectroscopy

Similar to solid-state ^{13}C NMR spectroscopy, ^{15}N NMR spectroscopy is a highly helpful approach, particularly for examining insoluble chitin, although it takes more time because the ^{15}N nuclide is not naturally abundant (<0.4%). The approach is quite costly because measurement periods are often of the order of 12–24 h, even when employing high-field NMR spectrometers with enormous volumes of material. Fungal chitin was the first 15N labeled by Watson et al. [43], making ^{15}N NMR spectroscopy considerably more useful. The solid-state ^{15}N NMR spectra for the samples of chitin tagged with ^{15}N and those with a natural abundance of ^{15}N required 60,000 and 2,000 repeats, respectively. Care should be taken to make sure that CP build-up is the same for both N-deacetylated nitrogen centers and N-acetylated, much like with spectroscopy of ^{13}NMR in solid-state.

The chitosan and chitin spectrum exhibits two distinct nitrogen atom signals, one of which is associated with the amine group and the other acetamide grouping.

8.2.5.1.4 ^{31}P NMR spectroscopy

With a 100% isotope abundance and a reasonably high magnetogyric ratio, ^{31}P is one of the more common NMR methods, making ^{31}P NMR spectroscopy one of the more common NMR procedures. Both natural and synthetic chitosans lack phosphorus atoms, making it unable to analyze them directly with solid-state ^{31}P NMR spectroscopy. However, both solid-state and liquid-state ^{31}P NMR spectroscopy may be used to analyze various soluble chitosan derivatives that include phosphorus atoms. Specific phosphorus compound types can be recognized in the spectrum by their distinctive chemical shift ranges.

8.2.6 Thermogravimetric Analysis (TGA)

The Mw of the polymer affects its thermal characteristics. Derivative thermogravimetric (DTG) curves for γ-, α-, and β-chitin are created, as shown in Fig. 8.8, a comparison and evaluation of their thermogravimetric profiles with relation to DD and thermal stability in two separate phases [21]. The evaporation by which water leaves hydrophilic groups in chitin chains is what causes the first loss of weight. The initial weight loss is seen for γ, α, and β-chitin at temperatures of ~130 °C (4.95%), ~205 °C (5.20%), and ~205 °C (7.93%), respectively. There are no obvious variations between any of the three chitin samples' earliest stages of breakdown. The second weight loss is due to the breakdown of the polymeric structure, namely the dehydration of saccharide rings and the deterioration of acetylated chitin units [21].

All three chitin isolates had DTGmax values of 381.99 °C (69.87%) for γ-chitin, 348.77 °C (72.56%) for β-chitin, and 391.09 °C (74.37%) for α-chitin. The findings showed that DTGmax for α-chitin was much greater than γ-chitin, but only modestly. Compared to α-chitin and β-chitin, the -chitin showed a minimal degradation percentage at its DTGmax temperature. This is explained by the structural arrangement of γ-chitin, which is made up of dense, spindly microfibers. The DTGmax values for α-chitin were reported to be between 350 °C and 400 °C in earlier publications [60]. *Periplaneta americana* wing chitin DTGmax values were recorded by Kaya et al. [61] to be 389 °C. Sajomsang et al. [62] noted that the DTGmax values for *Cicada sloughs* and *crab shells*, respectively, were 372 °C and 369 °C. Thermal stability for β-chitin was found to be between 250 °C and 350 °C [63]. In the lone publication on its characterization, the thermal stability of γ-chitin was quoted as 380.7 °C.

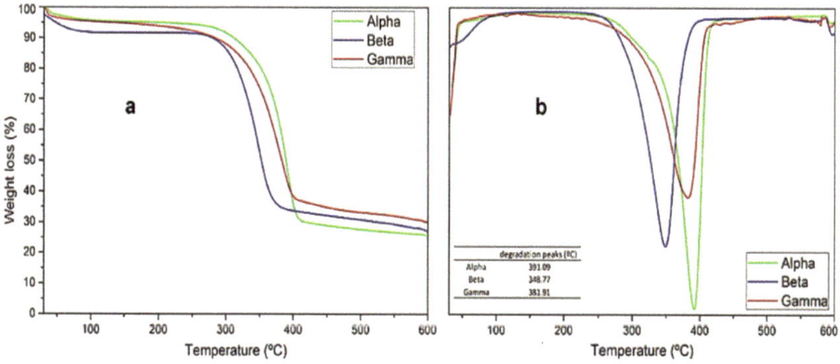

Figure 8.8 (a) TGA and (b) DTG of γ, β, and α-chitin studied (reproduced with permission from [21]).

8.2.7 Utilizing Energy-Dispersive Spectroscopy (EDS) for X-Ray Microanalysis

X-rays are used in the EDS analysis to hit a liquid nitrogen-cooled semiconductor detector, which produces an electrical signal proportional to the energy of the X-rays impacting it. These pulses are subsequently processed by a multichannel analyzer. The graph shows X-ray intensity (counts) against energy (keV). EDS microanalysis offers a simultaneous multielement analysis of the whole X-ray spectrum and is a quick method. The higher-order X-ray lines are not visible. The procedure for preparing samples for EDS analysis is identical to that for SEM analysis, with the exception that carbon is used for coating rather than gold. The same chitosan sample used for the SEM micrograph may be utilized for the EDS X-ray microanalysis.

EDS X-ray microscopy can be used for elemental analysis of a chitosan sample. In a study using Rafigh et al. [64], they found that the chitosan sample contained 48.18% C, 12.19% N, and 37.21% O elements estimated from X-ray microanalysis EDS.

8.2.8 Scanning Electron Microscope Analysis

SEM is a very helpful technique for visual validation of the surface's physical condition and morphology [65]. SEM has been used to study the surface morphology of chitin, fungus, and chitosan in crab shells, as well as deacetylated chitosan films and powder and chitosan and crosslinked chitin types before and after metal binding.

As shown in Fig. 8.9, SEM analysis is performed to look at the surface morphology of chitin and chitosan. Arthropod exoskeleton shape and the morphology of each reaction step involved in the production of chitosan are shown in Fig. 8.9 (a–c). The raw prawn shell has a varied morphology with a compact structure, white spots, and a well-defined shape, as shown in Fig. 8.9 (a–c) [66]. The uncooked prawn shell may include white spots that indicate the presence of $CaCO_3$. As a result of acid treatment after demineralization, $CaCO_3$ is removed, leaving behind white patches that resemble rounded holes, as shown in Fig. 8.9 (d–f).

The surface, however, transforms into a microfibrillar crystalline structure after acidification treatment [66].

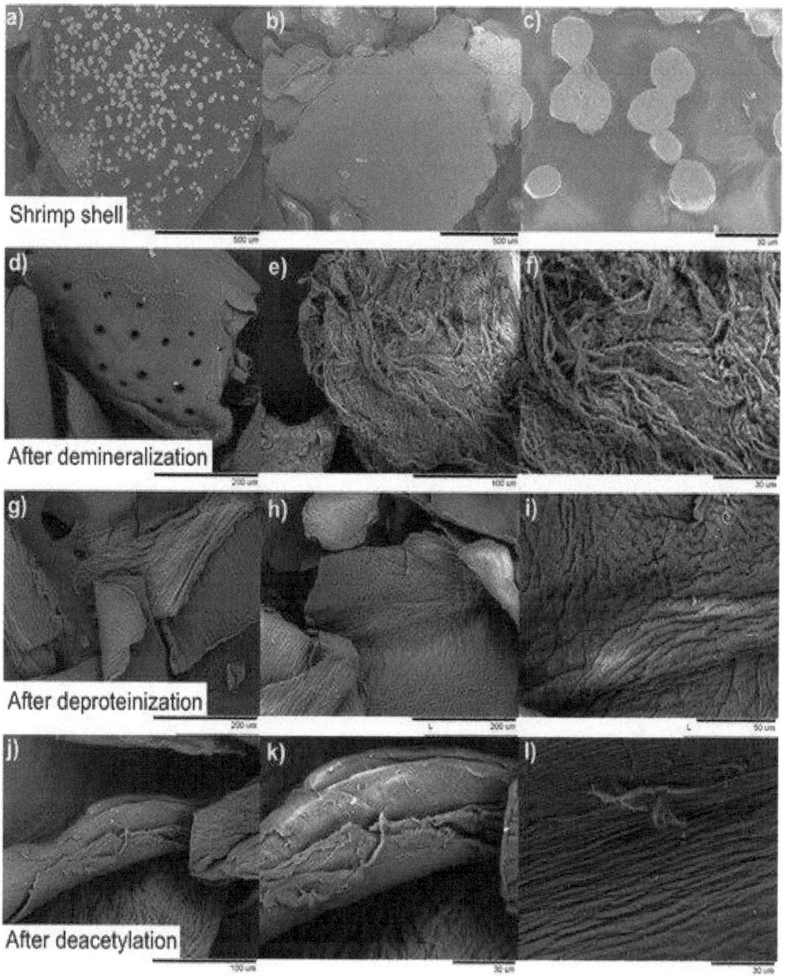

Figure 8.9 SEM images: (a) represent shrimp shells, (b) in the purification stage, (c) post-deproteinization stage, and (d) a post-deacetylation phase (reproduced with permission from [66]).

The surface morphologies of insect chitosan and chitin may generally be divided into the following groups: Nanofiber with nanopores (I), nanofiber alone (II), nanofiber alone (III), nanofiber alone (IV), smooth surface (V), and rough surface (VI). Both nanofiber and nanopore structures may be seen in the chitin of house crickets [67], grasshoppers [61], crickets [68], and orthopteran species. A nanofibrous structure may be seen in aquatic insects, water scavenger beetles, desert locusts [69], and Colorado potato beetle chitosan [70]. According to a few studies, the chitin of black army flies and cockroaches possessed nanopores without nanofibers [69] and nanofibers without nanopore structures. Additionally, the surface morphologies of the chitin from *Holotrichia parallel*, *Zophobas morio*, and *Catharsius molossus* for the extraction of chitosan were both rough and smooth. Anand et al. [71] noted that the shape of crab and squilla chitin resembled a sponge and a cauliflower leaf in this context. The scanning electron microscopy (SEM) analysis shed light on the microfibrillar crystalline structure as well as the porosity of the Conus chitin [72]. The chitin extracted from krill, prawns, and lobster shells likewise included the closely packed fibers [73]. Additionally, an SEM study of the surface morphologies of *P. monodon's* chitosan and chitin revealed microfibril and porous structures [74]. One of the essential characteristics that govern the efficient usage and application of chitosan and chitin is surface morphology. Chitosan's and chitin nanofiber and nanopore forms may find utility in the textile, food, and pharmaceutical industries.

8.2.9 Transmission Electron Micrograph (TEM)

The target material is traversed by the TEM's electron beam. The variations in sample density are what cause the contrast discrepancies in the picture. Chitosan sample preparation involves embedding in Epon-Spurr resin, heating it in a vacuum-sealed microwave for three minutes at 250 W, and then allowing it to polymerize for 24 hours at 60 °C [75]. The samples are then sliced at a thickness of 65–85 nm using a Leica Ultracut UCT ultramicrotome. Sections are then set up on copper grids with carbon coatings for TEM observation. At 300 keV, the Philips TEM is in operation.

8.3 Connection between Characterization Techniques and Physical Characteristics

The features of chitosan and chitin have a significant impact on their properties and, consequently, on their potential uses, as this chapter will demonstrate. In actuality, not all chitin or chitosan samples may be applied in the same ways. For this reason, a thorough characterization of the samples is required.

Chitin exists in three distinct crystal forms: γ, β, and α chitins. In accordance with its physical condition, chitosan is also crystalline and exhibits polymorphism. The residual crystallinity of a polymer might vary greatly depending on where it came from and how it was handled throughout the extraction process.

Both completely deacetylated chitosan (i.e., 100%) and chitin (DD = 0%) have the highest levels of crystallinity.

Additionally, it has been claimed that the relationship between the source and the material's surface area exists. The primary factors influencing the polymer characteristics are crystallinity, polydispersity, DD, and Mw. Ash content, moisture, and levels of metallic heavy, endotoxins, and proteins must all be assessed for uses that involve human consumption, like food and medicinal uses.

The DD has been cited as one of the most significant chemical properties, which may have an impact on how well chitosan functions in a variety of applications [76].

The biochemical and biopharmacological relevance of chitosan is significantly impacted by the average Mw's influence on the formation of viscosity in aqueous solutions [77]. It is significant to note that chitin's poor solubility makes it difficult to estimate Mw.

Numerous techniques have been documented for determining the properties of chitosan and chitin, as indicated in Table 8.1 [78–82]. Using techniques based on distinct concepts yields varied results. As a result, it is crucial to specify the characterization technique. Even the most well-defined chitosans now available on the market are often simply specified in terms of their average DD and average polymerization degree, their ash ratio, and lack of microorganisms that might cause contamination, with the polydispersity index being mentioned in certain cases. Moreover, to the characteristics listed above, the arrangement of the acetyl groups throughout the chitosan molecules' linear backbone may be essential in determining how the molecules interact with biological systems [83].

Table 8.1 Physicochemical properties of chitosan and chitin and methods for their determination [78–82]

DD	First derivative UV-spectrophotometry Infrared spectroscopy Conductometric titration Nuclear magnetic resonance spectroscopy Differential scanning calorimetry Potentiometric titration
Average Mw or distribution of Mw	Viscosimetry Light scattering Gel Permeation chromatography
Moisture content	Weighing analysis
Crystallinity	Radiation diffraction
Protein	Bradford technique
Ash content	Gravimetric method

8.4 Conclusion

This was the main focus of this chapter on the practical ramifications of using the most well-liked spectroscopic techniques for the structural study and physicochemical properties of chitosan and chitin. Additionally, the various spectroscopic methods are discussed. It should also be noted that the simultaneous use of at least two different spectroscopic methods is necessary for chitosan and chitin structural analysis. The main argument in favor of this is that it is challenging to collect varied but comparable data on the composition and physical and chemical properties of these molecules when employing several spectroscopic approaches. Spectrophotometric approaches have revolutionized the physical and chemical characterization of chitosan and chitin materials and are now extensively utilized by researchers working in the area as a result of continual developments in sampling procedures, analytical software, and instrumentation gear.

References

1. Piekarska K, Sikora M, Owczarek M, Jóźwik-Pruska J, Wiśniewska-Wrona M (2023) Chitin and chitosan as polymers of the future—obtaining, modification, life cycle assessment and main directions of application. *Polymers* 15: 793.
2. Shah AM, Qazi IH, Matra M, Wanapat M (2022) Role of chitin and chitosan in ruminant diets and their impact on digestibility, microbiota and performance of ruminants. *Fermentation* 8: 549.
3. Amor IB, Hemmami H, Laouini SE, Temam HB, Zaoui H, Barhoum A (2023) Biosynthesis MgO and ZnO nanoparticles using chitosan extracted from Pimelia Payraudi Latreille for antibacterial applications. *World Journal of Microbiology and Biotechnology* 39: 19.
4. Ben Amor I, Hemmami H, Laouini SE, Mahboub MS, Barhoum A (2022) Sol-gel synthesis of ZnO nanoparticles using different chitosan sources: effects on antibacterial activity and photocatalytic degradation of AZO Dye. *Catalysts* 12: 1611.
5. Kumirska J, Weinhold MX, Czerwicka M, Stepnowski P (2011) Influence of the chemical structure and physicochemical properties of chitin-and chitosan-based materials on their biomedical activity. *Biomedical Engineering, Trends in Materials Science*: 25–64.
6. Elieh-Ali-Komi D, Hamblin MR (2016) Chitin and chitosan: production and application of versatile biomedical nanomaterials. *International Journal of Advanced Research* 4: 411.
7. Kumirska J, Czerwicka M, Kaczyński Z, Bychowska A, Brzozowski K, Thöming J, Stepnowski P (2010) Application of spectroscopic methods for structural analysis of chitin and chitosan. *Marine Drugs* 8: 1567–1636.
8. Dutta PK, Dutta J, Tripathi V (2004) Chitin and chitosan: Chemistry, properties and applications. *Journal of Scientific and Industrial Research* 63: 20–31.
9. Duan C, Meng X, Meng J, Khan MIH, Dai L, Khan A, An X, Zhang J, Huq T, Ni Y (2019) Chitosan as a preservative for fruits and vegetables: A review on chemistry and antimicrobial properties. *Journal of Bioresources and Bioproducts* 4: 11–21.

10. Kumirska J, Weinhold MX, Thöming J, Stepnowski P (2011) Biomedical activity of chitin/chitosan based materials—influence of physicochemical properties apart from molecular weight and degree of N-acetylation. *Polymers* 3: 1875–1901.
11. El Knidri H, Belaabed R, Addaou A, Laajeb A, Lahsini A (2018) Extraction, chemical modification and characterization of chitin and chitosan. *International Journal of Biological Macromolecules* 120: 1181–1189.
12. Chen Q, Qi Y, Jiang Y, Quan W, Luo H, Wu K, Li S, Ouyang Q (2022) Progress in research of chitosan chemical modification technologies and their applications. *Marine Drugs* 20: 536.
13. Weinhold M (2010) Characterization of chitosan using triple detection size-exclusion chromatography and 13C-NMR spectroscopy. http://nbn-resolving.de/urn:nbn:de:gbv:46-00101909-11, Universität Bremen.
14. Szlachetko J, Sá J (2016) X-ray spectroscopy: The driving force to understand and develop catalysis. *Advanced Catalytic Materials: Photocatalysis and Other Current Trends.* IntechOpen.
15. Zimmermann P, Peredkov S, Abdala PM, DeBeer S, Tromp M, Müller C, van Bokhoven JA (2020) Modern X-ray spectroscopy: XAS and XES in the laboratory. *Coordination Chemistry Reviews* 423: 213466.
16. Sun X, Peng B, Ji Y, Chen J, Li D (2009) Chitosan (chitin)/cellulose composite biosorbents prepared using ionic liquid for heavy metal ions adsorption. *AIChE Journal* 55: 2062–2069.
17. Zhang H, Yang S, Fang J, Deng Y, Wang D, Zhao Y (2014) Optimization of the fermentation conditions of Rhizopus japonicus M193 for the production of chitin deacetylase and chitosan. *Carbohydrate Polymers* 101: 57–67.
18. Barwin Vino A, Ramasamy P, Vairamani S, Shanmugan A (2011) Physicochemical characterization of biopolymers chitin and chitosan extracted from squid Doryteuthis sibogae adam, 1954 pen. *International Journal of Pharmaceutical Research and Development* 2: 181–190.
19. Iwashita N (2016) X-ray powder diffraction. *Materials Science and Engineering of Carbon.* Elsevier, pp. 7–25.
20. S Silva D, AM Delezuk J, A La Porta F, Longo E, P Campana-Filho S (2015) Comparison of experimental and theoretical data on the structural and electronic characterization of chitin and chitosan. *Current Physical Chemistry* 5: 206–213.

21. Kaya M, Mujtaba M, Ehrlich H, Salaberria AM, Baran T, Amemiya CT, Galli R, Akyuz L, Sargin I, Labidi J (2017) On chemistry of γ-chitin. *Carbohydrate Polymers* 176: 177–186.

22. McMillan PF, Hofmeister A (2018) Infrared and Raman spectroscopy. *Spectroscopic Methods in Mineralogy and Geology*. De Gruyter, pp. 99–160.

23. Hajji S, Younes I, Ghorbel-Bellaaj O, Hajji R, Rinaudo M, Nasri M, Jellouli K (2014) Structural differences between chitin and chitosan extracted from three different marine sources. *International Journal of Biological Macromolecules* 65: 298–306.

24. Beil S, Schamberger A, Naumann W, Machill S, van Pée K-H (2012) Determination of the degree of N-acetylation (DA) of chitin and chitosan in the presence of water by first derivative ATR FTIR spectroscopy. *Carbohydrate Polymers* 87: 117–122.

25. Paulino AT, Simionato JI, Garcia JC, Nozaki J (2006) Characterization of chitosan and chitin produced from silkworm crysalides. *Carbohydrate Polymers* 64: 98–103.

26. Amor IB, Hemmami H, Laouini SE, Abdelaziz AG, Barhoum A (2023) Influence of chitosan source and degree of deacetylation on antibacterial activity and adsorption of AZO dye from water. *Biomass Conversion and Biorefinery*: 1–11.

27. Picollo M, Aceto M, Vitorino T (2018) UV-Vis spectroscopy. *Physical Sciences Reviews* 4: 20180008.

28. Perkampus H-H (2013) *UV-VIS Spectroscopy and its Applications*. Springer Science & Business Media.

29. Song Y, Cong Y, Wang B, Zhang N (2020) Applications of Fourier transform infrared spectroscopy to pharmaceutical preparations. *Expert Opinion on Drug Delivery* 17: 551–571.

30. Liu D, Wei Y, Yao P, Jiang L (2006) Determination of the degree of acetylation of chitosan by UV spectrophotometry using dual standards. *Carbohydrate Research* 341: 782–785.

31. Juárez-de la Rosa B, Ardisson P-L, Azamar-Barrios J, Quintana P, Alvarado-Gil J (2007) Optical, thermal, and structural characterization of the sclerotized skeleton of two antipatharian coral species. *Materials Science and Engineering: C* 27: 880–885.

32. McDonnell LA, Heeren RM (2007) Imaging mass spectrometry. *Mass Spectrometry Reviews* 26: 606–643.

33. Courant F, Antignac JP, Dervilly-Pinel G, Le Bizec B (2014) Basics of mass spectrometry based metabolomics. *Proteomics* 14: 2369–2388.

34. Siuzdak G (2004) An introduction to mass spectrometry ionization: an excerpt from *The Expanding Role of Mass Spectrometry in Biotechnology*; MCC Press: San Diego, 2005. *JALA: Journal of the Association for Laboratory Automation* 9(2): 50–63.

35. Forbes TP, Pettibone JM, Windsor E, Conny JM, Fletcher RA (2023) Rapid chemical screening of microplastics and nanoplastics by thermal desorption and pyrolysis mass spectrometry with unsupervised fuzzy clustering. *Analytical Chemistry* 2023, 95(33), 12373–12382.

36. Dunn WB, Broadhurst D, Begley P, Zelena E, Francis-McIntyre S, Anderson N, Brown M, Knowles JD, Halsall A, Haselden JN (2011) Procedures for large-scale metabolic profiling of serum and plasma using gas chromatography and liquid chromatography coupled to mass spectrometry. *Nature Protocols* 6: 1060–1083.

37. Vesentini D, Steward D, Singh AP, Ball R, Daniel G, Franich R (2007) Chitosan-mediated changes in cell wall composition, morphology and ultrastructure in two wood-inhabiting fungi. *Mycological Research* 111: 875–890.

38. Jones M, Weiland K, Kujundzic M, Theiner J, Kählig H, Kontturi E, John S, Bismarck A, Mautner A (2019) Waste-derived low-cost mycelium nanopapers with tunable mechanical and surface properties. *Biomacromolecules* 20: 3513–3523.

39. Fuchs B, Süß R, Nimptsch A, Schiller J (2009) MALDI-TOF-MS directly combined with TLC: a review of the current state. *Chromatographia* 69: 95–105.

40. Struzzi C, Scardamaglia M, Reckinger N, Sezen H, Amati M, Gregoratti L, Colomer J-F, Ewels C, Snyders R, Bittencourt C (2017) Probing plasma fluorinated graphene via spectromicroscopy. *Physical Chemistry Chemical Physics* 19: 31418–31428.

41. Cerqueira B, Vega F, Serra C, Silva L, Andrade M (2011) Time of flight secondary ion mass spectrometry and high-resolution transmission electron microscopy/energy dispersive spectroscopy: a preliminary study of the distribution of Cu^{2+} and Cu^{2+}/Pb^{2+} on a Bt horizon surfaces. *Journal of Hazardous Materials* 195: 422–431.

42. Chen X, Yang H, Zhong Z, Yan N (2017) Base-catalysed, one-step mechanochemical conversion of chitin and shrimp shells into low molecular weight chitosan. *Green Chemistry* 19: 2783–2792.

43. Watson HR, Apperley DC, Dixon DP, Edwards R, Hodgson DR (2009) An efficient method for 15N-labeling of chitin in fungi. *Biomacromolecules* 10: 793–797.

44. Peng L, Jayapalan S, Chankvetadze B, Farkas T (2010) Reversed-phase chiral HPLC and LC/MS analysis with tris (chloromethylphenylcarbamate) derivatives of cellulose and amylose as chiral stationary phases. *Journal of Chromatography A* 1217: 6942–6955.

45. Makino A, Kurosaki K, Ohmae M, Kobayashi S (2006) Chitinase-catalyzed synthesis of alternatingly N-deacetylated chitin: A chitin–chitosan hybrid polysaccharide. *Biomacromolecules* 7: 950–957.

46. Ahn S, Hong JY, Hong MK, Jang YP, Oh MS, Jung JH, Hong J (2009) Structural determination of sildenafil and its analogues in dietary supplements by fast-atom bombardment collision-induced dissociation tandem mass spectrometry. *Rapid Communications in Mass Spectrometry: An International Journal Devoted to the Rapid Dissemination of Up-to-the-Minute Research in Mass Spectrometry* 23: 3158–3166.

47. Sarode S, Upadhyay P, Khosa MA, Mak T, Shakir A, Song S, Ullah A (2019) Overview of wastewater treatment methods with special focus on biopolymer chitin-chitosan. *International Journal of Biological Macromolecules* 121: 1086–1100.

48. Nielson KE, Bowen GJ (2010) Hydrogen and oxygen in brine shrimp chitin reflect environmental water and dietary isotopic composition. *Geochimica et Cosmochimica Acta* 74: 1812–1822.

49. López F, Mercê A, Alguacil FJ, López-Delgado A (2008) A kinetic study on the thermal behaviour of chitosan. *Journal of Thermal Analysis and Calorimetry* 91: 633–639.

50. Schmitz C, González Auza L, Koberidze D, Rasche S, Fischer R, Bortesi L (2019) Conversion of chitin to defined chitosan oligomers: current status and future prospects. *Marine Drugs* 17: 452.

51. Yuen A, Chen T, Wang C, Wei W, Kabir I, Vargas J, Chan Q, Kook S, Yeoh G (2020) Utilising genetic algorithm to optimise pyrolysis kinetics for fire modelling and characterisation of chitosan/graphene oxide polyurethane composites. *Composites Part B: Engineering* 182: 107619.

52. Popa-Nita S, Lucas J-M, Ladavière C, David L, Domard A (2009) Mechanisms involved during the ultrasonically induced depolymerization of chitosan: characterization and control. *Biomacromolecules* 10: 1203–1211.

53. Kasaai MR (2010) Determination of the degree of N-acetylation for chitin and chitosan by various NMR spectroscopy techniques: A review. *Carbohydrate Polymers* 79: 801–810.

54. Bharti SK, Roy R (2012) Quantitative 1H NMR spectroscopy. *TrAC Trends in Analytical Chemistry* 35: 5–26.

55. Queiroz MF, Teodosio Melo KR, Sabry DA, Sassaki GL, Rocha HAO (2014) Does the use of chitosan contribute to oxalate kidney stone formation? *Marine Drugs* 13: 141–158.

56. Azaïs T, Tourné-Péteilh C, Aussenac F, Baccile N, Coelho C, Devoisselle J-M, Babonneau F (2006) Solid-state NMR study of ibuprofen confined in MCM-41 material. *Chemistry of Materials* 18: 6382–6390.

57. Stothers J (2012) *Carbon-13 NMR Spectroscopy: Organic Chemistry, A Series of Monographs*, Volume 24. Elsevier.

58. Knicker H (2011) Solid state CPMAS 13C and 15N NMR spectroscopy in organic geochemistry and how spin dynamics can either aggravate or improve spectra interpretation. *Organic Geochemistry* 42: 867–890.

59. Ghassemi N, Poulhazan A, Deligey F, Mentink-Vigier F, Marcotte I, Wang T (2021) Solid-state NMR investigations of extracellular matrixes and cell walls of algae, bacteria, fungi, and plants. *Chemical Reviews* 122: 10036–10086.

60. Ifuku S, Saimoto H (2012) Chitin nanofibers: preparations, modifications, and applications. *Nanoscale* 4: 3308–3318.

61. Kaya M, Baran T, Asan-Ozusaglam M, Cakmak YS, Tozak KO, Mol A, Mentes A, Sezen G (2015) Extraction and characterization of chitin and chitosan with antimicrobial and antioxidant activities from cosmopolitan Orthoptera species (Insecta). *Biotechnology and Bioprocess Engineering* 20: 168–179.

62. Sajomsang W, Gonil P (2010) Preparation and characterization of α-chitin from cicada sloughs. *Materials Science and Engineering: C* 30: 357–363.

63. Nagahama H, Maeda H, Kashiki T, Jayakumar R, Furuike T, Tamura H (2009) Preparation and characterization of novel chitosan/gelatin membranes using chitosan hydrogel. *Carbohydrate Polymers* 76: 255–260.

64. Rafigh S, Heydarinasab A (2017) Mesoporous chitosan-SiO_2 nanoparticles: synthesis, characterization, and CO_2 adsorption capacity. *ACS Sustainable Chemistry & Engineering* 5: 10379–10386.

65. Gniadek M, Dąbrowska A (2019) The marine nano-and microplastics characterisation by SEM-EDX: the potential of the method in comparison with various physical and chemical approaches. *Marine Pollution Bulletin* 148: 210–216.

66. de Queiroz Antonino RSCM, Lia Fook BRP, de Oliveira Lima VA, de Farias Rached RÍ, Lima EPN, da Silva Lima RJ, Peniche Covas CA, Lia Fook MV (2017) Preparation and characterization of chitosan obtained from shells of shrimp (Litopenaeus vannamei Boone). *Marine Drugs* 15: 141.

67. Ibitoye E, Lokman I, Hezmee M, Goh Y, Zuki A, Jimoh A (2018) Extraction and physicochemical characterization of chitin and chitosan isolated from house cricket. *Biomedical Materials* 13: 025009.

68. Kabalak M, Aracagök D, Torun M (2020) Extraction, characterization and comparison of chitins from large bodied four Coleoptera and Orthoptera species. *International Journal of Biological Macromolecules* 145: 402–409.

69. Kaya M, Baran T, Erdoðan S, Menteþ A, Özüsaðlam MA, Çakmak YS (2014) Physicochemical comparison of chitin and chitosan obtained from larvae and adult Colorado potato beetle (Leptinotarsa decemlineata). *Materials Science and Engineering: C* 45: 72–81.

70. Marei NH, Abd El-Samie E, Salah T, Saad GR, Elwahy AH (2016) Isolation and characterization of chitosan from different local insects in Egypt. *International Journal of Biological Macromolecules* 82: 871–877.

71. Anand M, Kalaivani R, Maruthupandy M, Kumaraguru A, Suresh S (2014) Extraction and characterization of chitosan from marine crab and squilla collected from the Gulf of Mannar Region, South India. *Journal of Chitin and Chitosan Science* 2: 280–287.

72. Mohan K, Ravichandran S, Muralisankar T, Uthayakumar V, Chandirasekar R, Rajeevgandhi C, Rajan DK, Seedevi P (2019) Extraction and characterization of chitin from sea snail Conus inscriptus (Reeve, 1843). *International Journal of Biological Macromolecules* 126: 555–560.

73. Wang Y, Chang Y, Yu L, Zhang C, Xu X, Xue Y, Li Z, Xue C (2013) Crystalline structure and thermal property characterization of chitin from Antarctic krill (Euphausia superba). *Carbohydrate Polymers* 92: 90–97.

74. Srinivasan H, Kanayairam V, Ravichandran R (2018) Chitin and chitosan preparation from shrimp shells Penaeus monodon and its human ovarian cancer cell line, PA-1. *International Journal of Biological Macromolecules* 107: 662–667.

75. Kremer A, Lippens S, Bartunkova S, Asselbergh B, Blanpain C, Fendrych M, Goossens A, Holt M, Janssens S, Krols M (2015) Developing 3D SEM in a broad biological context. *Journal of Microscopy* 259: 80–96.

76. Gharsallaoui A, Oulahal N, Joly C, Degraeve P (2016) Nisin as a food preservative: part 1: physicochemical properties, antimicrobial activity, and main uses. *Critical Reviews in Food Science and Nutrition* 56: 1262–1274.

77. Abd Elgadir M, Uddin MS, Ferdosh S, Adam A, Chowdhury AJK, Sarker MZI (2015) Impact of chitosan composites and chitosan nanoparticle composites on various drug delivery systems: A review. *Journal of Food and Drug Analysis* 23: 619–629.

78. Latańska I, Rosiak P, Paul P, Sujka W, Kolesińska B (2021) Modulating the physicochemical properties of chitin and chitosan as a method of obtaining new biological properties of biodegradable materials. *Chitin and Chitosan: Physicochemical Properties and Industrial Applications.* IntechOpen.

79. Aranaz I, Mengíbar M, Harris R, Miralles B, Acosta N, Calderón L, Sánchez Á, Heras Á (2014) Role of physicochemical properties of chitin and chitosan on their functionality. *Current Chemical Biology* 8: 27–42.

80. Li J, Zhao Y, Jiang X (2020) Quantitative analysis of protein in thermosensitive hydroxypropyl chitin for biomedical applications. *Analytical Biochemistry* 599: 113745.

81. Ioelovich M (2014) Crystallinity and hydrophility of chitin and chitosan. *Journal of Chemistry* 3: 7–14.

82. Kumari S, Rath P, Kumar ASH, Tiwari T (2015) Extraction and characterization of chitin and chitosan from fishery waste by chemical method. *Environmental Technology & Innovation* 3: 77–85.

83. Franca EF, Freitas LC, Lins RD (2011) Chitosan molecular structure as a function of N-acetylation. *Biopolymers* 95: 448–460.

Chapter 9

Application of Chitosan and Chitin in Food Packaging

Abdelmalek Zaater,[a,b] Saoussan Temmar,[c] Talha Bin Emran,[d,e] Abdelkrim Rebiai,[f] Asma Ben Amor,[c,f] Mohammed Tayeb Oucif Khaled,[c,f] and Ali Tliba[g]

[a]*Department of Agronomy, Faculty of Nature and Life Sciences, University of El Oued, El Oued, Algeria*
[b]*Biodiversity Laboratory and Application of Biotechnology in Agriculture, University of El Oued, El Oued, Algeria*
[c]*Department of Process Engineering and Petrochemical, Faculty of Technology, University of El Oued, El Oued, Algeria*
[d]*Department of Pharmacy, BGC Trust University Bangladesh, Chittagong, Bangladesh*
[e]*Department of Pharmacy, Faculty of Allied Health Sciences, Daffodil International University, Dhaka, Bangladesh*
[f]*Renewable Energy Development unit in Arid Zones, University of El Oued, El Oued, Algeria*
[g]*Laboratory Valorization and Technology of Saharan Resources, University of El Oued, El Oued, Algeria*

ttemmar8@gmail.com

The attention to biopolymers has risen as a result of the exhaustion of fossil fuel supply and the harm to the environment resulting from a buildup of plastic packaging that is not biodegradable. Several biopolymers have been produced from leftover food in

Chitin and Chitosan: Physical and Chemical Properties
Edited by Shakeel Ahmed, Soumeia Zeghoud, Hadia Hemmami, and Ilham Ben Amor

ISBN 978-981-5129-39-7 (Hardcover), 978-1-003-58977-8 (eBook)
www.jennystanford.com

order to decrease such waste while concurrently providing innovative materials used to package food. Chitosan serves as a substitute for artificial polymers and a supply of starting materials for new products. To decide if a material is suitable for use as a material for packing food, it is crucial to examine its permeability and mechanical properties. It is feasible to predict how films will react when handling, transferring, and storing packaged objects thanks to their mechanical properties. Pure chitosan films are the best choice for use in active and food packaging because of their mechanical and barrier characteristics. To alter these properties, chitosan can be mixed with plasticizers, other polysaccharides, lipids, and proteins, among other things. These mixtures maintain the biodegradability of the polymer and the food's quality attributes while modifying the final polymer's characteristics to meet food requirements and extend its useful life. In terms of their optical properties, pure chitosan films, which are optically transparent, demonstrate strong transmittance levels in the visible range. This is an important consideration in figuring out how well the client will respond to the videos. The remarkable UV absorbance of films based on chitosan also enables them to protect food against lipid oxidations brought on using ultraviolet rays.

9.1 Introduction

Along the food chain, food packaging is crucial for preventing chemical, physical, and biological dangers. To protect food from external factors, including dust, odors, light, shocks, humidity, and temperature, packaging is crucial [1]. Plastic packaging was made possible by the advent of petroleum-based polymers, including ethylene vinyl alcohol, polyester, and polypropylene, among others. Previously, packaging was made of metal, glass, or cardboard. Due to their high quality and reasonable pricing, products made from petroleum are the most often utilized components in the packaging of the food business [2].

Although they are not made from sustainable sources, petroleum-based products are neither recyclable, compostable, or biodegradable. Therefore their widespread usage has a detrimental effect on the environment [2]. Food packaging research needs to solve the environmental problems brought on by irresponsible handling of non-degradable materials and develop inventive replacement materials with fewer detrimental environmental consequences. In order to use them as alternatives to synthetics based on petroleum, a large variety of research is centered on biodegradable natural polymers. Chitosan [3], cellulose [4], starch [5], whey protein [6], and gelatin [7] are a few examples of biodegradable natural polymers that have been extensively researched for potential applications in the food packaging sector. In the last years, there has been a noticeable increase in articles on chitosan and the packaging business, as illustrated in Fig. 9.1.

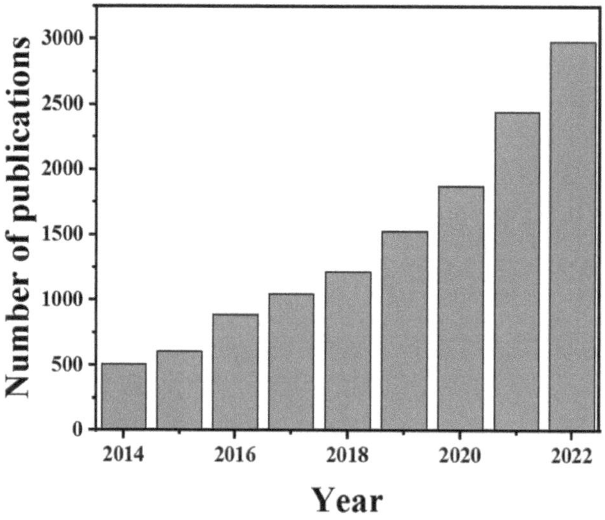

Figure 9.1 Publications regarding chitosan and food packaging in the last years (Source: ScienceDirect).

In addition to being biodegradable, nontoxic, and biocompatible, chitosan and its derivatives also exhibit antibacterial, antifungal, and metal-chelating properties [8]. Due to its remarkable characteristics that produce films, chitosan has been examined for application primarily in the chemical, pharmaceutical, and healthcare industries [9]. Chitosan is created by deacetylating chitin, a large natural polysaccharide that has been well-known since 1884. Chitin is produced using several organisms. It may be produced by several sustainable resources, chiefly by-products from the fishing industry. As a result, chitosan is a polysaccharide that is both inexpensive and convenient. Chitosan is an organic acid that is often soluble in dilute solutions like citric, malic, formic, acetic, lactic, or tartaric acid [10]. Chitosan is semi-crystalline in a solid state. Due to the aforementioned chitosan properties, it is becoming common to employ chitosan as a component of sheets or coatings for food packaging.

Recent shifts in consumer lifestyles (a lack of time for cooking and shopping), tendencies in manufacturing (tasty, fresh, extended validity period, and quality-monitored goods), retail, and food demand have resulted in the emergence of the beyond a device for physical obstacles and passive packaging new packaging approaches exist [11].

Innovative food packaging methods should concentrate on preventing the main reasons why food products deteriorate since doing so will raise food safety standards. The packaging sector is searching for answers so that it can have functional qualities like a particular gas barrier, moisture or gas absorbents, antioxidant activity, UV protection, antibacterial abilities, or the ability to reside and report on product quality [12]. These cutting-edge packaging choices will extend the shelf life of food goods and allow for longer-term product preservation. Two major cutting-edge technologies and packaging solutions that have surfaced to address this issue are active and smart packaging.

Energetic packaging uses products and components that protect, improve, or increase the food's shelf life that is being packed. Package of this kind preserves nutrients of food content prevents the growth of non-pathogenic pathogenic germs, or stops the migration of pollutants. Examples of illustrations of active packaging include scavengers of oxygen, ethylene absorbers, absorption of moisture, antioxidants, mechanisms for releasing or absorbing bait, ethanol emitters, and microbiological films [13].

In contrast, intelligent packaging includes parts and products that keep an eye on the condition of the food or the environment in which it is kept. These packets can communicate the state of the meal. But the food is unaffected by these containers. To enhance quality, extend shelf life, and anticipate potential problems, this pack performs smart functions, including detecting, registering, recording, and exchanging data.

Chitosan may be thought of as a biopolymer of great attention and suited for creating intelligent or biodegradable food packaging that is active based on the aforementioned traits. This chapter's goal is to supply a succinct analysis of the latest developments in chitosan-based intelligent and active packaging. This chapter will focus on how active additives affect the mechanical, functional, and barrier characteristics of active and smart films based on chitosan.

9.2 Techniques for Enhancing the Characteristics of Films Made with Chitosan

Chitosan's inherent disadvantages, including its high sensitivity to water, thermal stability, and low mechanical, result in a lower food shelf life than that of standard food packaging materials, which has limited its use in food packaging [14]. As a result, numerous strategies have been proposed to overcome these issues and improve the properties of products based on chitosan, such as coatings for surfaces [15], complexation, graft copolymerization [16], high-energy irradiation [17], and combining with additional biopolymers. When chitosan is combined with several other types of polymers to create a composite film, the benefits of the base polymers can be combined with the construction of a film that performs better than the combination of its components. Cellulose, pectin and its derivatives, starch, gelatin [18], polyvinyl alcohol [19], soy protein isolate, polylactic acid [20], and other natural synthetic and natural polymers have all been observed to combine with chitosan.

9.2.1 Mechanical Properties

Predicting the conduct and integrity displayed while handling, transport, and storage is made easier by studying the mechanical characteristics of biopolymers used as film for food packaging, such as percentage elongation and tensile strength. Additionally, the packaged product with protuberances must be considered.

Analysis of the mechanical axial parameters, such as deformation and puncture force, is equally crucial, even if it is uncommon.

Predicting the appropriateness and potential uses of the novel biopolymer involves comparing these values with the mechanical characteristics of available artificial polymers. The composition of the film, the nature of the ingredients, the measuring device, and the testing circumstances (temperature and humidity levels) all affect mechanical parameters like elongation ratio at fracture, tensile strength, puncture deformation, and puncture force. To assess the mechanical and hole characteristics of biodegradable films, a texturometer is often utilized [21].

The highest stress a film can withstand without breaking is known as its tensile strength (MPa) [22]. It stands for elongation resistance. The elongation at break percentage (%) represents the highest stretching of the film prior to rupture. Tensile power and biopolymer films' percentage of elongation at break often have an inverse relationship. When a film's strength increases (higher tensile strength), its extensibility decreases (low elongation at break). Typically, texturometer software's stress-strain curve is used to compute tensile ability and the elongation at break %. The test is conducted in accordance with the recommended procedure ASTM D-882 [23].

There are several documented mechanical characteristics data for pure chitosan films in the research, as shown in Table 9.1. The studies conducted indicate that these variations are caused by the various chitosan properties used (molecular weight (Mw), deacetylation degree (DD)), the solvent technique used to make the film, the amount of chitosan in the film, the time spent storing, and the test's pre- and post-measurement circumstances. Chitosan creates the film by creating hydrogen ties between the hydroxyl and amino groups. As the number of amino and hydroxyl groups rose during film production, the chitosan films' hydrogen bonding was also enhanced.

Because of this, the Mw of chitosan affects the kinds and quantities of interactions between polymers and solvents. High polymerization degree chains in high molecular weight chitosan encourage polymer-polymer interactions, which strengthen matrices and lower molecular weight. As can be observed in Table 9.1, when the reactions increase to raise the Mw, this leads to an increase in the tensile power and a reduction in the values of the percentage elongation.

Table 9.1 Mechanical characteristics of films made of pure chitosan under various circumstances

Expression	Acids used	Properties of chitosan	Conditions	Tensile strength (MPa)	Elongation ratio (%)	Ref.
(1%) Chitosan	Acetic	Mw150 kDa	20 °C 20% RH	87.68	-	[26]
(2%) Chitosan	Acetic	DD 95%	25 °C	68.8–50.2	7.6–5.5	[27]
	Citric	Mw 37–920 kDa	50% RH	6.7–17.4	41.9–117.8	
	Lactic			17.1–62.6	19.6–31.1	
	Malic			27.4–62.4	17.8–29.9	
(2%) Chitosan	Acetic	80%DD Mw100 kDa	25 °C 50% RH	49.6–59.4	5.40–8.35	[28]
(2.5%) Chitosan	Acetic	DD 85% High Mw	25 °C 60% RH	~60.0	~5.0	[29]
(2.5%) Chitosan	Acetic	DD 85% Few Mw	25 °C 60% RH	~10.0	~58.0	
(2.5%) Chitosan	Acetic	85% DD MW average	25 °C RH 60%	~18.0	~85.0	
(1%) Chitosan	Acetic	Mw is high	25 °C RH 50%	61.8	4.59	[30]
(1%) Chitosan	Acetic	Mw few	25 °C RH 50%	55.8	4.58	
(1%) Chitosan	Acetic	78% DD Mw 200 kDa	-	39.1	10.84	[31]
(2%) Chitosan	Acetic	DD 78% Mw 100 kDa	-	55.6	8	[32]

The mechanical characteristics of the films are impacted by storage time. Regardless of the molecular weight, an improvement in tensile strength was seen. The scientists attributed it to changes in chitosan macromolecule structure and a reduction in free volume during storage. The increase in tensile strength may be caused by these conformational changes. According to Kaewklin et al. [24], the solvent utilized affects how the behavior of the molecular weight as a function of elongation behaves. When examining the mechanical characteristics, higher tensile strength values were observed in relation to the solvent used, with acetic acid as the solvent, but resulted in larger percentages of elongation at break values [25]. The fact that chitosan forms dimers in an acetic acid solution, according to scientists, shows that the intermolecular connection is rather strong.

The values enable an analysis of the relationship between the film's characteristics and the deacetylation level, chitosan content, molecular weight, solvent technique, and measurement circumstances (RH: relative humidity, deacetylation degree, and Mw: molecular weight).

9.2.2 Barrier Properties

When used as food packaging, a polymeric film's barrier qualities are essential for determining or forecasting the product's shelf life. In an effort to keep the food's quality high, one of the major objectives of films for food packing is to delay the transfer of molecules between the environment and food. Knowing these qualities enables one to determine the permeability and the movement of specific molecules, such as gases like CO_2 or O_2, organic vapor, water vapor, or liquids, through the film. Small molecules can pass rather easily through plastics. They may pass through the polymer package wall from the external or internal environment, causing a product's quality to be constantly changing and shortening of shelf life [30].

Most research has been done on the permeability to water vapor of biodegradable films, mostly because of the critical function that water plays in degrading processes, keeping food crisp, fresh, or preventing dehydration, depending on the meal, and partially due to the measurement's simplicity. Conversely, oxygen permeability is a crucial metric but one that has received less attention due to the necessity for specialized measuring tools. Numerous food degradation processes involving oxygen include oxidation reactions, which alter food's flavor, aroma, and color, as well as microbe development, enzymatic browning, and vitamin loss. Additionally, fresh fruits and vegetables must be permeable to oxygen in order to breathe. According to several studies [30, 33], water-loving biopolymer films have strong oxygen barrier properties and increased permeability to water vapor.

The produced biofilm will be more suited for foods or applications can be determined based on these permeability metrics [30]. Water vapor permeability measures how much water permeates in terms of time and area (g/s m Pa) while taking into account the pressure difference and material thickness. The most popular technique for measuring the permeability to water vapor by gravimetry is typically the standard method ASTM E96.

Like a lot of other edible polysaccharide films, chitosan films have a strong hydrophilic character that makes them rather poor water barrier qualities [34], as seen in Table 9.2. Care must be used while comparing the figures listed in the literature. The Mw, DD, and chitosan concentration of films are among the factors that influence the water vapor permeability values. Additionally, a number of external factors, including the measurement procedure, the measurement environment (temperature and relative humidity), adjusting for the air gap effect, and the storage period and environment, all have an impact on these numbers.

In general, the permeability to water vapor values published [29, 30] is not considerably impacted by DD, Mw, or the kind of solvent used to create the dissolution. However, according to Kerch et al. [25], the permeability to water vapor increases as the Mw of the chitosan employed increases. It is impossible to determine whether DD, Mw, or both are to blame for the change in the permeability to water vapor since the authors of this work tested chitosan with varying degrees of deacetylation and molecular weights.

The permeability of the film varies while it is stored. Either high or low-weight chitosan films showed a rise in water vapor permeability.

According to the authors, during storage, chitosan macromolecules change their shape, and their free volume decreases. The macromolecules of chitosan' conformations change, going from erratic to much more spacious, which is what causes the rise in water vapor permeability, even as the free volume of the polymer decreases [25].

Table 9.2 Data on the permeability to water vapor of pure chitosan films under various circumstances

Expression	Acids used	Properties of chitosan	Conditions	Permeability to water vapor (g/m s Pa)	Ref.
(2%) Chitosan	Acetic	DD ~80%	38 °C	(4016–4500) g/m² day	[28]
		Mw 100 kDa	90% RH		
(2%) Chitosan	Acetic	DD 95%	25 °C	$(3.1–3.2)\cdot10^{-10}$	[27]
	Citric			$(4.0–5.1)\cdot10^{-10}$	
	Lactic	Mw 370–920 kDa	RH 50%	$(4.7–6.9)\cdot10^{-10}$	
	Malic			$(2.6–4.1)\cdot10^{-10}$	
(1%) Chitosan	Acetic	Mw ~150 kDa	25 °C	3.53 g mm/m² h kPa	[26]
		50% RH			
(1%) Chitosan	Acetic	DD 78%	90% RH	0.01322 g m/m² day kPa	[31]
		Mw 200 kDa			
(1%) Chitosan	Acetic	Mw few	38 °C	$8.07–10^{-13}$	[30]
			RH 90%		
(1%) Chitosan	Acetic	Mw is high	38 °C	$8.07–10^{-13}$	
			RH 90%		
Chitosan (2.5%)	Acetic	DD 85%	5 °C	$3.38–10^{-10}$	[29]
		Medium Mw	2000 Pa		
(2.5%) Chitosan	Acetic	DD 85%	5 °C	$4.14–10^{-10}$	[29]
		Mw Few	2000 Pa		
(2.5%) Chitosan	Acetic	DD 85%	5 °C	$4.55–10^{-10}$	
		High Mw	2000 Pa		

In accordance with the standard technique ASTM D-3935, oxygen permeability is often assessed [35]. Less often, it can also be determined using the ASTM D-1434 standard procedure [28]. Capillary processes are responsible for edible films' capacity to permeate oxygen. The chitosan films display good oxygen-barrier qualities and have low oxygen permeability, making them perfect for use in food packaging. According to Park et al. [27] and Valenzuela et al. [36], these values are comparable to those of ethylene vinyl alcohol copolymer films or commercial polyvinylidene chloride (PVC). As shown in Table 9.3, various oxygen permeability values have been reported in the research. Pure chitosan films' oxygen permeability is influenced by variables, including the chitosan's Mw and the solvent employed to make the film-forming solution [36].

Table 9.3 Information on the oxygen permeability of films made purely of chitosan in various circumstances

Expression	Acid used	Properties of chitosan	Conditions	Permeability to oxygen	Ref.
(1%) Chitosan	Acetic	Mw few	23 °C RH 50%	7.70 cm^3 μm/m^2 day kPa	[30]
(1%) Chitosan	Acetic	Mw is high	23 °C RH 50%	6.65 cm^3 μm/m^2 day kPa	
(2%) Chitosan	Acetic	DD ~80% Mw 100 kDa	27 °C RH 65%	(38.0–56.0) ml/m^2 day atm	[28]
(2%) Chitosan	Acetic	DD 95%	25 °C	(1.4–5.8) × 10^{-8} cc/m^2 day atm	[27]
	Citric			(3.3–3.0) × 10^{-8} cc/m^2 day atm	
	Lactic	Mw 370–920 kDa	RH 0%	(2.7–2.7) × 10^{-8} cc/m^2 day atm	
	Malic			(0.8–1.9) × 10^{-8} cc/m day atm	

The values enable an analysis of the relationship between the film's characteristics and the deacetylation level, chitosan content, molecular weight, solvent technique, and measurement circumstances (relative humidity, deacetylation degree, and Mw).

The values enable an analysis of the relationships between the film's permeability to oxygen and DD, Mw, solvent type, chitosan content, and measurement circumstances. RH stands for relative humidity.

9.2.3 Antimicrobial Characteristics

Less chemical use is required by consumers in order to prolong food's shelf-life goods. Because of this, there has been an increase in interest in recent years in creating materials with the ability to form films and possess antimicrobial qualities, which aid in increasing shelf life and food safety. In the creation of innovative materials for packaging food, the use of natural antimicrobial agents is gaining popularity as a more sustainable option for chemically synthesized antimicrobial polymers. Additionally, natural antimicrobial compounds work to increase the validity period and quality of food by preventing germs from growing on the food's surface as a result of direct contact with the coating or film [37]. Due to its ability to create films and coatings, chitosan is one of the most researched antibacterial agents for food packaging. Chitosan is more effective against yeasts than it is against a variety of foodborne filamentous fungi, bacteria, and yeast [38].

Chitosan's antibacterial activity mechanism is not completely understood. However, three primary theories about the mechanisms of chitosan's antibacterial effect have been put forth. The theory that altering the permeability of cells is caused by stacked electrostatically on the cell surface is the most widely accepted. Chitosan interacts by forming a polycationic structure with the negatively charged microbial cell membranes because the amino group at position C2, less than its pH = 6.3, is positively charged. Proteinaceous and other intracellular components leak due to this contact. Another theory is that chitosan's ability to penetrate the cell's nucleus because of its low Mw prevents the transcription of DNA into RNA because of adsorption molecules of DNA. Inhibiting the generation of toxins and microbial development, chitosan is able to function as an essential nutrition-chelating agent [37, 38]. This is the basis for the final suggested mechanism.

On the other hand, chitosan's antibacterial action is influenced by internal characteristics like the degree of polymerization and deacetylation. In this view, the amount of protonated amino groups (–NH_2) present in chitosan determines the antibacterial activity, which rises with larger degrees of deacetylation due to their charge density and increased solubility [39]. Because of this, regardless of the kind of acid used for solubility, chitosan with a high amount of DD>85% and Mw range from 28 to 1671 kDa has demonstrated the greatest antibacterial activity in water-based solutions (resistance to bacteria). The suggested processes are supported by findings from experiments in which the activity was investigated by adding microorganisms to a solid or liquid culture medium containing chitosan. The inhibitory mechanism for films has not been clarified. Asphyxia and other anaerobic circumstances are additional external variables that need to be taken into account [40]. Additionally, it has been noted that the antimicrobial influence is influenced by the food composition (oil starch, NaCl, and proteins) and the force of interaction between the antimicrobial compound and the polymer matrix, such as the interaction between chitosan and starch.

Chitosan's antimicrobial properties films have been studied using several groups of (–) and (+) gram, bacteria, yeasts, and fungi. These groups are summarized in Table 9.4.

Inhibition is the development of a wide range of bacteria and fungi; chitosan has been shown in several studies to have antibacterial activity and a method of action that prevents food spoiling. Numerous variables, including Mw, DD, temperature, salinity, and pH, among other things, have an impact on both impacts. These characteristics significantly contribute to the antibacterial action. Due to these differences, it is extremely difficult to compare them and determine the minimal inhibitory concentration for every bacterium [41].

For instance, chitosan has greater action against a number of bacteria when tested at low pH levels [42]. The antibacterial action of chitosan also has an impact on temperature, which is a key element in the development of bacteria or fungi. For various temperatures, a difference in the lowest inhibitory concentration value was noted [42]. For chitosan samples of various molecular weights, the impact of temperature and pH was examined.

The findings demonstrated that antibacterial activity rises as temperature and pH rise. The impact of molecular weight was also influenced by pH. These findings demonstrated the significance of reporting the pH, temperature, DD, and Mw in order to compare identical methods.

Table 9.4 Use of (+) and (–) gram, fungi, bacteria, and yeast populations to investigate the antibacterial properties of chitosan membranes

	Microorganism	Ref.
Fungi	*Cryptococcus humiculus*	[43]
	Sclerotinia sclerotium	[40]
	Botrytis cinerea	[44]
	Monilinia fructicola	[40]
	Rhizopus stolonifera	[40]
	Aspergillus niger	[40]
Yeast	*Candida lambica*	[45]
Gram-positive	*Staphylococcus aureus*	[40]
	Brochothrix thermosphacta	[46]
	Lactobacillus plantarum	[47]
	Lactobacillus curvatus	
	Pediococcus acidilactici	
	Lactobacillus sakei	
	Bacillus cereus	
	Listeria monocytogenes	
Gram-negative	*Pseudomonas fluorescens*	[47]
	Salmonella typhimurium	[41]
	Pseudomonas aeruginosa	[41]
	Photobacterium phosphoreum	[47]
	Escherichia coli	[41]
	Enterobacter aeromonas	[47]

9.2.4 Optical Properties

The ability to shield food from the negative traces of light, particularly UV radiation, is one of the desired properties of materials for packaging. Since ultraviolet light with a wavelength between 200 nm and 280 nm is one of the most frequent oxidation initiators in the food chain, UV radiation is to blame for causing lipid oxidation. Unwanted tastes and scents that occur in food are the result of these oxidations. In order to increase the food product's validity period while maintaining food quality and appropriate organoleptic qualities, active films with good UV-light barrier capabilities must be developed [48].

The transparency of an optical property is also a crucial factor since it directly affects customer acceptance. According to Vilela et al. [49], transparent packaging is preferred by consumers. So, they may view the product content and evaluate the visual appeal. As a result, it is critical to create films with sufficient transparency and effective UV radiation protection barriers.

Being optically transparent films, the UV-visible spectrum of films made only of chitosan displays transmittance between 88% and 93% in the visible spectrum [49]. Additionally, films made of chitosan have exceptional UV absorption. Unsaturated bonds, namely C-O bonds found in chitosan, account for the absorption capability by rays detected in the band between 270 nm and 280 nm [50]. To create novel polymers with suitable optical characteristics, these features must be taken into account in all potential chitosan and other component combinations.

9.3 Chitosan Blends

In order to improve the physical qualities of each component, a polymer blend is created by mixing at least two polymers or copolymers in a phase-separated or suitable combination [51]. The goal of polymer blending is to create composite materials that combine component features, potentially improving their useful qualities and minimizing their drawbacks [52]. The vast range of physical, mechanical, thermal, optical, and barrier qualities that come from polymer mixing as a method to enhance packaging materials is crucial to its effectiveness. In order to build blends that are appropriate for a certain application, it is crucial to analyze these qualities.

Numerous research has been published that examine the creation and qualities of films made from chitosan combined with additional synthetic and natural polymers as a consequence of the increased interest in employing chitosan for packaging applications [53]. The two primary categories of chitosan blend in this study are chitosan-synthetic polymers and chitosan-natural biopolymers blend.

9.3.1 Blends of Natural Biopolymers with Chitosan

It is feasible to improve the useful characteristics of films made from chitosan mixing them with other natural biopolymers like proteins and polysaccharides, as well as its derivatives [51]. The compatibility of chitosan with other biopolymers depends on the ability to interact electrostatically because of the cationic nature of chitosan at appropriate pH levels and the presence of highly polar groups that may establish dipole associations or intermolecular hydrogen interactions with the appropriate functional groups of other various biopolymers [18]. Starch, pectin (from corn, rice, cassava, potatoes, etc.), carrageenan, alginate, xylan, xanthan gum, kefiran, glucose, and their derivatives may all be combined with chitosan, according to a paper by Wang et al. [54].

Animal and non-animal protein-based films (such as those made from soy protein isolate, corn zein, wheat gluten, and quinoa protein) are highly valued for their wide availability, excellent mechanical properties, and potential for producing biodegradable films. These films also exhibit good gas barrier qualities against non-condensable gases like carbon dioxide, nitrogen, and oxygen, as well as aromas [55]. In comparison to single proteins and chitosan film, the chitosan-protein mix film may have higher functional qualities, which would encourage its use in food packaging [54].

9.3.2 Blends of Chitosan-Synthetic Polymers

In-depth research has been done on the advantages of adding chitosan to synthetic polymers, like polylactic acid, polyvinyl pyrrolidone, and polyvinyl alcohol, on the physical, biological, and mechanical aspects of composite films. Because they have a wider variety of mechanical qualities, greater chemical resistance, and lower production costs than natural polymers, synthetic polymers have a better chance of succeeding as biodegradable materials [56]. Chitosan is possibly miscible with certain synthetic polymers, mostly because hydroxyl groups from the synthetic polymer and amine and hydroxyl groups from chitosan generate intermolecular hydrogen bonds [57]. Chitosan and synthetic biopolymers can sometimes be combined to improve the mechanical and water barrier qualities of films, depending on how the various polymer components interact.

9.4 Examples of Films Based on Chitosan

9.4.1 Examples of Applications for Active Films

In several cuisines, the use of active film-based chitosan has been investigated. The chitosan-based active films with antioxidant potential have received the greatest attention in the literature. According to de Ramabu et al. [58], the antioxidant films based on chitosan enhanced with olive pomace and mango leaf extract were both effectively protective films, evaluated for 31 and 28 days as highly protective films against nuts oxidation, respectively. The nut oil's peroxide content was examined by the authors to confirm the potent antioxidant action of the produced chitosan active films [59]. Tests on the physical characteristics of fresh fruit and vegetable storage, such as color, weight loss, firmness, or respiration rate, also point to the beneficial effects of chitosan antioxidant films. For example, essential oil, films with cloves added, banana peels extract, ascorbic acid, aloe vera, and procyanidins enhance some storage time factors of apple, blueberry, pomegranate arils, litchi fruit, and mango.

Similar outcomes were seen when films were treated with citric acid to keep the color and texture of green chilies or when mushrooms were treated with -tocopherol to maintain their firmness and reduce browning [60].

Additionally, meat products have had the efficiency of chitosan active antioxidant films assessed. A fast rise in thiobarbituric acid (used to detect oxidation of lipids) was seen in mortadella sausages wrapped with neat chitosan films during chilled storage, according to Moradi et al. [61]. However, over the first six days, oxidation in the samples wrapped in chitosan films containing Zataria Multiflora Boiss essential oil and grape seed extract dramatically decreased [61]. The ability of thyme, prickly ash, cloves, geranium, fennel, and testing has been done with cinnamon extracts on chitosan-based films to increase the shelf life of chilled pork. The findings show that the active film allows for a reduction in the total number of bacteria that are alive after 7 days and successfully prevents meat discoloration. Additionally, syringic acid-enriched active chitosan films have a preservation effect on quail eggs, showing a slower decline in yolk index, weight loss, and Haugh unit values compared to uncoated eggs [62].

In many of these studies, oils' additional antibacterial characteristics have demonstrated a growth inhibition impact on harmful microorganisms like *L. monocytogenes* and a delay in the oxidation of lipids found in these meals [61]. The use of chitosan/ anise (*Pimpinella anisum L.*) essential oil films has improved the chemical characteristics of chicken burgers and delayed the oxidation of fats [63].

9.4.2 Examples of Intelligent Films

As Table 9.5 summarizes, chitosan has been assessed as a possible source of content for intelligent films. Films with an intelligent pH pointer are used to assess food freshness. When the product's pH changes, the indicator's color changes. This allows customers to determine whether the food they are going to eat is in good shape right away. A lot of studies have been done on chitosan-based films with sophisticated pH indicators during the past few years. Fishery products have been the subject of research. For instance, fish were used to study the effects of adding alizarin to chitosan film. When the pH of the solution changed from 4 to 10, it was seen that the color of the film dramatically changed from minor yellow to purple [64]. When employed to preserve fish, butterfly pudding extract caused the color of chitosan films to shift from purple-blue to dark green [65]. After 7 days of storage, the film's color changed from light blue to yellow due to further extracts, this time from red cabbage. This suggests that there have been pH fluctuations and that the fish fillets have been contaminated [66]. Using a sticker sensor, a pH-touchy chitosan-based film with colors derived from Bauhinia Blakeana Dunn flowers has also been used to gauge fish's freshness and pH fluctuations. The color changes from purple to green when the meal's volatile amines are released.

Table 9.5 Applications for films made of chitosan that are active and intelligent

Active ingredients	Function	Foodstuff	Ref.
	(A) Active		
Olive pomace	Antioxidant	Nuts	[59]
Mango leaf extract	Antioxidant	Cashew nuts	[58]
Banana peels extract	Antioxidant	Apple	[60]
Thymus vulgaris EO	Antimicrobial/flavor absorber	Cooked pork	[68]
Aloe vera gel	Absorber/ethylene	Mango	[69]
TiO_2	Ethylene absorber	Tomato	[70]
Neat chitosan	Oxygen scavenger	Sliced sucuk	[71]
Ginger EO	Antimicrobial/absorber moisture	Barracuda fish	[72]
	(B) Intelligent		
Butterfly pudding extract	pH index	Fish	[65]
Alizarin	pH	Fish	[64]
Carrot anthocyanins	pH index	Pasteurized milk	[73]
Black rice bran anthocyanins	Freshness indicator	Pomfret and shrimp	[74]
Purple sweet and jabuticaba fruit potato anthocyanins	Time-temperature indicator	Meat	[75]
Red cabbage extracts	pH index	Fish	[66]

Chitosan purple and black rice extract samples have been produced for the meat products, and these may be used to calculate the pH change in the examined pork samples [62]. Due to the moderate anthocyanin concentration, the results showed that chitosan-purple rice extracts are more suited for preventing the degradation of pork than chitosan-black rice extracts [62]. Pork belly slices can be wrapped with a material made of red cabbage anthocyanin, poly(vinyl alcohol), and chitosan. When in a mildly alkaline range, the film takes on a yellowish hue and a faint green tint. According to Zhang et al. [67], the sensitive chitosan-Bauhinia Blakeana Dunn film is also effective in identifying pig meat contamination. According to observations, the detection film's color changes from brown to green after 48 hours, clearly showing that the sample's pH has changed.

9.5 Conclusions and Future Views

During marketing, packaging is an essential element that protects the product and ensures food safety. Food production has been highlighted as a cause of social issues and pollution because of the non-biodegradability of traditional plastic packaging and the insufficient waste management system. Biobased and biodegradable materials have lately gained a lot of interest as solutions to these problems. Using chitosan as a packaging material may assist in reducing the environmental issue, despite a number of drawbacks in terms of thermal stability, mechanical characteristics, barrier, and price of production. The combination of chitosan with several natural and synthetic polymers successfully overcome these limitations, allowing the films to be utilized for specific applications. Given that it will enable a partial replacement of the current synthetic plastic packaging materials that are freely accessible on the market, this technology seems to have a promising future for creative food packaging design.

Application possibilities for chitosan blend-based films include fresh foods (vegetables, seafood, and meat) and foods with a short- to medium-term shelf life. Films could present a creative replacement for traditional plastic film, for recycling is either impossible or made difficult by the presence of food leftovers.

They offer biodegradability/degradability features as a workable end-of-life solution in certain circumstances. Undoubtedly, nanotechnologies offer a promising complementary tool for improving the mechanical and barrier qualities of chitosan-based films as well as for the addition of other functionalities (antioxidant capacity and antimicrobial), but due to toxicological concerns, the development and implementation of these confront the opposition. A barrier and mechanical strength may be provided by biobased nanoparticles that are incorporated into nanocomposite composites, like nano-cellulose, while yet preserving the entire biodegradability of the material. Last but not least, despite contributions made in the literature over the time period under discussion, the subject of biodegradability is still generally disregarded. This exclusion is the result of two things: first, developing food packaging from a truly comprehensive and interdisciplinary perspective is difficult, and second, the synthesis of a new material using biodegradable components would make a new biodegradable substance. This must be verified since antimicrobial agents have been included in films. Thus, it cannot be assumed. In fact, compounds that stop microbial food deterioration may be harmful to the composting process. Therefore, biodegradability should be one of the desirable properties in future studies on films made with chitosan and generally on sustainable materials.

References

1. Kalpana S, Priyadarshini S, Leena MM, Moses J, Anandharamakrishnan C (2019) Intelligent packaging: Trends and applications in food systems. *Trends in Food Science & Technology* 93: 145–157.
2. Ludwicka K, Kaczmarek M, Białkowska A (2020) Bacterial nanocellulose—A biobased polymer for active and intelligent food packaging applications: Recent advances and developments. *Polymers* 12: 2209.
3. Priyadarshi R, Rhim J-W (2020) Chitosan-based biodegradable functional films for food packaging applications. *Innovative Food Science & Emerging Technologies* 62: 102346.
4. Cazón P, Vázquez M (2021) Bacterial cellulose as a biodegradable food packaging material: A review. *Food Hydrocolloids* 113: 106530.
5. Fang Y, Fu J, Tao C, Liu P, Cui B (2020) Mechanical properties and antibacterial activities of novel starch-based composite films

incorporated with salicylic acid. *International Journal of Biological Macromolecules* 155: 1350–1358.

6. Schmid M, Müller K (2019) Whey protein-based packaging films and coatings. *Whey Proteins*. Elsevier, pp. 407–437.
7. Ramos M, Valdés A, Beltrán A, Garrigós MC (2016) Gelatin-based films and coatings for food packaging applications. *Coatings* 6: 41.
8. Bakshi PS, Selvakumar D, Kadirvelu K, Kumar N (2020) Chitosan as an environment friendly biomaterial–a review on recent modifications and applications. *International Journal of Biological Macromolecules* 150: 1072–1083.
9. Siripatrawan U, Harte BR (2010) Physical properties and antioxidant activity of an active film from chitosan incorporated with green tea extract. *Food Hydrocolloids* 24: 770–775.
10. Cazón P, Vázquez M (2019) Applications of chitosan as food packaging materials. *Sustainable Agriculture Reviews 36: Chitin and Chitosan: Applications in Food, Agriculture, Pharmacy, Medicine and Wastewater Treatment*: 81–123.
11. Dainelli D, Gontard N, Spyropoulos D, Zondervan-van den Beuken E, Tobback P (2008) Active and intelligent food packaging: legal aspects and safety concerns. *Trends in Food Science & Technology* 19: S103–S112.
12. Wyrwa J, Barska A (2017) Innovations in the food packaging market: Active packaging. *European Food Research and Technology* 243: 1681–1692.
13. Nura A (2018) Advances in food packaging technology-a review. *Journal of Postharvest Technology* 6: 55–64.
14. Al-Tayyar NA, Youssef AM, Al-Hindi R (2020) Antimicrobial food packaging based on sustainable bio-based materials for reducing foodborne pathogens: A review. *Food Chemistry* 310: 125915.
15. Khwaldia K, Basta AH, Aloui H, El-Saied H (2014) Chitosan–caseinate bilayer coatings for paper packaging materials. *Carbohydrate Polymers* 99: 508–516.
16. Wang Y, Du H, Xie M, Ma G, Yang W, Hu Q, Pei F (2019) Characterization of the physical properties and biological activity of chitosan films grafted with gallic acid and caffeic acid: A comparison study. *Food Packaging and Shelf Life* 22: 100401.
17. Shahbazi M, Rajabzadeh G, Ahmadi SJ (2017) Characterization of nanocomposite film based on chitosan intercalated in clay platelets by electron beam irradiation. *Carbohydrate Polymers* 157: 226–235.

18. Bonilla J, Poloni T, Lourenço RV, Sobral PJ (2018) Antioxidant potential of eugenol and ginger essential oils with gelatin/chitosan films. *Food Bioscience* 23: 107–114.

19. Yoon S-D, Kim Y-M, Kim BI, Je J-Y (2017) Preparation and antibacterial activities of chitosan-gallic acid/polyvinyl alcohol blend film by LED-UV irradiation. *Journal of Photochemistry and Photobiology B: Biology* 176: 145–149.

20. Li K, Jin S, Liu X, Chen H, He J, Li J (2017) Preparation and characterization of chitosan/soy protein isolate nanocomposite film reinforced by Cu nanoclusters. *Polymers* 9: 247.

21. Ghasemlou M, Khodaiyan F, Oromiehie A (2011) Physical, mechanical, barrier, and thermal properties of polyol-plasticized biodegradable edible film made from kefiran. *Carbohydrate Polymers* 84: 477–483.

22. Takeuchi Y, Ikeda N, Tahara K, Takeuchi H (2020) Mechanical characteristics of orally disintegrating films: Comparison of folding endurance and tensile properties. *International Journal of Pharmaceutics* 589: 119876.

23. Mesa Vargas DF, Vidal-Lesso A, Alfaro Ayala JA (2018) A material model fitting for recycled polyethylene terephthalate implemented in the finite element modelling. In: *ASME International Mechanical Engineering Congress and Exposition*. American Society of Mechanical Engineers, p. V005T007A006.

24. Um IC, Park YH (2007) The effect of casting solvent on the structural characteristics and miscibility of regenerated silk fibroin/poly (vinyl alcohol) blends. *Fibers and Polymers* 8: 579–585.

25. Kerch G, Korkhov V (2011) Effect of storage time and temperature on structure, mechanical and barrier properties of chitosan-based films. *European Food Research and Technology* 232: 17–22.

26. Li J, Zivanovic S, Davidson Pa, Kit K (2010) Characterization and comparison of chitosan/PVP and chitosan/PEO blend films. *Carbohydrate Polymers* 79: 786–791.

27. Park S, Marsh K, Rhim J (2002) Characteristics of different molecular weight chitosan films affected by the type of organic solvents. *Journal of Food Science* 67: 194–197.

28. Srinivasa P, Ramesh M, Kumar K, Tharanathan R (2004) Properties of chitosan films prepared under different drying conditions. *Journal of Food Engineering* 63: 79–85.

29. Bof MJ, Bordagaray VC, Locaso DE, García MA (2015) Chitosan molecular weight effect on starch-composite film properties. *Food Hydrocolloids* 51: 281–294.

30. Leceta I, Guerrero P, De La Caba K (2013) Functional properties of chitosan-based films. *Carbohydrate Polymers* 93: 339–346.
31. Srinivasa P, Ramesh M, Kumar K, Tharanathan R (2003) Properties and sorption studies of chitosan–polyvinyl alcohol blend films. *Carbohydrate Polymers* 53: 431–438.
32. Srinivasa P, Ramesh M, Kumar K, Tharanathan R (2003) Properties and sorption studies of chitosan–polyvinyl alcohol blend films. *Carbohydrate Polymers*, 53, 431–438.
33. Rao M, Kanatt S, Chawla S, Sharma A (2010) Chitosan and guar gum composite films: Preparation, physical, mechanical and antimicrobial properties. *Carbohydrate Polymers* 82: 1243–1247.
34. Aider M (2010) Chitosan application for active bio-based film production and potential in the food industry. *LWT-Food Science and Technology* 43: 837–842.
35. Souza MP, Vaz AF, Silva HD, Cerqueira MA, Vicente AA, Carneiro-da-Cunha MG (2015) Development and characterization of an active chitosan-based film containing quercetin. *Food and Bioprocess Technology* 8: 2183–2191.
36. Valenzuela C, Abugoch L, Tapia C (2013) Quinoa protein–chitosan–sunflower oil edible film: Mechanical, barrier and structural properties. *LWT-Food Science and Technology* 50: 531–537.
37. van den Broek LA, Knoop RJ, Kappen FH, Boeriu CG (2015) Chitosan films and blends for packaging material. *Carbohydrate Polymers* 116: 237–242.
38. Suresh P, Raj KR, Nidheesh T, Pal GK, Sakhare P (2015) Application of chitosan for improvement of quality and shelf life of table eggs under tropical room conditions. *Journal of Food Science and Technology* 52: 6345–6354.
39. Moreira MdR, Roura SI, Ponce A (2011) Effectiveness of chitosan edible coatings to improve microbiological and sensory quality of fresh cut broccoli. *LWT-Food Science and Technology* 44: 2335–2341.
40. Martínez-Camacho A, Cortez-Rocha M, Ezquerra-Brauer J, Graciano-Verdugo A, Rodriguez-Félix F, Castillo-Ortega M, Yépiz-Gómez M, Plascencia-Jatomea M (2010) Chitosan composite films: Thermal, structural, mechanical and antifungal properties. *Carbohydrate Polymers* 82: 305–315.
41. Helander I, Nurmiaho-Lassila E-L, Ahvenainen R, Rhoades J, Roller S (2001) Chitosan disrupts the barrier properties of the outer membrane of Gram-negative bacteria. *International Journal of Food Microbiology* 71: 235–244.

42. Tayel AA, Moussa S, Opwis K, Knittel D, Schollmeyer E, Nickisch-Hartfiel A (2010) Inhibition of microbial pathogens by fungal chitosan. *International Journal of Biological Macromolecules* 47: 10–14.

43. Baker LG, Specht CA, Donlin MJ, Lodge JK (2007) Chitosan, the deacetylated form of chitin, is necessary for cell wall integrity in Cryptococcus neoformans. *Eukaryotic Cell* 6: 855–867.

44. Romanazzi G, Nigro F, Ippolito A, DiVenere D, Salerno M (2002) Effects of pre-and postharvest chitosan treatments to control storage grey mold of table grapes. *Journal of Food Science* 67: 1862–1867.

45. Shih P-Y, Liao Y-T, Tseng Y-K, Deng F-S, Lin C-H (2019) A potential antifungal effect of chitosan against Candida albicans is mediated via the inhibition of SAGA complex component expression and the subsequent alteration of cell surface integrity. *Frontiers in Microbiology* 10: 602.

46. Giatrakou V, Ntzimani A, Savvaidis I (2010) Effect of chitosan and thyme oil on a ready to cook chicken product. *Food Microbiology* 27: 132–136.

47. Devlieghere F, Vermeulen A, Debevere J (2004) Chitosan: antimicrobial activity, interactions with food components and applicability as a coating on fruit and vegetables. *Food Microbiology* 21: 703–714.

48. Cazón P, Vázquez M, Velazquez G (2018) Composite films of regenerate cellulose with chitosan and polyvinyl alcohol: Evaluation of water adsorption, mechanical and optical properties. *International Journal of Biological Macromolecules* 117: 235–246.

49. Vilela C, Pinto RJ, Coelho J, Domingues MR, Daina S, Sadocco P, Santos SA, Freire CS (2017) Bioactive chitosan/ellagic acid films with UV-light protection for active food packaging. *Food Hydrocolloids* 73: 120–128.

50. Hajji S, Chaker A, Jridi M, Maalej H, Jellouli K, Boufi S, Nasri M (2016) Structural analysis, and antioxidant and antibacterial properties of chitosan-poly (vinyl alcohol) biodegradable films. *Environmental Science and Pollution Research* 23: 15310–15320.

51. Cazón P, Vázquez M (2020) Mechanical and barrier properties of chitosan combined with other components as food packaging film. *Environmental Chemistry Letters* 18: 257–267.

52. Parameswaranpillai J, Thomas S, Grohens Y (2014) Polymer blends: state of the art, new challenges, and opportunities. *Characterization of Polymer Blends*: 1–6.

53. Kumar S, Mukherjee A, Dutta J (2020) Chitosan based nanocomposite films and coatings: Emerging antimicrobial food packaging alternatives. *Trends in Food Science & Technology* 97: 196–209.

54. Wang H, Qian J, Ding F (2018) Emerging chitosan-based films for food packaging applications. *Journal of Agricultural and Food Chemistry* 66: 395–413.

55. Arfat YA, Ahmed J, Hiremath N, Auras R, Joseph A (2017) Thermo-mechanical, rheological, structural and antimicrobial properties of bionanocomposite films based on fish skin gelatin and silver-copper nanoparticles. *Food Hydrocolloids* 62: 191–202.

56. El Bourakadi K, Merghoub N, Fardioui M, Mekhzoum MEM, Kadmiri IM, Essassi EM, Bouhfid R (2019) Chitosan/polyvinyl alcohol/thiabendazoluim-montmorillonite bio-nanocomposite films: Mechanical, morphological and antimicrobial properties. *Composites Part B: Engineering* 172: 103–110.

57. Bonilla J, Fortunati E, Atarés L, Chiralt A, Kenny JM (2014) Physical, structural and antimicrobial properties of poly vinyl alcohol–chitosan biodegradable films. *Food Hydrocolloids* 35: 463–470.

58. Rambabu K, Bharath G, Banat F, Show PL, Cocoletzi HH (2019) Mango leaf extract incorporated chitosan antioxidant film for active food packaging. *International Journal of Biological Macromolecules* 126: 1234–1243.

59. de Moraes Crizel T, de Oliveira Rios A, Alves VD, Bandarra N, Moldão-Martins M, Flôres SH (2018) Active food packaging prepared with chitosan and olive pomace. *Food Hydrocolloids* 74: 139–150.

60. Zhang W, Li X, Jiang W (2020) Development of antioxidant chitosan film with banana peels extract and its application as coating in maintaining the storage quality of apple. *International Journal of Biological Macromolecules* 154: 1205–1214.

61. Moradi M, Tajik H, Razavi Rohani SM, Oromiehie AR (2011) Effectiveness of Zataria multiflora Boiss essential oil and grape seed extract impregnated chitosan film on ready-to-eat mortadella-type sausages during refrigerated storage. *Journal of the Science of Food and Agriculture* 91: 2850–2857.

62. Yang K, Dang H, Liu L, Hu X, Li X, Ma Z, Wang X, Ren T (2019) Effect of syringic acid incorporation on the physical, mechanical, structural and antibacterial properties of chitosan film for quail eggs preservation. *International Journal of Biological Macromolecules* 141: 876–884.

63. Mahdavi V, Hosseini SE, Sharifan A (2018) Effect of edible chitosan film enriched with anise (Pimpinella anisum L.) essential oil on shelf life and quality of the chicken burger. *Food Science & Nutrition* 6: 269–279.

64. Ezati P, Rhim J-W (2020) pH-responsive chitosan-based film incorporated with alizarin for intelligent packaging applications. *Food Hydrocolloids* 102: 105629.

65. Yan J, Cui R, Qin Y, Li L, Yuan M (2021) A pH indicator film based on chitosan and butterfly pudding extract for monitoring fish freshness. *International Journal of Biological Macromolecules* 177: 328–336.

66. Silva-Pereira MC, Teixeira JA, Pereira-Júnior VA, Stefani R (2015) Chitosan/corn starch blend films with extract from Brassica oleraceae (red cabbage) as a visual indicator of fish deterioration. *LWT-Food Science and Technology* 61: 258–262.

67. Zhang X, Lu S, Chen X (2014) A visual pH sensing film using natural dyes from Bauhinia blakeana Dunn. *Sensors and actuators B: Chemical* 198: 268–273.

68. Quesada J, Sendra E, Navarro C, Sayas-Barberá E (2016) Antimicrobial active packaging including chitosan films with Thymus vulgaris L. essential oil for ready-to-eat meat. *Foods* 5: 57.

69. Shah S, Hashmi MS (2020) Chitosan–aloe vera gel coating delays postharvest decay of mango fruit. *Horticulture, environment, and Biotechnology* 61: 279–289.

70. Kaewklin P, Siripatrawan U, Suwanagul A, Lee YS (2018) Active packaging from chitosan-titanium dioxide nanocomposite film for prolonging storage life of tomato fruit. *International Journal of Biological Macromolecules* 112: 523–529.

71. Şahin A, Çarkcıoğlu E, Demirhan B, Candoğan K (2017) Chitosan edible coating and oxygen scavenger effects on modified atmosphere packaged sliced sucuk. *Journal of Food Processing and Preservation* 41: e13213.

72. Remya S, Mohan C, Bindu J, Sivaraman G, Venkateshwarlu G, Ravishankar C (2016) Effect of chitosan based active packaging film on the keeping quality of chilled stored barracuda fish. *Journal of Food Science and Technology* 53: 685–693.

73. Tirtashi FE, Moradi M, Tajik H, Forough M, Ezati P, Kuswandi B (2019) Cellulose/chitosan pH-responsive indicator incorporated with carrot anthocyanins for intelligent food packaging. *International Journal of Biological Macromolecules* 136: 920–926.

74. Capello C, Trevisol TC, Pelicioli J, Terrazas MB, Monteiro AR, Valencia GA (2021) Preparation and characterization of colorimetric indicator films based on chitosan/polyvinyl alcohol and anthocyanins from agri-food wastes. *Journal of Polymers and the Environment* 29: 1616–1629.

75. Wu C, Sun J, Zheng P, Kang X, Chen M, Li Y, Ge Y, Hu Y, Pang J (2019) Preparation of an intelligent film based on chitosan/oxidized chitin nanocrystals incorporating black rice bran anthocyanins for seafood spoilage monitoring. *Carbohydrate Polymers* 222: 115006.

Chapter 10

Chitin and Chitosan Applications in Medication Delivery Systems

Mohamed Sharaf[a,b] **and Chen-Guang Liu**[b]

[a]*Biochemistry Department, Faculty of Agriculture, Al-Azhar University, Nasr City, Cairo, Egypt*

[b]*Biochemistry and Molecular Biology Department, College of Marine Life Sciences, Ocean University of China, Qingdao, PR China*

mohamedkamel@azhar.edu.com

For drug encapsulation and release, systems for delivering drugs employ polymers. Compared to synthetic polymers, natural polymers often provide benefits such as biodegradability, biocompatibility, and recognizable physiological moieties, which aid cellular functions. Chitosan has several qualities, including drug release kinetics, modifiability, cross-linking potential with various polymers, antibacterial activity, gel-forming capacity, bioadhesive properties, immunostimulatory effects, macrophage activation, and gas permeability. This chapter discusses how chitosan is used for drug delivery, as well as drug release methods. Chitosan, a biodegradable, biocompatible polysaccharide produced from chitin, shows potential for medication delivery and biological uses. Various chitosan and chitin extraction methods provide materials with distinct characteristics that may be modified to increase bioactivity. Oral, ophthalmic, transdermal,

Chitin and Chitosan: Physical and Chemical Properties
Edited by Shakeel Ahmed, Soumeia Zeghoud, Hadia Hemmami, and Ilham Ben Amor

ISBN 978-981-5129-39-7 (Hardcover), 978-1-003-58977-8 (eBook)
www.jennystanford.com

vaginal, and nasal chitosan-based drug delivery devices target and prolong drug release. Chitosan has been utilized to repair bone, cartilage, cardiac, ophthalmic, wound healing, and periodontal tissue, among other biological processes. Chitosan has also been employed in applications for gene transfer, vaccination, bio-imaging, and cosmetics. The biocompatibility and characteristics of modified chitosan derivatives have been improved, creating novel materials with exciting potentials in a range of applications in biomedicine. The most recent research on chitosan and its use in biomedical science and medication delivery are summarized in this chapter.

10.1 Introduction

Drug delivery systems consist of pharmacologically active drugs that are coupled with a suitable carrier, forming the fundamental components. The fundamental criteria for ideal delivery systems include the ability to selectively direct the active factor exclusively to the intended location of action inside the body while ensuring the provision of an acceptable quantity of the active agent at a consistent and reasonable degree over a predetermined duration [1]. Drug carriers, often composed of polymers, are used to fulfill the primary essential attribute of systems delivery, which is to induce plasma realty concentration within the therapeutic ambit [2].

The idea of controlled distribution systems has been established by this feature. The creation of specialized delivery systems is facilitated by the second characteristic. This involves encapsulating the active component inside a delivery system that can carry the medication to the desired place and calls for careful consideration of the carrier, administration route, and target of release [3]. With the effectiveness of delivery systems or the accomplishment of the therapeutic goal, the outcome is contingent upon several variables, including the characteristics of the drug and the polymers, the condition being treated, the dose form, and the mode of administration. Delivery systems enable drug bioavailability and duration of pharmacological action to be prolonged, drug dose and frequency to be decreased, plasma level fluctuations to be reduced, side effects to be decreased, and patient compliance to be increased [3].

Polymers are commonly employed as carriers for the delivery of active chemicals. A range of delivery methods may be generated by their use, including micelles, tablets, hydrogels, capsules, and particle systems (beads, micro- and nanoparticles). Toxicology, biodegradability, and biocompatibility are factors to consider before choosing an acceptable polymer because the latter two can result in the desired drug release and enhance the elimination of carriers subsequent to medication delivery. Their three-dimensional structure has the capacity to both encapsulate the medicine and regulate its release based on the situation. Moreover, they possess the capacity to adapt the pharmacokinetics of the energetic compound, thereby enhancing the curative indicator through the mitigation of adverse effects and subsequently impacting the overall efficacy of the system [4, 5]. Both natural and synthetic polymers are commonly employed in the design and development of delivery methods. However, natural polymers, which are derived from different sources like plants, fungi, animals, and bacteria, possess notable advantages over synthetic polymers. These advantages include lower toxicity, greater biocompatibility, enhanced biodegradability, and increased accessibility. As cited in reference [6], chitosan, alginate, and cellulose are frequently employed natural polymers in the field of medicine.

Chitin was found in the cell walls of fungi, green algae, yeast, crab and insect cuticles, and prawn shells, in contrast to chitosan, which is far less frequent in the environment and only found in the cell walls of select fungi [7].

Chitosan and chitin are widely used in practically all sectors, such as medicine, dentistry, environmental protection, food processing chemistry, agriculture, and biotechnology, as a result of the aforementioned outstanding qualities. Additionally, it has bioactivities that have led to its usage as an anti-HIV-1, antifungal, antibacterial, and anticancer agent. Chitosan is currently used in the medical and pharmaceutical fields, specifically for wound healing, tissue engineering, drug delivery, cardiovascular disease, and age-related illnesses [8]. This is possible due to the fact that chitosan can be created in a variety of shapes, including tablets, hydrogels, capsules, microsponges, particle systems, films, membranes, and nanocomposites, allowing the production of flexible devices that may be tailored to the needs of end users. This chapter, however, focuses on the characteristics of chitosan that make it appropriate for drug administration, various chitosan-based dosage forms and potential drug-releasing mechanisms, and the use of chitosan in different delivery of drug methods. Drug delivery and the biomedical industry have both indicated potential uses for chitosan and chitosan-based systems (Fig. 10.1).

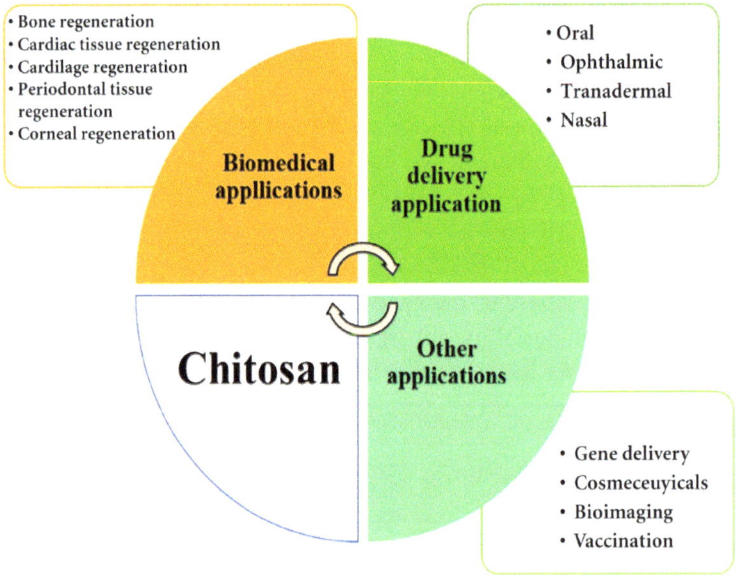

Figure 10.1 Several applications of chitosan.

Modern medicine places a high priority on drug delivery because it affects a medicinal substance's effectiveness and safety. Controlled drug distribution throughout the human body is a challenging procedure that necessitates overcoming numerous obstacles, including the liver's and kidneys' clearance of medications, the bloodstream's rapid drug degradation, and the limited permeability of biological membranes. To prevent the medicine from degrading, lengthen its life in circulation, and improve its localization at the target site, an appropriate carrier or vehicle is required. A number of essential qualities, including biodegradability, biocompatibility, and controlled release properties, should be present in the optimum drug delivery device [9]. After its intended purpose, the means of transportation may be effectively metabolized and expelled from the body thanks to the virtue of biodegradability. On the other hand, biocompatibility refers to the absence of any adverse effects caused by the delivery vehicle on the body. Pharmaceutical substances possessing controlled release properties exhibit the ability to gradually release the active drug compound over a predetermined duration, thereby sustaining therapeutic concentrations and reducing the necessity for frequent administration. Chitosan has demonstrated considerable potential in the field of medication delivery due to its unique characteristics. The Drug Administration and Food (FDA) has classified this polymer as "Generally Recognised as Safe" (GRAS) and has authorized its utilization in the fields of engineering with tissues and pharmaceutical delivery applications, due to its remarkable biocompatibility and biodegradability [10]. The LD_{50} of chitosan following swallowing on rats has been found to exceed 16 g/kg, as indicated by various acute toxicity tests. These tests have consistently shown that chitosan is safe for systemic use [11]. The primary amino group of the recurring glycosidic residue in chitosan functions as a location for chemical attachment and imparts a cationic charge to the molecule. The positive charge possessed by chitosan renders it pH-sensitive, thereby facilitating the formation of ionic compounds with pharmaceutical agents and governing the controlled administration of medications. One of the primary advantages of chitosan is its versatility in being able to undergo transformation into various dosage forms, each possessing distinct characteristics and applications. The subsequent sections (Table 10.1) discuss several commonly employed chitosan-based delivery methods.

Table 10.1 Discuss several commonly employed chitosan-based delivery methods

System type	Overview	Procedure for preparation	Key attributes/features	Ref.
Powder/ micro-granule system	Polymeric chitosan is used to bind irregularly shaped micron-sized drug aggregates, forming a subset of solid dosage forms.	• The processes of spray drying • The processes of salt-/organic solvent-induced precipitation • The processes of gelation	• Useful for large-dose, bulky medications • A powder or granule's small particle size makes it easier to dissolve quickly in the body, enhancing its bioavailability • Increases the medicines' chemical stability when integrated • Ease of use, handling, and preparation	[12]
Hydrogel system	A three-dimensional network structure made up of interconnecting polymer chains can carry a lot of water. The molecular-level manipulation of gelation chemistry allows for the development of hydrogels with tailored physicochemical characteristics, rendering them suitable for many biological applications.	• The processes of physical cross-linking • The processes of enzymatic cross-linking • The processes of photo-crosslinking • The processes of chemical cross-linking	• Excellent biocompatibility and injectability • A hydrating effect • Possible to include in situ forming qualities	[13]
Nanofiber system	The cutting-edge platform is a ground-breaking method for encapsulating or attaching drugs to nanometer-sized fibers.	• The processes of templating • The processes of electrospinning • The processes of melt/solution blowing	• Bilayered or trilayered nanofibers can be prepared to combine several medications	[14]

System type	Overview	Procedure for preparation	Key attributes/features	Ref.
	Nanofibers possess inherent advantages for applications involving controlled drug release and tissue regeneration owing to their elevated surface-to-volume ratio.		• Heavy drug use and effective drug release are made possible by their large surface area compared to their volume • Modifying the composition and concentration of the polymer can result in modified medication release	
Microsphere system	The diameters of these spherical particles vary between 10 nm and 1000 nm. The release profile may be modulated by the use of several microsphere variations, including fibrous microspheres, hollows, and core shells.	• The processes of spray drying • The processes of emulsification • The processes of thermal cross-linking • The processes of ionic gelation • The processes of coagulation/ precipitation	• Flexibility in the administration method • High trapping and drug loading effectiveness • Physically stabilizes entrapped biomolecules • Sustained drug release	[15]
Nanoparticle system	Nanoparticles are used as particulate systems because of their range of sizes (1 to 100 nm) and their capacity for surface alterations. This characteristic renders them adaptable transporters for the delivery of pharmaceuticals, proteins, and genetic material.	• The process of evaporation of an emulsion-solvent • The phenomenon of reverse micellization • The processes of emulsification and cross-linking • The process of polyelectrolyte complexation • The technique of desolvation	• Enhanced permeability and retention effects (in cancer) or high site-specific drug localization with the use of targeting ligands • Can be changed to produce stimuli-triggered medication release (pH, redox, temperature) • It can make it easier for drugs to be delivered together	[16]

10.2 Oral Drug Delivery System (ODDS)

The term "oral drug delivery" pertains to the method of administering a drug by mouth, commonly in the shape of liquids, tablets, or capsules. These dosage forms are consumed by the patient, they pass through several stages, starting from oral intake and ending with absorption into the bloodstream. Given that it is non-invasive and convenient, it is the most popular and practical method of medication delivery. Since it may be administered by patients themselves and does not require medical assistance, it is frequently the most convenient and preferred method. Additionally, oral drug delivery enables the prolonged sustained release of medication, which can enhance therapeutic effects and patient compliance. The gastrointestinal tract's large mucosal surface area enhances medication absorption, making it a highly preferred route for drug administration. Due to their vulnerability to gastrointestinal secretions, particularly enzymes, and the fluctuating pH levels, many medications cannot be taken orally. When a medicine is taken orally, physiological variables, such as transit time and the availability of food, blood perfusion, and the fact that some drugs have a significant role in several biological processes and very little absorption windows may make it much harder to absorb [17]. Chitosan has advantages for usage in systems that dispense oral medications. Due to its cationic properties, it can easily interact with negatively charged cell membranes, facilitating cellular absorption. Chitosan may partially dissolve in water, unlike many other insoluble in water polymers, and can form a hydrogel in wet settings. Its exceptional ability to stick to mucosal surfaces makes it a fantastic option for oral medication delivery. The P-glycoprotein efflux pump can also be inhibited by various chitosan derivatives, like glycol chitosan solution or nanoparticles made of glycol chitosan [17]. This increases the possibility of absorption into the bloodstream. The uses of chitosan (ODDS) are addressed below. Furthermore, Arif and his colleagues presented the invention of chitosan-PMLA nanoparticulate systems as carriers for antibacterial glycolipids, which was motivated by the necessity to discover substitutes to manage *Helicobacter pylori* infections [18, 19]. The search conducted using Ren et al. [20] examined the potential of cinnamyl-chitosan, a novel excipient, which is N-cinnamyl substituted and O-amine functionalized. The formulation of directly compressed tablets involved the utilization of microcrystalline cellulose as a binder/filler, stearate of magnesium as a form of lubricant, and acetaminophen as the active pharmaceutical ingredient. Additionally, either chitosan or the altered cinnamyl-chitosan was incorporated into the tablets. The sole factor contributing to the differences observed in the ability to flow, compressive behavior, and disintegrating properties of the tableting blends was the variation in chitosan type, as the quantities of the other ingredients remained consistent across all formulations. The 20% cinnamyl-chitosan mix exhibited plastic behavior that was comparable to that of the control group, but the force-displacement profile analysis showed that they were more elastic due to their greater particle size. Cinnamyl-chitosan was easily deformed as a result, allowing for high compaction of the mixture at only moderately high compressive pressures. Cinnamyl-chitosan combinations with larger particle sizes exhibit good flowability and compaction properties with no tableting problems. They additionally demonstrate excellent acetaminophen-releasing characteristics as excipients in contrast to MCC. The cinnamyl-chitosan compound, after undergoing chemical modification, exhibited enhanced antibacterial and antioxidative characteristics, along with improved mechanical strength and efficacy in the delivery of medication. Consequently, it shows promise as a potential direct compression excipient replacement. In work conducted by Yasufuku et al. [21], the authors utilized the antioxidant properties of modified lower molecular weight chitosan to develop chitosan tablets with oral administration and extended-release abilities. The chitosan displayed scavenging capability due to the oxidative stress induced by hydroxyl and superoxide radicals. This activity enabled the chitosan to resist disintegration and achieve a controlled release of theophylline, a representative medication, for a duration exceeding 12 hours.

This approach is promising for ODDS, particularly for treating stomach wounds. A chitosan-catechol adhesive patch (Chitoral) designed after mussels that can be used as an oral mucosal medication delivery device was reported by Ryu et al. [22]. The podium is built on a chitosan basis that has been carboxyl-terminated catechol hydro-caffeic acid modified. When this modified chitosan solution is freeze-dried, it forms a porous, sponge-like sticky substance that can hold medicinal medicines inside the polymeric network. The platform dissolves instantly when it gets formed, interacts with saliva, and immediately transforms into an insoluble in water, sticky hydrogel-like substance via covalent cross-linking and physical entanglement. It also forms inter-molecular complexes with oral mucins. Pig tongue and labial mucosa lap-shear adhesion tests showed a separation stress of 10.3 6.1 kPa. By enabling a sustained release of encapsulated medications, triamcinolone acetonide-loaded chitoral has been shown to speed the healing of severe oral ulcers.

10.3 Colon-Specific Drug Delivery System (CDDS)

The advantages associated with the colon encompass its approximate neutrality in terms of pH, limited presence of digestive enzymes, and extended duration of transit. The objective of colon-specific drug delivery is to address both systemic conditions, like nocturnal asthma and angina, and local conditions, including colorectal cancer, ulcerative colitis, Crohn's disease, and irritable bowel syndrome. Furthermore, the potential adverse impacts and optimal dosage of the medication could potentially be mitigated through the implementation of localized therapeutic interventions [23]. The present technique of drug administration presents a novel pathway for pharmaceuticals that exhibit instability or limited absorption within the upper gastrointestinal tract, particularly proteins and peptides. Various methods currently exist for targeting the colon, which include bacteria-responsive, pressure-responsive, time-responsive, and pH-responsive approaches, according to the specific triggering mechanism. Due to its resistance to enzymatic degradation in the upper part of the gastrointestinal tract, chitosan exhibits heightened susceptibility to bacterial enzymes found in the colon, resulting in the cleavage of its glycosidic bond [24]. Mucoadhesive micro- and nanoparticles were frequently employed in the treatment of colonic diseases. In a particular study, the process of encapsulating egg yolk immunoglobulin within microbeads composed of chitosan-Ca-pectinate was effectively carried out. The study demonstrated that the release of the medication exhibited a slow rate within the small intestine and stomach, though it exhibited a fast rate within the colon. According to Xing et al. [25], these microbeads have the ability to stop Clostridium difficile poison from adhering to the intestinal mucosa. By the double ionic gelation method, in their study, Lanjhiyana et al. [26] created tinidazole cross-linked microbeads using sodium alginate and chitosan. The researchers successfully demonstrated the ability of these microbeads to maintain the discharge of the medicine. The enhanced mechanical strength of the microbeads can be attributed to the process of twofold ionic gelation.

The main therapy strategy for conditions like Crohn's disease and ulcerative colitis primarily centers on the amelioration of symptomatic pain. Non-steroidal anti-inflammatory medications (NSAIDs) can accomplish this goal when administered locally. To administer budesonide specifically to the ileum and ascending colon, in their study, Wakure et al. [27] developed a chitosan capsule that exhibits pH responsiveness, incorporating an enteric coating composed of Eudragit S100/L100 in a 1:1 ratio. The enteric-coated chitosan capsule was designed to release the medication specifically after the ileocecal junction in the gastrointestinal tract. Subsequently, the chitosan in the capsule underwent biodegradation by colonic bacteria. This phenomenon was confirmed through the utilization of scintigraphic imaging technique. The present study focuses on the advancement of novel delivery techniques for 5-aminosalicylic acid through the utilization of drug intercalation within a double-layered Mg_2 Al hydroxide structure, subsequently incorporating it into a chitosan matrix. Furthermore, the application of pectin to the aforementioned substance was employed as a means to establish a barrier that would defend against the acidic conditions present in the stomach. Based on the findings of Ribeiro et al. [28], it has been found that the aforementioned bio-nanocomposite beads have the ability to release 5-aminosalicylic acid, specifically in the colon. Consequently, these beads hold potential for therapeutic applications in the treatment of Crohn's disease and ulcerative colitis. The successful delivery of paracetamol to the intestines and colon was achieved using a pH-sensitive hydrogel composed of sodium alginate and hydroxyethyl acrylic chitosan [29]. Colorectal cancer, ranking third in terms of mortality attributed to cancer, is predominantly managed through the administration of oral medications. This approach facilitates targeted drug distribution within the colon. Chitosan has been employed in the development of diverse pharmaceutical formulations aimed at facilitating the specific delivery of various medications to the colon. In a particular investigation, a combinatorial technique was utilized to incorporate both 5-fluorouracil and curcumin within nanoparticles composed of N,O-carboxymethylchitosan. The nanoparticles exhibited enhanced anticancer properties and demonstrated continued release of medication over a duration of four days in an in vitro research [30]. The researchers employed a combinatorial approach to generate folate-chitosan nanoparticles that were loaded with both 5-fluorouracil and leucoverin. These nanoparticles were specifically designed for targeted delivery to the colon, with the aim of treating colon cancer. The application of Eudragit S-100 coating to these nanoparticles resulted in a discernible sustained release of pharmaceutical agents, specifically in the colon, while no such release was observed in simulated gastric or intestinal fluid. In their study, Jyoti et al. [31] employed a two-step process to generate curcumin microspheres. Initially, curcumin was blended with chitosan microspheres in a 1:1 proportion, followed by complexation with 2-HP-CD to form curcumin-2-HP-CD. The study found that the distribution of curcumin in the colon was approximately 8.36 times higher in the microspheres compared to the curcumin suspension when both were exposed to simulated colonic fluid with a pH range of 7.0–8.0.

10.4 Ophthalmic Delivery System (OPDDS)

Ocular medicine administration is a significant research issue in the pharmaceutical business due to the distinct pharmacodynamic and pharmacokinetic characteristics of the eye. The unique structure of the eye makes it difficult to sustain an appropriate medication concentration at the necessary site of activity because it inhibits the entry of molecules of medicine. In order to treat ophthalmic problems orally, significant doses of active medicines that may have serious side effects are needed to reach therapeutic concentrations. The blood-aqueous, blood-retinal, and blood-vitreous barriers are further obstacles that prevent systemic entry [32]. As a result, the only viable method of administration for ophthalmic dose forms is topical. Conjunctivitis, blepharitis, and keratitis sicca are examples of conditions that can be treated with drops, gels, and creams. Other conditions that can be treated by intraocular therapy through the cornea include glaucoma and uveitis.

The ocular bioavailability of topical therapies is restricted to a value below 5% [33, 34] as a result of inadequate pre-corneal retention, limited corneal penetration, and the presence of effective mechanisms for drug elimination, like protein binding, tear drainage, and induced tear formation. Enhancing corneal residence time has been used to address bioavailability and the requirement for frequent medication delivery. Mucoadhesive polymers, viscosity enhancers, and in situ, gel-forming technologies accomplish this. Chitosan naturally has these properties. Due to its biocompatibility, antibacterial properties, pH neutrality, and capacity to break in the presence of lysozyme (muramidase, an enzyme present in tears and innate immune system component) [35], this material exhibits potential for the development of controlled release platforms in ocular applications.

Nanoparticles have been the subject of extensive research as the primary chitosan-based delivery strategy for ocular applications. In the ionic gelation process commonly employed for their production, tripolyphosphate (TPP) is typically utilized as a cross-linking agent. The utilization of the amino groups that carry a positive charge in this reaction is attributed to their interaction with the TPP groups that possess a negative charge, resulting in a decrease in the mucoadhesive capacity [36], for the purpose of improving azithromycin's precorneal residence time, ocular penetration, and bioavailability (to treat bacterial conjunctivitis). Eid et al. [37] synthesized chitosan-coated niosomes. A modified thin-film hydration technique with chitosan solution coating produced drug-loaded niosomes. The optimized particles had 376 nm diameters, 32.1 mV surface charges, 74.2% entrapment efficiency, and 3114 dyne/cm^2 mucoadhesion forces. Due to the higher permeability factor of chitosan-coated niosomes and the continuous dissolution of the encapsulated drug, the platform increased azithromycin concentration in the eyes of rabbits three-fold compared to commercial drops.

In situ, hydrogels and ocular delivery methods based on films or membranes provide superior therapeutic results. Typically, an in situ gelling method is liquid when it is first injected into the eye. However, it quickly changes into a viscoelastic gel inside the conjunctiva sac of the eye in response to diverse stimuli. As a result, the gel's residence duration is greatly increased, and the medications that are contained in it are gradually released over an extended period of time. As a consequence, this prolonged release increases the drug's bioavailability and lowers the frequency of delivery. Numerous kinds of in situ gelling systems, such as those that are temperature, pH, and ionic-activated, have been suggested as innovative ophthalmic formulations [38]. Hexanoyl glycol chitosan (H-GCS), which gels in situ and is temperature-sensitive, has been reported by Shi et al., as a technique for improving ocular bioavailability. Levofloxacin appears to have no effect on the aqueous H-GCS solution's ability to gel at 32 °C, where it goes through a conventional sol-gel transition. After 24 hours of incubation, L-929 and HCEC cells exhibit only modest cytotoxicity to H-GCS at doses below 0.8 mg/mL. The ocular application of a hydrogel containing 2% weight of H-GCS did not induce any ocular discomfort in rabbits after a single injection. Following topical administration, the application of a 2 wt% H-GCS hydrogel has been seen to considerably prolong the absorption of medications in the precorneal region, as compared to an aqueous solution. This extended absorption period leads to an increase in ocular bioavailability [39].

Li et al. [40] synthesized a chitosan solution that was air-dried to make a drug-containing film to increase the drug's insoluble aqueous solubility. The chitosan solution included 1 mg/mL brimonidine tartrate, an antiglaucoma drug. The transparent, mucoadhesive film was stable. The loaded drug formed minute particles in the chitosan film. The chitosan film's brimonidine tartrate ocular permeability was measured using a Franz diffusion cell. More than 75% of the drug was released in less than 6 minutes and 90% in 30 minutes. The drug's corneal permeability was 1.62 10^5 cm/s, far greater than previous systems.

10.5 Buccal Drug Delivery System (BDDS)

The buccal mucosa is considered the most favorable route for drug administration due to several advantageous factors. Firstly, the area's high level of vascularization ensures a high bioavailability of drugs. Additionally, the relatively immobile mucosa layer results in low enzymatic activity, further enhancing drug absorption. Moreover, the buccal mucosa allows for adaptability in dosage form design, enabling the administration of drugs for each of the local and systemic effects. These factors contribute to the attractiveness of the buccal mucosa as a drug delivery route [41]. Nevertheless, the barrier function of the buccal mucosa, inadvertent swallowing of the delivery system, constraints on oral intake, and enzymatic salivary scavenging are recognized as the limitations associated with buccal drug administration [42]. The utilization of mucoadhesive polymers, which possess the ability to retain dosage forms on the mucosa layer for an extended duration, thereby facilitating diffusion, seems to offer greater feasibility compared to alternative approaches for augmenting transbuccal drug delivery. This is primarily due to the prevention of lasting damage to the mucosa membrane caused by penetration-boosting agents [42]. Therefore, chitosan might be seen as a potential polymer for the development of buccal medication delivery systems as a result of its strong mucoadhesive properties combined with its capacity to increase absorption. The highest common oral disease, oral candidiasis, is brought on using the opportunistic yeast pathogen Candida albicans. This infection is typically seen in people who are immune-impaired. In addition to increasing the likelihood of various adverse reactions and interactions between drugs, oral antifungal therapy also increases the risk of Candida strain resistance due to increased concentrations. So, local delivery of antifungal drugs to treat oral candidiasis seems to represent the best approach to avoid the aforementioned consequences [43].

The emulsification/internal gelation technique was employed to encapsulate nystatin, an antifungal medication, within three distinct varieties of mucoadhesive microspheres composed of hydrogel, chitosan-coated, and alginate materials. The controlled release of Nystatin was achieved through the use of chitosan-coated microspheres, resulting in a concentration gradient that effectively suppressed the growth of *Candida albicans*. Based on the findings of an in vivo study, it can be concluded that the absence of nystatin in the circulatory system indicates the absence of any adverse effects on the aforementioned systems [44].

The chitosan-based gels containing celecoxib exhibited favorable physiochemical and rheological properties, resulting in enhanced mucosal retention. Azone was employed as a substance to enhance the permeation of other compounds. Consequently, the gel that is generated has the potential to be applied via the buccal cavity for the purpose of treating malignancies [45].

Insulin is delivered buccally to avoid oral and parenteral issues. One study used dimethyl ethyl and thiolated chitosan to make insulin nanoparticles. Chitosan (17.1%), dimethyl ethyl (67.89%), and the thiolated derivative showed the highest insulin penetration in rabbit buccal mucosa ex vivo. Compared to chitosan and dimethyl ethyl chitosan, thiolated derivative-based nanoparticles were the most water-soluble [46].

The development of a novel hydrogel incorporating lidocaine involved a two-step process. Initially, catechol groups that function were chemically linked to the chitosan polymer's main structure. Subsequently, genipin was employed as a cross-linking agent to bind the previously modified chitosan molecules together. In comparison to a chitosan hydrogel with a contact time of only 1.5 h, it has been reported that a cross-linked catechol-chitosan-based hydrogel demonstrates in vitro adhesion for a duration exceeding 6 h. A concentration of 1 μg/ml of lidocaine was observed in the serum of bunnies following an in vivo investigation utilizing the novel hydrogel [41].

10.6 Transdermal Drug Delivery System (TDDS)

The skin is a particularly beneficial route for giving many drugs since it is the biggest and most expansive organ in the body. The transdermal route of administration has many benefits in comparison to intravenous and oral delivery methods. These advantages include the ability to bypass metabolism in its first pass, maintain steady blood levels, and achieve both localized and systemic distribution. Moreover, it may be used to effectively manage other medical diseases, such as chronic pain, hormone imbalances, and cardiovascular issues [47]. The stratum corneum, however, frequently interferes with transdermal delivery's ability to work. A multilamellar lipid bilayer that serves as a physical barrier to medication molecules covers keratin filaments that are encased in a cornified envelope. Even though a variety of transdermal therapies are commercially available (including ointments, sprays, spreads, and patches), limitations such as inadequate bioavailability, limited drug diffusion, unfavorable lateral effects (like skin sensitivity), and low patient compliance are still issues [48]. Due to their exceptional stability and adhesion, synthetic high-molecular-weight polymers like PLA and PEG are being investigated by researchers as possible solutions to these problems, but their uses are restricted by their inertness to degradation, cytotoxicity, and immunogenicity. Chitosan (and its derivatives) can be used in certain circumstances to help increase medication penetration [49]. The underlying method involves changing the secondary molecular makeup of stratum corneum proteins, redistributing transmembrane proteins to open closed junctions in the granular layer, and increasing stratum corneum content of water (to assist the administration of hydrophilic medicines) [50].

Shekh et al. [51] created a scaffold carrier with a prolonged and regulated release profile for transdermal distribution by functionalizing polyacrylonitrile nanofibers with oxidized chitosan. Chitosan's oxidized state was validated by SEM, which also showed an increase in surface roughness and average nanofiber diameter from 218 nm to 354 nm. Acyclovir nanofibers treated with oxidized chitosan in an in vitro drug release research had a more sustained release curve than untreated ones (as predicted by the Korsmeyer-Peppas model by a Fickian dispersal mechanism). This action may be brought about by Schiff bases between the molecules of drugs on nanofibers and oxidized chitosan. Long-acting antivirals like acyclovir may be less hazardous by lowering the drug's initial high local concentration thanks to oxidized chitosan-modified nanofibers. Human adipose-derived stem cells and blood compatibility experiment revealed that the scaffold was hemocompatible and cell biocompatible. Indulekha et al. [52] created a transdermal delivery system for drugs that exhibits responsiveness to changes in heat and pH, thereby facilitating the administration of medication as per individual requirements. In order to produce a gel-like material, it was necessary to graft thermosensitive poly (N-vinyl caprolactam) onto pH-responsive chitosan using a specific technique. The objective of this study was to design and implement a system that is easily navigable for patients, facilitating the self-administration of medications through the application of a heating pad to the transdermal system for drug delivery. The investigators performed in vitro tests for drug release on two model pharmaceuticals, namely acetaminophen and etoricoxib, at three discrete temperatures (32 °C, 25 °C, and 39 °C) and two distinct pH levels (7 and 5.5), with the aim of assessing the efficacy of the system. Based on the findings, it was noted that the maximum drug release was recorded for both drugs at a pH of 5.5 and a temperature of 39 °C. Moreover, it was discovered in the course of in vitro skin permeation tests employing a Franz diffusion cell that both treatments exhibited enhanced drug release when the skin was exposed to elevated temperatures. Radwan-Pragowska et al. [53] conducted a study wherein they developed chitosan-based patches that were cross-linked and functionalized with zinc oxide nanoparticles (ZnO NPs). The objective of this study was to enhance the controlled distribution of cannabidiol. The completed composites were subjected to SEM and fluorescence microscopy analyses, which revealed remarkable levels of porosity and favorable swelling possessions. The incorporation of ZnO NPs led to improvements in the physicochemical properties of the patches, such as increased mechanical strength, enhanced drug loading capacity, and the capability to release pharmaceuticals over an extended duration through a first-order kinetics process. The nanocomposites demonstrated conductive characteristics, rendering them suitable for use in iontophoresis. The study employed both indirect and direct procedures, specifically the XTT analyses and morphological analysis, to evaluate the cytotoxicity of the produced materials. The results obtained from both approaches indicated that the transdermal systems exhibited a high level of biocompatibility with skin cells, specifically L929 mouse fibroblasts.

Chitosan nanoparticles with highly controlled size and surface charge have a high specific surface area and excellent skin permeability. Nanoparticle-based skin-treatment systems have been studied. Riezk et al. [54] found that chitosan nanoparticles can transport medicine for cutaneous leishmaniasis. The researchers generated positively and negatively charged chitosan nanoparticles using tripolyphosphate sodium and dextran sulfate, respectively. When loaded with Amphotericin B (AmB), both nanoparticles killed Leishmania's major intracellular amastigotes in vitro, much as unencapsulated AmB. Both nanoparticles were less damaging to KB cells and red blood cells. Fereig et al. [55] provided a description of chitosan nanoparticles loaded with tacrolimus, which can be utilized for the medical treatment of psoriasis. For the inclusion of hydrophobic medicines in chitosan, the scientists proposed a scalable, hazardous, solvent-free, and improved ion gelation method. The particles' size after drug loading was 140.8 nm, and their entrapment competence was 65.5%. In an in vitro investigation on skin infusion, the investigators discovered that tacrolimus in chitosan nanoparticles had a permeability of 24% after 24 hours, compared to tacrolimus cream's permeability of 61%. This suggests that the tacrolimus release might be greatly delayed by the chitosan nanoparticles, thereby lowering the drug's systemic toxicity. Additionally, tacrolimus cream deposition rates over 24 hours were 11.4% against tacrolimus chitosan nanoparticle deposition rates of 75%, showing the possibility of the nanoparticles for tacrolimus delivery to the skin. Skin sclerosis at the laceration site is brought on by the condition psoriasis, which significantly hinders the transdermal administration of medication. Tacrolimus' transdermal permeability was shown to be enhanced by chitosan nanoparticles, which also delayed the release of the medication from the cortex, increased its act time at target areas, and lessened the drug's overall influence.

10.7 Periodontal Drug Delivery System (PDDS)

Periodontal illness is a pathological case that develops as a result of the build-up of microaerophilic bacterial plaque and anaerobic [56]. It begins with inflammation and eventually leads to the irreparable loss of periodontium, which comprises the cementum, alveolar bone, gingiva, and periodontal ligament. Scaling and root planing are currently being used as mechanical treatments to halt the loss of teeth and dental bone. However, the prescription of local or systemic antibiotics is crucial to halting the spread of illness. Antibiotic resistance may result from systemic delivery of the antibiotic, which necessitates a greater dose to obtain the target concentration in the gingival crevicular fluid [56]. Due to their low dosage required, topical treatments, including toothpaste, gels, and mouthwash, have been utilized successfully [56, 57]. The biggest difficulty, however, is retaining these local formulations in the right place due to their repeated applications and greater initial concentration requirements [58]. Designing tools that can be inserted into the dental pocket (intrapocket) is the best line of action. Chitosan is the best material to use in the device since it has the mucoadhesive and biodegradable properties that are required.

The major goals of contemporary periodontal treatment are to heal the inflammatory tissue and lower the bacteria burden. Numerous studies have put different antibiotics into intrapocket films or chips or in situ/thermoreversible gel. In this context, minocycline was combined with chitosan and -glycerophosphate to create a thermoresponsive gel [59]. Gel-treated rats had considerably lower gingival indices and probing depths than the ointment-treated rats, showing that the produced gel could maintain the drug release in periodontitis. Sheshala et al. [60] conducted an examination of in situ gels containing moxifloxacin. These gels were created using different amounts of poloxamer, both alone and in combination with chitosan and glycerophosphate. The formulations with poloxamers, glycerophosphate, hydroxypropyl methylcellulose, and chitosan showed sustained drug release for 8 hours, it was reported. They found too the aforementioned two formulations also showed antibacterial activity against *Aggregatibacter actinomycetemcomitans* and *Streptococcus mutans* [61].

The main benefit of intrapocket films/chips over thermo-reversible gels is retention in the tooth socket for a greater amount length of time, leading to the prolonged release of the encapsulated medicine without any systemic effects. Moxifloxacin was added to intrapocket films of the single-layer, triple-layer, and double-layer varieties at a concentration of 30% w/w chitosan. Single-layer films in an in vitro drug release research showed burst release on the initial day, while double-layer and triple-layer films were able to regulate this initial burst release. Satranidazole with chitosan (5%) subgingival films that were developed exhibited an initial burst release of the medication followed by a steady release for 96 hours. The number of gram-positive bacteria was shown to be decreased by the movies [62].

10.8 Nasal Drug Delivery System (NDDS)

The term "nasal delivery of drugs" describes the practice of giving medicine through the nasal passages. This approach capitalizes on the abundant vascularization in the mucosal membranes of the nasal cavity to expedite the absorption and systemic dissemination of pharmaceutical substances. The nasal administration of topical medications has long been used to treat localized conditions of the upper respiratory tract, such as infections, congestion, and allergies. Small molecular weight medications are now more frequently administered via the nasal route for treating chronic illnesses, including diabetes and obesity. Researchers are particularly interested in the nasal route's potential for the treatment of long-term neurodegenerative disorders of the central nervous system (CNS), such as Alzheimer's disease and Parkinson's disease, as well as for the quicker management of conditions like migraine and cramps [63]. The nasal route offers several advantages compared to the parenteral route, such as the potential for self-administration and the avoidance of first-pass metabolism. These benefits contribute to enhanced patient convenience. The way of the nose is not without its difficulties, though. The volume of a given formulation is frequently constrained due to the high nasal secretion rate, which can also reduce the amount of time that medications can stay in the nasal cavity. Additionally, prolonged targeted therapy may harm the nasal cavity. Additionally, when delivered nasally, only highly lipophilic compounds had the best absorption [64]. Researchers are working on methods to enhance nasal adsorption, prolong the residence duration of medications in the nose, and change the physicochemical characteristics of pharmaceuticals in order to solve these problems. One potential strategy involves the use of chitosan, a substance containing amino groups that possess a positive charge. These positively charged amino groups can engage in interactions with their negatively charged counterparts found in the mucosal layers, including sialic acid. Chitosan may be able to hold a formulation in the nasal cavity for extended periods of time as a result of this interaction's potential to influence the epithelial membrane's permeability and facilitate the passage across the membrane, hydrophilic compounds [65]. Additionally, chitosan's mucoadhesion and modulatory influence on mucociliary clearance has sparked the creation of a number of intriguing nasal administration platforms.

Ibuprofen (IBU) was delivered from the nose to the brain using a thermosensitive hydrogel made of chitosan and -glycerophosphate as a gelling agent. It has been demonstrated that the non-steroidal anti-inflammatory medicine IBU delays the onset of Alzheimer's disease. However, because of the brain-blood barrier and IBU's poor solubility, there is a limit to the quantity of IBU that may enter the brain, whether either orally or through a vein. The researchers devised a simple intranasal solution that undergoes a transition from a liquid state to a semi-solid state, exhibiting mucoadhesive characteristics when exposed to physiological temperatures, in order to overcome these limitations. The gel undergoes a process of solidification within a time frame of 4–7 minutes when exposed to a temperature range of 30–35 °C. This solidification process occurs at a pace that is 2.9 times quicker than the rate at which the mucociliary clearance of the nasal cavity takes place, which typically requires 20–25 minutes. The phenomenon of shear-thinning, observed at room temperature, facilitates the generation of small droplets by nasal spray dispenser. The utilization of laser diffraction resulted in the generation of an ideal spray characterized by a reduced droplet size (Dv50), enlarged spray area, and enhanced surface area coverage while employing chitosan with a low molecular weight range of 110–150 kDa. The composition boosts ibuprofen's water solubility by 100 without chemical solvents. Reversible tight junction manipulation may accelerate IBU transport into nasal epithelial cells in vitro [65].

Trapani et al. [66] synthesized nanoparticles using a mixture of N-O-carboxymethyl chitosan and dopamine amide in order to target Parkinson's disease. The nanoparticles were synthesized utilizing the nanoprecipitation method and exhibited a zeta potential of 32.4 ± 1.60 mV.

They were spherical, had a bimodal size distribution, and were 289.0 nm in size. The research showed a dopamine encapsulation effectiveness of 89.0%. The particles were shown to be amorphous and thermally stable by solid-state examination utilizing FT-IR, DSC, and thermogravimetric techniques, making them potentially appropriate for use in pharmaceutical formulations. The amide connection between carboxymethyl chitosan and dopamine was validated by X-ray photoelectron spectroscopy analysis, which also showed that there was no free dopamine on the nanoparticle's surface. Additionally, at the greatest dopamine concentration examined (75 M), olfactory ensheathing cells showed very mild cytotoxicity in cell survival experiments. Results using epifluorescence microscopy showed that mucin was essential for nanoparticle absorption, probably as a result of adsorption on the surface, the experiment resulted in the formation of a rigid corona architecture that exclusively binds to certain cell surface mucin receptors (Fig. 10.2).

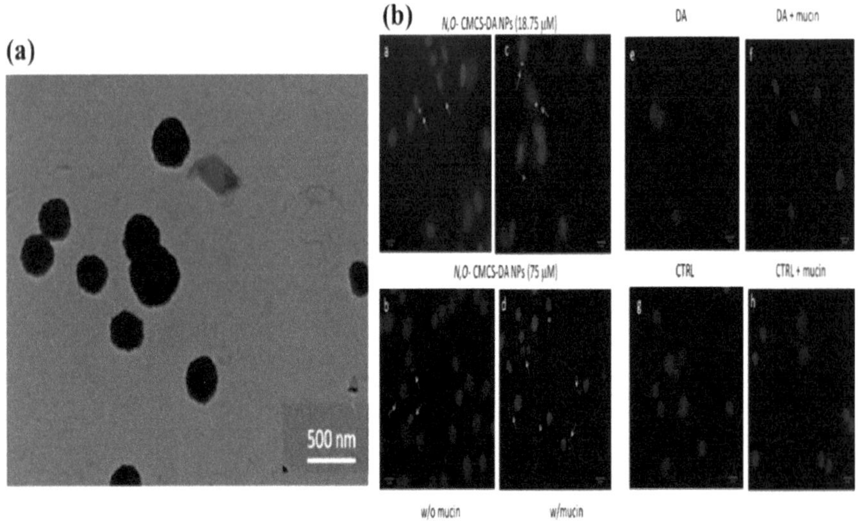

Figure 10.2 Epifluorescence microscopy (a) and cryo-TEM images (b) of nanoparticles conjugated with N-O-carboxymethyl chitosan-dopamine amide (Copyright MDPI 2019. Reproduced with permission from [66]).

Chitosomes, luteolin-loaded nanoparticles coated in chitosan, were created by Abbas et al. [67] for Alzheimer's disease-related cognitive impairment. 7.5:1.5:6.7 phosphatidylcholine, phosphatidylserine, and cholesterol were dissolved in 1.5 mL of 100% ethanol. The phospholipids were dissolved in ethanol and added dropwise to 5 mL of 0.1 M acetate buffer (pH 4.4) magnetically agitated at 800 rpm at 25 °C for 1 h. The empty liposomal solution (F1) was stabilized overnight at 4 °C. Before adding phospholipids to the aqueous medium, luteolin concentrations (40 mg%, F2; 20 mg% w/v, F1) were dissolved in the ethanolic solution. Chitosan coatings were created by titration (F4–F6). The entrapment efficiencies for luteolin loadings of 20 and 40 mg% were found to be 80.6% ± 1.28% and 77.9% ± 1.33%, respectively. The best chitosomes were 412.8 nm particle size, 0.378 PDI, and 37.3 zeta potential after coating with 4 mg% chitosan solution. Chitosomes released approximately 90% of imprisoned medication in 24 hours at pH 6.4, which mimics the nasal environment. The mucin solution's sialic groups and the chitosomes' chitosan covering formed a gel-like structure that maintained the nanoparticles in the nasal cavity longer. After chitosome therapy, mice had better short- and long-term spatial memory. The biochemical investigation demonstrated better antioxidant effects and reduced pro-inflammatory mediator levels, while histological analysis showed enhanced neuronal survival and fewer amyloid plaques. Chitosomes reduced Alzheimer's disease deterioration by 50% by decreasing hyperphosphorylated tau protein and aggregation (Fig. 10.3).

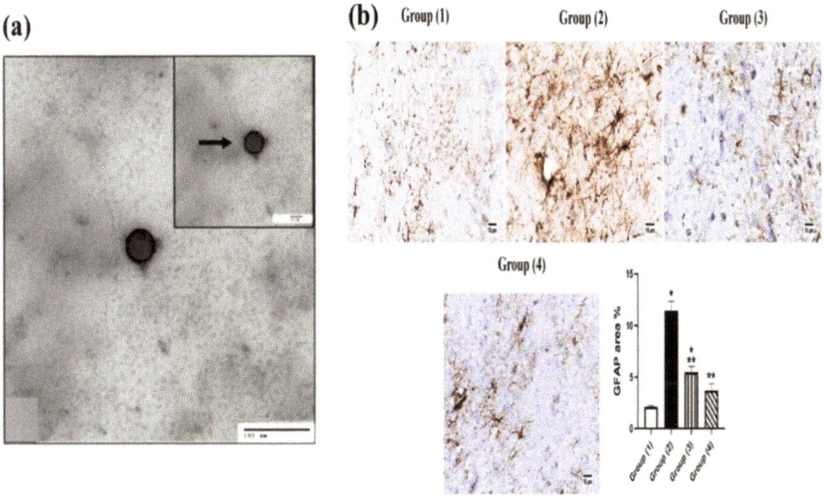

Figure 10.3 TEM picture of luteolin-loaded chitosomes at a magnification of 150,000. The layer of chitosan coating is shown by arrows. Photomicrograph immunohistochemistry of the expression of GFAP in brain tissue (a). GFAP is clearly expressed in Group 2 (the disease control), but there is no morphological difference between the two treated groups (b). (Copyright MDPI 2019. Reproduced with permission from [67]).

10.9 Vaginal Drug Delivery System (VDDS)

The vagina has been recognized as a possible channel for medication delivery in addition to its function as a genital organ involved in reproduction. Numerous clinical diseases, including bacterial, protozoal, fungal, and viral infections that might affect the vaginal lumen, can be treated by administering antimicrobial/ antiviral medications locally only to the cervicovaginal area [68].

In comparison to oral administration, this mode of administration provides a number of benefits, including the avoidance of first-pass metabolism, a decrease in gastrointestinal adverse effects, and convenience of use. Any vaginal medication delivery system must be developed with consideration for the specific vaginal milieu, including the vagina's pH range (usually between 5 and 4), bacteria, and cyclical fluctuations [69]. Although topical vulvovaginal treatments with commercial goods have a low patient acceptance due to their pain and short drug retention times, vaginal administration of therapeutic agents for local infections has shown equal effectiveness to administration by mouth [70]. To maximize the acceptability and therapeutic effectiveness of such medicines, mucoadhesive systems must be developed. Chitosan exhibits antibacterial properties and possesses mucoadhesive and mucopenetrating characteristics, rendering it a viable choice for the advancement of vaginal medication delivery systems. Its usage in such systems may boost patient compliance and extend the duration that medications are retained [71]. Additionally, chitosan-based formulations have been demonstrated to be secure, effective, and well-tolerated, underscoring its promise as a reliable method of vaginal medication administration.

In their study, Campos et al. [72] developed a gynecological gel utilizing a chitosan matrix and including a methanolic extract derived from Mitracarpus frigidus (MFM) for the purpose of addressing vulvovaginal candidiasis (VVC), a condition characterized by the proliferation of Candida albicans, an opportunistic pathogen. Local VVC antifungal therapy typically induces fungal drug resistance and hypersensitivity. MFM, an antioxidant and antifungal, was employed to build a novel medication to circumvent these constraints. Scientists tested MFM-chitosan-gel in an experimental VVC model. As the extract concentration grew, the gel became pseudoplastic and stretchable. MFM chitosan-gel influenced mouse *C. albicans*-VVC model concentration-dependently. 2.5% and 5% MFM chitosan-gel reduced vaginal fungal load by 80.69% and 89.43%, respectively, whereas 10.0% MFM completely removed it on the final day. The tissue organization of the vaginal mucosa in infected animals that were treated with MFM chitosan-gel showed similarities to non-infected animals, as seen in Fig. 10.4. Additionally, there was a significant decrease in the presence of inflammatory cells, which were predominantly observed in the deep lamina propria layers. MFM chitosan-gel-treated mice had less apparent vulvovaginal mucosa epithelial cell death and vacuolization.

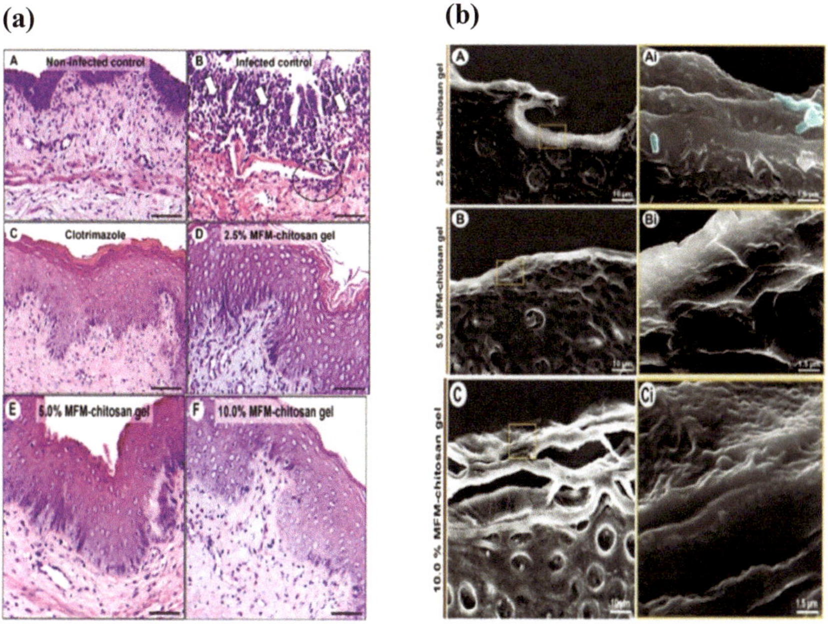

Figure 10.4 Sections of the vulvovaginal histology from the experimental animal groups. Female rats were either given clotrimazole or MFM-chitosan gel therapy, or neither, and were either intravaginally infected with *C. albicans* or not (a); after topical vaginal therapy with chitosan gel with MFM in different concentrations: 2.5%, 5.0% and 10.0%, SEM images were taken of the vulvovaginal epithelium in rat models of vulvovaginal candidiasis (b). Reproduced from [72], with permission; 2020 Elsevier copyright.

A chitosan/anion polyelectrolyte complex (PEC) with significant mucoadhesion and regulated release was created by Darwesh et al. for how fluconazole vaginal implants are made. Utilizing mannitol, the vaginal inserts were made via lyophilization. In employing a factorial design, the objective is to examine the influence of the anion type on a certain phenomenon, and the chitosan/anion ratio on the insert's ability to release drugs and adhere to mucous membranes was investigated. Compritol® 888, a release retardant, was added to the optimized insert to modify its release profile, which was based on a 5:5 ratio of chitosan to sodium alginate. With regard to friability percentages, drug content, and swelling indices at six hours, fluconazole inserts displayed satisfactory results. Statistical analysis revealed that the elements under study had a substantial impact on the detachment force and release characteristics. According to histological analysis, the platform demonstrated improved antifungal effectiveness against Candida albicans both in vivo and in vitro with fewer inflammatory cells (Fig. 10.5) [73].

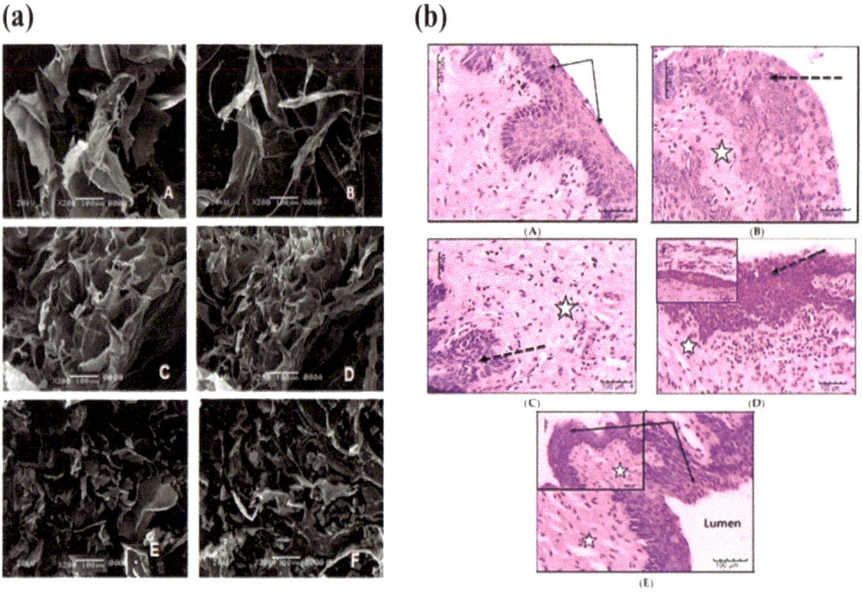

Figure 10.5 Imaging using scanning electron microscopy of (A) 5:5 CH/Alg PEC-based insert, unloaded (B) 5:5 CH/Alg PEC Fluconazole (C) 5:5 CH/XG PEC-based insert, unloaded (D) 5:5 CH/XG PEC (E) Fluconazole 5:5 CH/Carb PEC-based unloaded insert (F) 5:5 CH/PEC/fluconazole for carp (200) (a). Candida-infected vaginal tissue treated with a fluconazole PEC-based vaginal insert and fluconazole solution was examined histologically (b). Reproduced from [73], with permission; © MDPI 2018.

10.10 Conclusion and Outlook

In conclusion, chitin and chitosan are promising materials in the domains of drug delivery due to their varied characteristics and uses. Through a variety of methods, chitosan may be extracted from chitin, and the resultant chitosan can be altered to improve its bioactivities and characteristics. Researchers have been able to modify chitosan to make it more suitable for particular purposes, leading to the creation of brand-new chitin/chitosan-based products. For oral, ocular, nasal, transdermal, and vaginal modes of administration, chitin/chitosan-based drug delivery systems have been created, offering targeted and prolonged drug release. Chitosan has also been utilized as a material for wound dressings because of its antibacterial and hemostatic qualities, which speed up the growth of new tissues and prevent infections. The use of chitosan as a scaffold material in tissue engineering has been driven by its biocompatibility and its capacity to facilitate cell growth and differentiation. As a result, the review gives information on the biological uses of chitosan, like bone regeneration, periodontal tissue regeneration, heart muscle regeneration, cartilage tissue regeneration, wound healing, and corneal regeneration. The flexibility of chitosan is demonstrated by its usage in cosmeceuticals, bioimaging, and gene delivery. These applications have demonstrated tremendous promise. Their employment in diverse applications holds enormous promise for the future, and their potential in biological research and medication delivery is immense. Despite the substantial advancements made in chitin/chitosan-based materials, issues like repeatability and stability still need to be resolved. To fully explore the potential of chitin/chitosan-based materials in diverse biomedical applications and to analyze the potential of chitosan in conjunction with other materials, more research is necessary. Overall, chitosan's adaptability, in addition to its biodegradability, biocompatibility, and special features, makes it a promising substance for the creation of novel biomedical applications for medication delivery that might fundamentally alter the medical sector.

Acknowledgment

I express appreciation and thanks to Prof. Mohammed Gaber Taha and Prof. Mohammed Mabrouk Aldanasory, Biochemistry Department, Faculty of Agriculture, Al Azhar University, Egypt, who supported and encouraged me throughout this work.

References

1. Cirillo, G., Hampel, S., Spizzirri, U. G., Parisi, O. I., Picci, N., and Iemma, F. (2014). Carbon nanotubes hybrid hydrogels in drug delivery: A perspective review. *BioMed Research International*, 2014(1), 825017.
2. Coelho, J. F., Ferreira, P. C., Alves, P., Cordeiro, R., Fonseca, A. C., Góis, J. R., and Gil, M. H. (2010). Drug delivery systems: Advanced technologies potentially applicable in personalized treatments. *EPMA Journal*, 1, 164-209.
3. Liechty, W. B., Kryscio, D. R., Slaughter, B. V., and Peppas, N. A. (2010). Polymers for drug delivery systems. *Annual Review of Chemical and Biomolecular Engineering*, 1, 149–173.
4. Riva, R., Ragelle, H., des Rieux, A., Duhem, N., Jérôme, C., and Préat, V. (2011). Chitosan and chitosan derivatives in drug delivery and tissue engineering. *Chitosan for Biomaterials II*, 19–44.
5. Khan, S., Sharaf, M., Ahmed, I., Khan, T. U., Shabana, S., Arif, M., Kazmi, S., and Liu, C. (2022). Potential utility of nano-based treatment approaches to address the risk of Helicobacter pylori. *Expert Review of Anti-Infective Therapy*, 20(3), 407–424.
6. Ahmed, T. A., and Aljaeid, B. M. (2016). Preparation, characterization, and potential application of chitosan, chitosan derivatives, and chitosan metal nanoparticles in pharmaceutical drug delivery. *Drug Design, Development and Therapy*, 483–507.
7. Amor, I. B., Hemmami, H., Laouini, S. E., Abdelaziz, A. G., and Barhoum, A. (2023) Influence of chitosan source and degree of deacetylation on antibacterial activity and adsorption of AZO dye from water. *Biomass Conversion and Biorefinery*, 1–11.
8. Cheung, R. C. F., Ng, T. B., Wong, J. H., and Chan, W. Y. (2015). Chitosan: An update on potential biomedical and pharmaceutical applications. *Marine Drugs*, 13(8), 5156–5186.

9. Homayun, B., Lin, X., and Choi, H. J. (2019). Challenges and recent progress in oral drug delivery systems for biopharmaceuticals. *Pharmaceutics*, 11(3), 129.
10. Kantak, M. N., and Bharate, S. S. (2022). Analysis of clinical trials on biomaterial and therapeutic applications of chitosan: A review. *Carbohydrate Polymers*, 278, 118999.
11. Kean, T., and Thanou, M. (2010). Biodegradation, biodistribution and toxicity of chitosan. *Advanced Drug Delivery Reviews*, 62(1), 3–11.
12. Singh, D. J., Lohade, A. A., Parmar, J. J., Hegde, D. D., Soni, P., Samad, A., and Menon, M. D. (2012). Development of chitosan-based dry powder inhalation system of cisplatin for lung cancer. *Indian Journal of Pharmaceutical Sciences*, 74(6), 521.
13. Michalik, R., and Wandzik, I. (2020). A mini-review on chitosan-based hydrogels with potential for sustainable agricultural applications. *Polymers*, 12(10), 2425.
14. Al-Jbour, N. D., Beg, M. D., Gimbun, J., and Alam, A. M. (2019). An overview of chitosan nanofibers and their applications in the drug delivery process. *Current Drug Delivery*, 16(4), 272–294.
15. Saranya, T. S., Rajan, V. K., Biswas, R., Jayakumar, R., and Sathianarayanan, S. (2018). Synthesis, characterisation and biomedical applications of curcumin conjugated chitosan microspheres. *International Journal of Biological Macromolecules*, 110, 227–233.
16. Garg, U., Chauhan, S., Nagaich, U., and Jain, N. (2019). Current advances in chitosan nanoparticles based drug delivery and targeting. *Advanced Pharmaceutical Bulletin*, 9(2), 195.
17. Desai, N., Rana, D., Salave, S., Gupta, R., Patel, P., Karunakaran, B., Kommineni, N., et al. (2023). Chitosan: A potential biopolymer in drug delivery and biomedical applications. *Pharmaceutics*, 15(4), 1313.
18. Arif, M., Sharaf, M., Samreen, K. S., Chi, Z., and Liu, C.-G., (2021) Chitosan-based nanoparticles as delivery-carrier for promising antimicrobial glycolipid biosurfactant to improve the eradication rate of Helicobacter pylori biofilm. *Journal of Biomaterials Science, Polymer Edition*, 32, 813–832.
19. Arif, M., Sharaf, M., Samreen, Dong, Q., Wang, L., Chi, Z., and Liu, C. G. (2021). Bacteria-targeting chitosan/carbon dots nanocomposite with membrane disruptive properties improve eradication rate of Helicobacter pylori. *Journal of Biomaterials Science Polymer Edition*, 32(18), 2423–2447.

20. Ren, G., Clancy, C., Tamer, T. M., Schaller, B., Walker, G. M., and Collins, M. N. (2019). Cinnamyl O-amine functionalized chitosan as a new excipient in direct compressed tablets with improved drug delivery. *International Journal of Biological Macromolecules*, 141, 936–946.

21. Yasufuku, T., Anraku, M., Kondo, Y., Hata, T., Hirose, J., Kobayashi, N., and Tomida, H. (2010). Useful extend-release chitosan tablets with high antioxidant activity. *Pharmaceutics*, 2(2), 245–257.

22. Ryu, J. H., Choi, J. S., Park, E., Eom, M. R., Jo, S., Lee, M. S., Kwon, S. K., and Lee, H. (2020) Chitosan oral patches inspired by mussel adhesion. *Journal of Controlled Release*, 317, 57–66.

23. Shukla, R. K., and Tiwari, A. (2012). Carbohydrate polymers: Applications and recent advances in delivering drugs to the colon. *Carbohydrate Polymers*, 88(2), 399–416.

24. Kean, T., and Thanou, M. (2011) *Chitin and Chitosan: Sources, Production and Medical Applications*, 292–318.

25. Xing, P., Shi, Y., Dong, C., Liu, H., Cheng, Y., Sun, J., Li, D., Li, M., Sun, K., and Feng, D. (2017). Colon-targeted delivery of IgY against Clostridium difficile toxin A and B by encapsulation in chitosan-Ca pectinate microbeads. *AAPS PharmSciTech*, 18, 1095–1103.

26. Lanjhiyana, S. K., Bajpayee, P., Kesavan, K., Lanjhiyana, S., and Muthu, M. S. (2013). Chitosan–sodium alginate blended polyelectrolyte complexes as potential multiparticulate carrier system: Colon-targeted delivery and gamma scintigraphic imaging. *Expert Opinion on Drug Delivery*, 10(1), 5–15.

27. Wakure, B. S., Yadav, A. V., Bhatia, N. M., and Salunke, M. A. (2013). Development of chitosan capsule for colon specific delivery of budesonide. *International Journal of Pharmaceutical Sciences and Research*, 4(8), 3239.

28. Ribeiro, L. N., Alcântara, A. C., Darder, M., Aranda, P., Araújo-Moreira, F. M., and Ruiz-Hitzky, E. (2014). Pectin-coated chitosan–LDH bionanocomposite beads as potential systems for colon-targeted drug delivery. *International Journal of Pharmaceutics*, 463(1), 1–9.

29. Treenate, P., and Monvisade, P. (2017). In vitro drug release profiles of pH-sensitive hydroxyethylacryl chitosan/sodium alginate hydrogels using paracetamol as a soluble model drug. *International Journal of Biological Macromolecules*, 99, 71–78.

30. Anitha, A., Sreeranganathan, M., Chennazhi, K. P., Lakshmanan, V. K., and Jayakumar, R. (2014). In vitro combinatorial anticancer effects of 5-fluorouracil and curcumin loaded N,O-carboxymethyl chitosan

nanoparticles toward colon cancer and in vivo pharmacokinetic studies. *European Journal of Pharmaceutics and Biopharmaceutics*, 88(1), 238–251.

31. Jyoti, K., Bhatia, R. K., Martis, E. A., Coutinho, E. C., Jain, U. K., Chandra, R., and Madan, J. (2016). Soluble curcumin amalgamated chitosan microspheres augmented drug delivery and cytotoxicity in colon cancer cells: In vitro and in vivo study. *Colloids and Surfaces B: Biointerfaces*, 148, 674–683.
32. Gote, V., Sikder, S., Sicotte, J., and Pal, D. (2019). Ocular drug delivery: Present innovations and future challenges. *Journal of Pharmacology and Experimental Therapeutics*, 370(3), 602–624.
33. Vasdev, N., Chaudhari, N., Polaka, S., Rajpoot, K., Gondaliya, P., Sayyed, A. A., Sengupta, P., and Tekade, R. K. (2023). Current progress in preservative-free topical ophthalmic formulations. *Journal of Drug Delivery Science and Technology*, 79, 103996.
34. Soni, V., Pandey, V., Tiwari, R., Asati, S., and Tekade, R. K. (2019). Design and evaluation of ophthalmic delivery formulations. In *Basic Fundamentals of Drug Delivery*, . Academic Press, pp. 473–538.
35. Zamboulis, A., Nanaki, S., Michailidou, G., Koumentakou, I., Lazaridou, M., Ainali, N. M., Xanthopoulou, E., and Bikiaris, D. N. (2020) Chitosan and its derivatives for ocular delivery formulations: Recent advances and developments. *Polymers*, 12(7), 1519.
36. Algharib, S. A., Dawood, A., Zhou, K., Chen, D., Li, C., Meng, K., Zhang, A., Luo, W., Ahmed, S., Huang, L., and Xie, S. (2022). Preparation of chitosan nanoparticles by ionotropic gelation technique: Effects of formulation parameters and in vitro characterization. *Journal of Molecular Structure*, 1252, 132129.
37. Eid, H. M., Naguib, I. A., Alsantali, R. I., Alsalahat, I., and Hegazy, A. M. (2021). Novel chitosan-coated niosomal formulation for improved management of bacterial conjunctivitis: A highly permeable and efficient ocular nanocarrier for azithromycin. *Journal of Pharmaceutical Sciences*, 110(8), 3027–3036.
38. Fang, G., Yang, X., Wang, Q., Zhang, A., and Tang, B. (2021). Hydrogels-based ophthalmic drug delivery systems for treatment of ocular diseases. *Materials Science and Engineering: C*, 127, 112212.
39. Shi, H., Wang, Y., Bao, Z., Lin, D., Liu, H., Yu, A., Lei, L., Li, X., and Xu, X. (2019). Thermosensitive glycol chitosan-based hydrogel as a topical ocular drug delivery system for enhanced ocular bioavailability. *International Journal of Pharmaceutics*, 570, 118688.

40. Li, B., Wang, J., Gui, Q., and Yang, H. (2020). Drug-loaded chitosan film prepared via facile solution casting and air-drying of plain water-based chitosan solution for ocular drug delivery. *Bioactive Materials*, 5(3), 577–583.

41. Xu, J., Strandman, S., Zhu, J. X., Barralet, J., and Cerruti, M. (2015). Genipin-crosslinked catechol-chitosan mucoadhesive hydrogels for buccal drug delivery. *Biomaterials*, 37, 395–404.

42. Lancina III, M. G., Shankar, R. K., and Yang, H. (2017). Chitosan nanofibers for transbuccal insulin delivery. *Journal of Biomedical Materials Research Part A*, 105(5), 1252–1259.

43. Cossu, A., Wang, M. S., Chaudhari, A., and Nitin, N. (2015). Antifungal activity against Candida albicans of starch Pickering emulsion with thymol or amphotericin B in suspension and calcium alginate films. *International Journal of Pharmaceutics*, 493(1–2), 233–242.

44. Aguilar-Pérez, K., Avilés-Castrillo, J., Medina, D. I., Parra-Saldivar, R., and Iqbal, H. M. (2020) Insight into nanoliposomes as smart nanocarriers for greening the twenty-first century biomedical settings. *Frontiers in Bioengineering and Biotechnology*, 8, 579536.

45. Cid, Y. P., Pedrazzi, V., de Sousa, V. P., and Pierre, M. B. R. (2012). In vitro characterization of chitosan gels for buccal delivery of celecoxib: Influence of a penetration enhancer. *AAPS PharmSciTech*, 13, 101–111.

46. Mortazavian, E., Dorkoosh, F. A., and Rafiee-Tehrani, M. (2014). Design, characterization and ex vivo evaluation of chitosan film integrating of insulin nanoparticles composed of thiolated chitosan derivative for buccal delivery of insulin. *Drug Development and Industrial Pharmacy*, 40(5), 691–698.

47. Jeong, W. Y., Kwon, M., Choi, H. E., and Kim, K. S. (2021). Recent advances in transdermal drug delivery systems: A review. *Biomaterials Research*, 25(1), 24.

48. Phatale, V., Vaiphei, K. K., Jha, S., Patil, D., Agrawal, M., and Alexander, A. (2022). Overcoming skin barriers through advanced transdermal drug delivery approaches. *Journal of Controlled Release*, 351, 361–380.

49. Jana, S., and Jana, S., (eds.). (2020). *Functional Chitosan: Drug Delivery and Biomedical Applications*. Springer Nature.

50. Ma, J., Wang, Y., and Lu, R. (2022). Mechanism and application of chitosan and its derivatives in promoting permeation in transdermal drug delivery systems: A review. *Pharmaceuticals*, 15(4), 459.

51. Shekh, M. I., Amirian, J., Stadler, F. J., Du, B., and Zhu, Y. (2020). Oxidized chitosan modified electrospun scaffolds for controllable release of acyclovir. *International Journal of Biological Macromolecules*, 151, 787–796.
52. Indulekha, S., Arunkumar, P., Bahadur, D., and Srivastava, R. (2016). Thermoresponsive polymeric gel as an on-demand transdermal drug delivery system for pain management. *Materials Science and Engineering: C*, 62, 113–122.
53. Radwan-Pragłowska, J., Janus, Ł., Piątkowski, M., Sierakowska, A., Szajna, E., Matýsek, D., and Bogdał, D. (2021). Development of stimuli-responsive chitosan/ZnO NPs transdermal systems for controlled cannabidiol delivery. *Polymers*, 13(2), 211.
54. Riezk, A., Van Bocxlaer, K., Yardley, V., Murdan, S., and Croft, S. L. (2020). Activity of amphotericin B-loaded chitosan nanoparticles against experimental cutaneous leishmaniasis. *Molecules*, 25(17), 4002.
55. Fereig, S. A., El-Zaafarany, G. M., Arafa, M. G., and Abdel-Mottaleb, M. M. (2021). Tacrolimus-loaded chitosan nanoparticles for enhanced skin deposition and management of plaque psoriasis. *Carbohydrate Polymers*, 268, 118238.
56. Parhi, R. (2020). Drug delivery applications of chitin and chitosan: A review. *Environmental Chemistry Letters*, 18(3), 577–594.
57. Parhi, R. (2019). Chitin and chitosan in drug delivery. *Sustainable Agriculture Reviews 36: Chitin and Chitosan: Applications in Food, Agriculture, Pharmacy, Medicine and Wastewater Treatment*, pp. 175–239.
58. Jain, K. K. (2008). Drug delivery systems-an overview. *Drug Delivery Systems*, 1–50.
59. Ruan, H., Yu, Y., Liu, Y., Ding, X., Guo, X., and Jiang, Q. (2016). Preparation and characteristics of thermoresponsive gel of minocycline hydrochloride and evaluation of its effect on experimental periodontitis models. *Drug Delivery*, 23(2), 525–531.
60. Sheshala, R., Anuar, N. K., Abu Samah, N. H., and Wong, T. W. (2019). In vitro drug dissolution/permeation testing of nanocarriers for skin application: A comprehensive review. *AAPS PharmSciTech*, 20, 1–28.
61. Sheshala, R., Quah, S. Y., Tan, G. C., Meka, V. S., Jnanendrappa, N., and Sahu, P. S. (2019). Investigation on solution-to-gel characteristic of thermosensitive and mucoadhesive biopolymers for the

development of moxifloxacin-loaded sustained release periodontal in situ gels. *Drug Delivery and Translational Research*, 9, 434–443.

62. Nair, S. C., and Anoop, K. R. (2014). Design and in vitro evaluation of controlled release satranidazole subgingival films for periodontitis therapy. *International Journal of Pharmaceutical Sciences Review and Research*, 24(1), 8–14.
63. Liu, X., Yan, S., Li, M., Zhang, S., Guo, G., and Yin, Q. (2022). *Spray Dried Levodopa-Doped Powder Potentially for Intranasal Delivery*, vol. 14, p. 1384.
64. Keller, L. A., Merkel, O., and Popp, A. (2022). Intranasal drug delivery: Opportunities and toxicologic challenges during drug development. *Drug Delivery and Translational Research*, 1–23.
65. Gholizadeh, H., Cheng, S., Pozzoli, M., Messerotti, E., Traini, D., Young, P., Kourmatzis, A., and Ong, H. X. (2019). Smart thermosensitive chitosan hydrogel for nasal delivery of ibuprofen to treat neurological disorders. *Expert Opinion on Drug Delivery*, 16(4), 453–466.
66. Trapani, A., Cometa, S., De Giglio, E., Corbo, F., Cassano, R., Di Gioia, M. L., Trombino, S., Hossain, M. N., Di Gioia, S., and Trapani, G. J. P. (2022). Novel nanoparticles based on N,O-carboxymethyl chitosan-dopamine amide conjugate for nose-to-brain delivery. *Pharmaceutics*, 14(1), 147.
67. Kiss, T., Ambrus, R., Abdelghafour, M. M., Zeiringer, S., Selmani, A., Roblegg, E., Budai-Szűcs, M., Janovák, L., Lőrinczi, B., Deák, Á., and Katona, G. (2022). Preparation and detailed characterization of the thiomer chitosan–cysteine as a suitable mucoadhesive excipient for nasal powders. *International Journal of Pharmaceutics*, 626, 122188.
68. Osmałek, T., Froelich, A., Jadach, B., Tatarek, A., Gadziński, P., Falana, A., Gralińska, K., Ekert, M., Puri, V., and Wrotyńska-Barczyńska, J. J. P. (2021). Recent advances in polymer-based vaginal drug delivery systems. *Pharmaceutics*, 13(6), 884.
69. Shapiro, R. L., DeLong, K., Zulfiqar, F., Carter, D., Better, M., and Ensign, L. M. (2022). In vitro and ex vivo models for evaluating vaginal drug delivery systems. *Advanced Drug Delivery Reviews*, 191, 114543.
70. Pandey, M., Choudhury, H., Abdul-Aziz, A., Bhattamisra, S. K., Gorain, B., Carine, T., Wee Toong, T., Yi, N. J., and Win Yi, L. J. P. (2020). Promising drug delivery approaches to treat microbial infections in the vagina: A recent update. *Polymers*, 13(1), 26.

71. Araujo, V. H. S., de Souza, M. P. C., Carvalho, G. C., Duarte, J. L., and Chorilli, M. (2021). Chitosan-based systems aimed at local application for vaginal infections. *Carbohydrate Polymers*, 261, 117919.

72. Campos, L. M., de Oliveira Lemos, A. S., da Cruz, L. F., de Freitas Araújo, M. G., de Mello Botti, G. C. R., Júnior, J. L. R., Pinto, P. F., et al. (2020). Development and in vivo evaluation of chitosan-gel containing Mitracarpus frigidus methanolic extract for vulvovaginal candidiasis treatment. *Biomedicine and Pharmacotherapy*, 130, 110609.

73. Darwesh, B., Aldawsari, H. M., and Badr-Eldin, S. M. (2018). Optimized chitosan/anion polyelectrolyte complex based inserts for vaginal delivery of fluconazole: In vitro/in vivo evaluation. *Pharmaceutics*, 10(4), 227.

Chapter 11

Application of Chitosan in Textiles

Deepak Chandran[a*], Sureshkumar Rajalingam[b], Marthandan Viswanathan[b], Pran Mohankumar[b], Deepika Krishnan[b], Ankitha Indu Jisha[b], Abhirami Vinu Nair[b], and Aishwarya Prashanth[b]

[a]*Department of Animal Husbandry, Government of Kerala, India*
[b]*Amrita School of Agricultural Sciences, Amrita Vishwa Vidyapeetham University, Coimbatore, Tamil Nadu, India 64210*

drdeepakchandran24@gmail.com

The use of chitosan-based products at several stages of textile production seems promising. Due to a number of its intrinsic qualities, biopolymer chitosan has been studied for usage in the textile industry. These characteristics include non-toxicity, antibacterial activity, biocompatibility, and biodegradability. Chitosan is used as a replacement for synthetic polymers in a variety of textile industry end uses. The reaction processes and applications of this material are explored.

Chitin and Chitosan: Physical and Chemical Properties
Edited by Shakeel Ahmed, Soumeia Zeghoud, Hadia Hemmami, and Ilham Ben Amor

ISBN 978-981-5129-39-7 (Hardcover), 978-1-003-58977-8 (eBook)
www.jennystanford.com

11.1 Introduction

The carbohydrate chitin, which is a renewable resource, is deacetylated to produce chitosan. In order to extract chitin, also known as polyβ-(1,4)-2-acetoamido2-deoxy-D-glucose, from the shells of crustaceans (such as crab and prawns), calcium carbonate must first be dissolved in an acid solution [1]. Next, proteins must be solubilized in an alkaline solution. Chitosan comes from a variety of sources since chitin is the second most common natural substance on earth after cellulose [1–3]. Due to its unusual mix of qualities, including biodegradability, moisture permeability, biocompatibility, non-toxicity, adsorption, reactivity, and antibacterial activity, chitosan has long been attractive to researchers as a natural material [4, 5]. Its uses in the biological, engineering, pharmaceutical, food, and other industries have all been thoroughly studied. Chitosan has recently been used in textile materials and textiles, and because of its many benefits for clothing, including its antibacterial properties, high absorption rate, and capacity to stop bleeding, this sector has made considerable advancements [6, 7].

Chitosan is a poly-(1-4)-2-amido-2-deoxy-D-glucose linear polyamine. The sole variation between its molecular structure (Fig. 11.1) and those of cellulose and chitin is the amino group ($-NH_2$) in lieu of the hydroxyl group (OH) on the C-2 position [8, 9]. Chitosan may be protonated into a polyelectrolyte in aqueous acidic environments because it contains $-NH_2$ groups. As a result, chitosan exhibits solubility in an aqueous acidic environment, and almost all of its properties rely on the solution's pH level. This is also the reason chitosan has antibacterial qualities and prevents a wide range of yeasts, bacteria, and fungi from growing [10]. The many reactive hydroxyl and amino groups found in chitosan, a pseudonatural cationic polymer, allow it to take part in a variety of reactions, including ionic, graft, and chelation processes [11]. Because of its properties, chitosan may be used to treat textiles in a variety of ways that increase their antibacterial activity, colorfastness, crease recovery, and other properties.

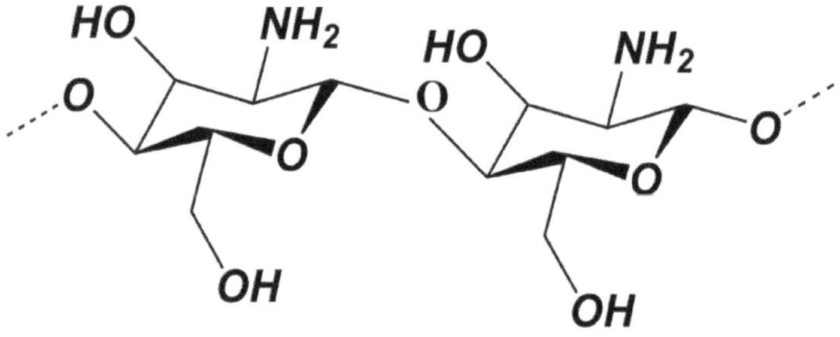

Figure 11.1 Chitosan's molecular structure.

The textile industry already uses chitosan for a wide range of purposes, including the production of chitosan fibers, chitosan nonwoven fabrics, textiles' initial treatment, yarn, dyeing, functional fabric finishing, wastewater treatment, and printing [12]. This chapter attempts to outline chitosan's numerous uses in the textile industry and to highlight its adaptable features in relation to textile finishing.

11.2 Production of Fiber Using Chitosan

11.2.1 Spinning Technique

11.2.1.1 Wet spinning

An acetylated polysaccharide is chitin. In comparison to semi-crystallized chitosan fibers, the addition of acetyl groups improves interchain forces and the percentage of crystallization, which results in greater dry and wet strength. Additionally, chitosan fibers with a higher moisture-regaining capacity perform poorly in terms of building the necessary strength [13].

Wet spinning is the most widely used technique for producing chitin and chitosan fib [14]. This procedure (Fig. 11.2) uses the extrusion process in a coagulation bath to create chitosan fibers from a viscous chitosan solution. When chitosan is dissolved in 1–10% acetic acid at a pH below 6, a viscous dope is produced. A candle system filtering unit removes contaminants, and then the liquid is degassed in a reservoir for five hours at a pressure of roughly 30 mbar to remove any remaining air bubbles. After that, the extrusion procedure begins to produce fibers, with a metering pump, a reservoir, and a spinneret making up the extrusion chamber. Under 1.5 bar of pressure, the chitosan dope is extruded into a coagulation bath. A coagulant solution in water, such as cupric ammonia, KOH, NaOH-AcONa, NaOH-Na_2SO_4, 40% NaOH, methanol, $CuSO_4$-concentrated ammonia, $CuSO_4$-NH_4OH, etc., is present in the coagulation bath [15]. The compounds in the tailors that were previously described serve as coagulation inhibitors. The take-up rollers, drying rollers, drawing, and filtering system are optimized in this process based on the production rate and necessary fiber qualities. Aqueous methanol and ethanol or distilled water are used to wash the extra coagulant from the fiber. Predrying bath is another name for this washing bath. To improve the intended mechanical property in drying procedures, certain extra physical and chemical processes are used [16].

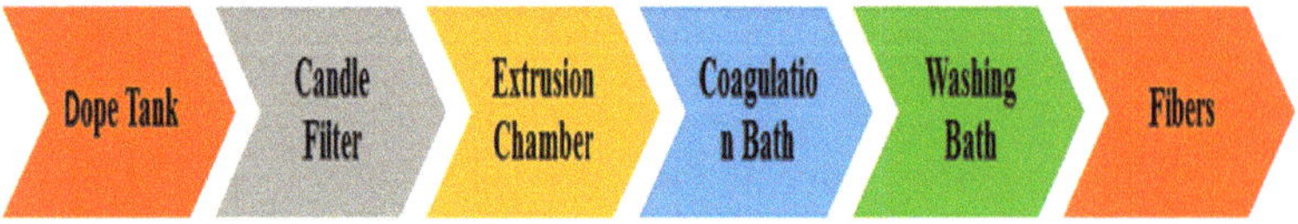

Figure 11.2 Typical wet spinning block diagram.

11.2.1.2 Solvent system

The appropriate fiber shapes and qualities are produced with the aid of a well-chosen solvent solution and coagulation bath. A variety of solvent systems, including trichloroacetic acid, dimethylacetamide-lithium chloride, dichloroacetic acid-isopropyl ether-formic acid, and a combination of trichloroacetic acid-chloral hydrate-dichloromethane (DCM), have been explored by several researchers (Table 11.1). Moreover, the pH-sensitive properties of chitosan are quite advantageous in the fiber processing industry. Chitosan, a straightforward solvent system, may be easily dissolved by acetic acid, especially at pH 6. Moreover, the utilization of a coagulation bath including sodium hydroxide (NaOH) has the advantage of simplicity in production and rapid enhancement of the chitosan fibers.

Table 11.1 Melt chitin and chitosan using a solvent system

Chitin dissolution	Chitosan dissolution
Chloral hydrate (40%), trichloroacetic acid (40%), methylene chloride (20%)	Watered-down organic acid (formic acid, acetic acid)
Lithium thiocyanate	p-Toluene sulfonic acid
$Ca(CNS)_2/LiCNS/CaI_2/CaBr_2$	N,N-dimethylformamide (DMF)/ dimethylsulfoxide
Xanthate [O(sodium thio) thiocarbonyl cellulose] -sodium N-acylchitosan xanthate [O-(sodium thio) thiocarbonyl N-acylchitosan]	1-butyl-3-methylimidazolium bromide ([BMIM][Br]), ionic liquid
Isopropyl ether, dichloroacetic acid, and formic acid	Urea, guanidine hydrochloric acid

11.2.2 Dry-Spinning Technique

The initial concentration of the chitosan used for dry yarn [17] should be more than or equal to 0.5% (w/w), which is the critical concentration for chain tangle. Chitosan is typically dispersed with acetic acid. The extrusion process comes after the procedure outlined above in a similar fashion. During the coagulation process, the chitosan monofilament comes into contact with 125 ml of gaseous ammonia, which has a purity of 20% w/w. The flow rate of gaseous ammonia may be estimated to be around 0.24 m^3/h for a duration of 7.5 seconds based on the observed relationship between the speed of the extruder and the first roller, which is roughly 3.4 m/min. Stretching is achieved by maintaining the speed of the first roller at 14% of its maximum, which leads to the formation of a straight fiber line throughout the coagulation process. In order to mitigate the occurrence of un-stretching, it is important to maintain a consistent speed. To keep the stretch ratio 1:12 for the manufactured monofilament, a secondary roller with increased speed is utilized. The oven, about 65 cm in size, is subjected to a consistent temperature of around 110 °C for the entirety of the drying process (Fig. 11.3). A flow rate of 1.5 cubic meters per hour is utilized for the air, and the monofilament is subjected to the oven for approximately 10 seconds. The chitosan fiber is thereafter subjected to a one-week treatment in a humid environment with a circulating temperature of 25 °C and a water concentration of 55% w/w [18].

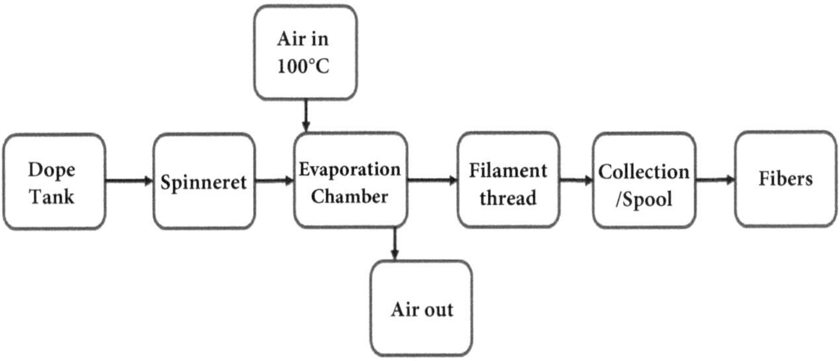

Figure 11.3 A common dry spinning block diagram.

11.2.3 Electrospun

Antonin Formhals submitted a patent application in 1934 for the production of textile yarn. Electrostatic spinning, often known as electrospinning, is the manufacturing process. The primary concept behind this technique is to create nanofibers by dissolving a polymer solution and decreasing surface tension using an electric potential [19]. In this technique, the requisite electric potential is provided by a high-voltage external source. The chitosan solution is prepared in the first phase of the procedure (Fig. 11.4) by being placed in a pipette that has been charged by a source-side positive electrode. The pendant polymer droplet located at the tip of the pipette undergoes a process in which it divides into a charged jet due to the emergence of a repulsive force. The phenomenon of radial-charge repulsion, wherein electric forces propel charged particles toward the collector and then neutralize them, leads to the fragmentation of a singular jet into several jets. When the mixture exits the pipette tip, the multifilament fluid jet dries and the solvent evaporates. In the collector, the solidified polymers are placed [20].

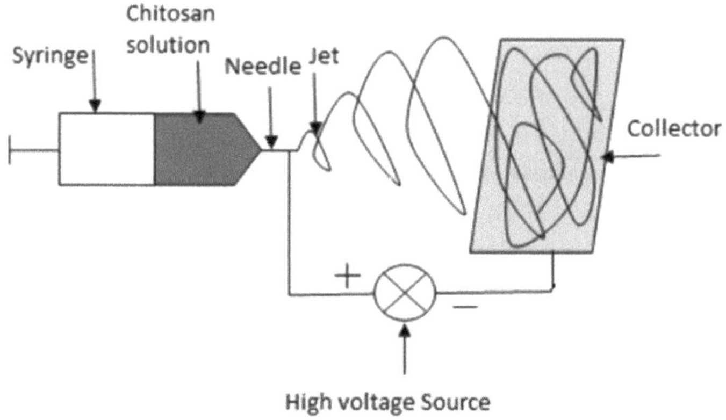

Figure 11.4 A typical electrospinning process diagram.

The key factor affecting the spinnability of the chitosan in this procedure is the solvent. The rheology of chitosan fibers is influenced by a variety of chemical and physical characteristics of chitosan, including conductivity and solution viscosity. Numerous experiments have been done to determine the best solvents for dissolving chitosan, including diluted hydrochloric acid, dichloroacetic acid, acetic acid, trifluoroacetic acid (TFA), formic acid, etc. The correct choice of the solvent would be aided by knowledge of the interactions between the solvent and the chitosan, as well as the impact of the solvent's ionic strength, viscosity, and pH [21].

Chitosan's greater viscosity at increasing concentrations is the key issue. For instance, chitosan becomes excessively viscous over 2% (w/w) and it can no longer flow. Contrarily, chitosan at low concentrations does not contain enough material to provide stable and solid fibers. Every chitosan molecule comprises OH and NH_2 groups, which interact quickly at a high viscosity level. As a result, at increasing concentrations, the chitosan chains interact more with one another, forming a three-dimensional network and very sticky gel.

Trifluoroacetic acid (TFA) is a suggested solvent that has been shown to completely dissolve 8% of pure chitosan. TFA and dichloromethane are combined to produce fibers with a diameter of 200–650 nm [22].

11.3 Fibers' Mechanical Characteristics

Nonwoven materials may be made with nanofibers thanks to electrospinning. To verify a method, it is crucial to consider the diameter of the fibers, homogeneous deposition at the nanoscale, the thickness of the fibers, and bead forming. Tensile power and tenacity are essential factors to consider when evaluating the quality of chitosan fibers in relation to these morphological and structural characteristics. At a final dry plane expansion rate of 3.7, the chitosan fiber was produced with an 82 g/denier elasticity modulus and 2 g/denier tenacity [23].

The polymorphism of chitin also affects the fibers; α-chitin is extremely hard, whereas γ- and β-chitin exhibit better qualities in toughness and elasticity [24]. The chitosan fibers, which are produced after spinning and exist in a wet state, have very low tensile strength. For example, glutaraldehyde in post-treatment, phthalate and phosphate ions in the solution, epichlorohydrin in the coagulation bath, and other methods are used to increase mechanical strength. Chitosan nanofibers are made possible by the physical, chemical, and mechanical properties of chitosan fibers as well as their capacity to be constructed in a variety of shapes. Examples include the removal of metals and colors from water [25], the capture of viruses in the air, and other filtering processes. Additionally, the technical advancements in textile processing related to chitin and chitosan would affect vascular grafts [26], tissue engineering [27], transporting and delivering medications [28], dressing for wound skin [29], and reinforcing microfiber composite [30].

11.4 Examples of Textile Applications of Chitosan

11.4.1 Yarn and Fibers

Chitosan is a tractable substance that may be used to create fibers. Wet spinning and electrospinning are the main methods used to manufacture chitosan solutions for use in fiber preparation.

After the chitosan has been dissolved in an acidic water solution, deaeration and filtering are usually followed by spinning. The chitosan fibers are refined, dried, and subjected to appropriate post-treatment after spinning [31]. In the textile business, chitosan fibers are typically used in conjunction with secondary fibers due to their drawbacks, which include strong electrostatic, high cost, and poor mechanical qualities. Preserving the scientifically proven qualities of chitosan in textiles can increase the economic value of the chitosan fibers and get the greatest performance from chitosan fibers [31]. On the other hand, 100% chitosan has drawbacks as well, including the inability to be washed, poor elasticity and softness, and the inability to be stabilized in place due to its thinness [5].

Because cellulose and chitosan have somewhat similar chemical structures, it is simple to combine the two to create fibers. Over time, these mixed fibers maintain both the physical and excellent characteristics of cellulosic fiber and chitosan. Chitosan and cellulose share a similar chemical structure, which makes these fibers extremely dyeable with both reactive and direct dyes and soft to the touch.

Several research has been conducted to investigate the impact of various elements, such as chitosan fiber properties, blending ratios, blending procedures, and other variables, on the performance of yarn samples. The objective of these studies is to address the limitations of chitosan fibers and improve the functionality of mixed yarns [4].

The fiber-blending approach gives more uniformity of fiber distribution, while the sliver-blending technique provides superior tensile qualities of yarn samples. According to Lam et al. [31], chitosan concentrations of 50% or more in the blend ratio diminish the stiffness of the fabric and provide the inner surface of the chitosan/cotton blend fabric with softer handling. This aided in the creation of chitosan fibers for use in medical textiles [4]. The air-jet spinning technique is used to create a yarn blend of ramie and chitosan that has particular antibacterial qualities and a good yarn structure. Chitosan is combined with other fibers in addition to cellulosic fibers to create high-quality yarns.

11.4.2 Nonwoven Textiles

The cost of chitosan fibers is much higher than that of other fibers [5, 31]. Chitosan fibers still have several significant limitations, such as excessive electrostatic and poor mechanical qualities, which prevent chitosan fiber use. This is true even though many researchers have tried to enhance the performance of chitosan fibers using a variety of ways [5, 31]. Chitosan utilization in the textile sector can be facilitated by nonwoven technology, cutting production costs and facilitating prototype manufacture [31]. Nonwoven textiles often have a highly branching, asymmetrical structure and uncontrollably brittle fibers. Consequently, it is necessary to employ both a wet-laid and a plasticizer mat over the course of the method. However, the process of spinning pure chitosan is hindered due to the limited solubility of chitosan in aqueous acids and the presence of substantial hydrogen bonding inside and between chains. Furthermore, the chitosan material with a purity of 100% exhibits inadequate thickness for effective stabilization in wound healing applications. It also has limited flexibility, low softness, and is not washable [32]. Because of its antimicrobial properties and resistance to assault from urine, juice, and bile from the pancreas as absorbable sutures, chitosan's textile applications are beneficial for wound healing [5]. As a result, practically all nonwoven chitosan textiles also require support from supporting components [33, 34].

11.4.3 Printing and Dyeing

Chitosan has a high number of amino groups, which provide anionic dyes (such as acid, direct, and reactive dyes) additional dye sites to bind to through electrostatic attraction and van der Waals forces when applied to textiles [32]. Fabrics made of cotton and wool that have been pretreated with chitosan have an excellent affinity for anionic dyes, high dye absorption, and strong color. Because of van der Waals forces, chitosan exhibits high adsorbability to cotton fibers because of structural similarities between chitosan and cellulose. Additionally, chitosan can bind to cellulose via a Schiff base reaction involving the amino group of chitosan and the reducing end (-CHO) of cellulose. Hydrogen bonding is also crucial for the incorporation of chitosan into cellulose [32].

Therefore, when used to functionally finish cotton textiles, chitosan has high fastness. Because chitosan may cover the neps generated in developing cotton fibers, cotton textiles coated with chitosan exhibit consistent depths of shade after dyeing [35]. Chitosan's application in fabric pretreatment has drawbacks, including stiffness, poor colorfastness, and ring dyeing.

This simultaneously resolves ring dyeing problems and improves the fastness and stiffness of materials. According to Srisuk's research, chitosan with a low molecular weight (Mw) might lessen these coloring flaws [36]. Chitosan can be used in printing to increase dye absorption and provide fabrics with a similar level of colorfastness. Chitosan can be used as a thickening and binder in textile printing because it has the ability to bind to water and is reactive owing to the presence of $-NH_2$ and -OH groups [37]. Chitosan improves the hand of printed fabrics after being combined with other suitable auxiliaries. Chitosan and gelatinized starch were combined by Abdou et al. for screen printing textiles [38].

The color fastness and color strength of printed textiles are both improved. This is because printing pastes' pseudoplastic characteristic is caused by the homogenous molecular structure of starch and chitosan. The edible coating exhibits enhanced film-forming and mechanical capabilities due to the interaction between the starch's hydroxyl groups and the chitosan's amino groups [38]. Chitosan has the potential to be utilized in digital inkjet printing as an alternative to sodium alginate, the conventional thickener, for fabric pretreatment. This substitution aims to enhance the color fastness and precision of printed materials, particularly by enhancing the vividness of colors and the clarity of outlines [39].

11.4.4 Functional Finishing

11.4.4.1 Antimicrobial finishing

The presence of amino groups causes chitosan to have a positive charge that is beneficial for cell development and adhesion.

Since bacteria are typically thought of as being negatively charged, coming into touch with chitosan can have a positive inhibitory impact [40].

The cationic amino groups present in chitosan interact with the anionic groups found in bacteria, thereby interfering with the bacteria's physiological processes and impeding their growth. Additionally, the chitosan located on the surface of the bacterial cell forms a polymer membrane, which hinders the passage of nutrients into the cell. Furthermore, the chitosan with a lower Mw is capable of penetrating the bacterial cell. These are the three main antimicrobial mechanisms of chitosan [5, 41]. The amount of chitosan that has been deacetylated does not directly correlate with the antibacterial activity of cloth treated with chitosan. But, one drawback of using chitosan in textiles is that it negatively affects the hand of the fabric, especially when high Mw chitosan is utilized.

Chitosan's Mw and antibacterial activity have been investigated in connection with one another [42]. Research demonstrates that Mw of chitosan and the tested bacterial strains are both related to the antibacterial activity of chitosan. As the Mw of chitosan grew, its antibacterial action against Staphylococcus aureus became stronger. In contrast, when the Mw of chitosan increases, the antibacterial efficacy against *Escherichia coli* is reduced. When using modest treatment doses, the Mw impact was more pronounced. Therefore, it's crucial to strike a balance between a fabric's hand quality and antibacterial activity by selecting the right chitosan Mw for the antimicrobial finishing procedure.

In addition to molecular weight, chitosan concentration, and deacetylation level also affect the antibacterial efficacy. That is, when chitosan concentration rises, its Mw drops below 305 kDa, the level of deacetylation rises, and antibacterial action becomes more potent [43]. Poor laundry duration is another drawback of fabrics treated with chitosan. Crosslinkers or binders can boost the percent add-on. The crosslinker or binder might, however, react with the amino groups in chitosan and conceal them, which would reduce the antibacterial effectiveness. The inclusion of additional auxiliaries can also help to improve the fabric's bad hand following chitosan treatment. However, the chitosan's antibacterial properties must be unaffected by the addition of chemicals [44]. Fabrics coated with chitosan were treated to UV light by Ferrero et al. [45], and the outcome was strong antibacterial and good hand qualities. Chitosan's fastness can be increased by dilution with an acetic acid solution before application to textiles and by allowing adequate time for the chitosan to infiltrate fibers before curing (for example, 12 hours at room temperature or 1 hour at 50 °C).

Habiba et al. [46] studied the use of chitosan-bound cotton fibers as a means to develop antibacterial textile products. The outcomes demonstrated that there was considerable contact between the chitosan particle and the cellulose matrix. It has been demonstrated that cotton treated with chitosan has remarkable antibacterial action against *E. coli* and *S. aureus*. The usage of cotton cloth with chitosan coating for surgical materials and very durable antibacterial finishing textile goods was therefore determined. For wholesome and sanitary textiles, Li et al. [47] created chitosan fiber with strong antioxidant and acid-resistant capabilities. Chitosan fiber was first colored with natural lac dye made of compounds with polyphenolic anthraquinones and then cross-linked using a water-soluble aziridine crosslinker. The good natural antibacterial activity of chitosan fiber was said to be unaffected by cross-linking and dyeing, making the cross-linked chitosan fibers appropriate for bioactive, healthful, and sanitary medical textile goods.

The sonochemical coating of textiles with hybrid antibacterial ZnO/chitosan nanoparticles was explored by Petkova et al. [48]. The present study employed a one-step simultaneous sonochemical deposition technique to fabricate hybrid antibacterial coatings on cotton fibers, utilizing a combination of chitosan and ZnO NPs. The combined application of ZnO nanoparticles and chitosan coatings demonstrated enhanced antibacterial efficacy against *S. aureus* and *E. coli*, surpassing the solo application of ZnO and chitosan coatings. This improvement in efficacy was evident even at low concentrations of ZnO. It was also mentioned that the antibacterial action of the treated materials held up to several washing cycles at 75 °C.

Chitosan nanoparticles coated with silver were used to modify the surface of wool fibers, according to Sahan et al. [49].

Regarding antibacterial effectiveness, hydrophilicity, coloring, surface morphology, air permeability, and tensile strength, the impacts of surface modification on wool materials were examined. Data from the investigation into the effects of atmospheric plasma and enzyme treatments on wool textiles are shown in Table 11.2.

Table 11.2 Antimicrobial effectiveness (%) (reproduced with permission from [49])

	Reduce bacteria (%)		
	Types of bacteria		
Treatment	*K. pneumoniae*	*E. coli*	*S. aureus*
Untreated	–	–	–
Plasma + chitosan	40.42	83.16	77.02
Chitosan	33.73	81.83	73.15
Enzyme + chitosan-silver nanoparticles + plasma	84.24	99.99	99.27
Enzyme + nanochitosan + plasma	68.62	93.98	90.35
Chitosan-silver nanoparticles	66.62	97.66	96.38
Plasma + nanochitosan	60.66	91.55	86.99
Nanochitosan	51.16	89.56	84.86

Table 11.2 demonstrates that the antibacterial activity is more significantly influenced by chitosan and nano chitosan particles. It was discovered that chitosan, plasma, and enzyme treatments all significantly influenced the characteristics of wool. The tensile strength, fastness, hydrophilicity, antibacterial, dyeability, and air permeability of wool textiles were among these qualities. The combined effects of plasma, enzyme, and nano chitosan treatments during the dyeing process produced 2.5 times more K/S values for the treated cloth compared to the untreated fabric.

The primary objective of the research has consistently been the development of chitosan-derived microfiber yarn that is both biocompatible and biodegradable, with a specific focus on its potential uses in the medical field [50]. According to Chellamani et al. [51], a comprehensive review has been conducted on the use of chitosan-coated textile substrates in the context of wound care applications. Shabunin et al. [52] provided a description of a bilayer wound dressing composed of aliphatic polyamide nanofibers and a chitosan-based composite layer. Chitosan nanofibers and chitin nanofibrils were used as the filler to create the chitosan-based composite. In 28 days, experimental tests showed that the third-degree burn wound surface had epithelized to a maximum of 97.8% [52]. The information we gathered revealed that burn wounds might be effectively treated using composite chitosan-copolyamide dressings.

11.4.5 Agent for Shrink-Proof Finishing

Wool fibers feel and shrink readily under mechanical action in the presence of water due to structural features. Due to the scales on the wool fibers, interlocking and hooking of adjacent fibers cause felting shrinkage. Chitosan has effective abilities to make films. As a result, when chitosan is applied to wool fibers, a thin membrane is created that protects the fibers and reduces the interlocking and hooking between them as well as the directional friction effect of wool [53]. Since chitosan and wool both have positive charges in acidic circumstances, there is difficult sorption between the two substances. Chitosan may better adhere to wool fibers and distribute itself uniformly due to the fibers' improved wettability and anionic nature. Wool fibers can develop sulfonate and carboxylate groups by oxidative (i.e., air or oxygen) plasma treatments, which offer anionic groups and increase wettability [54]. Hence, the utilization of plasma pretreatment has the potential to enhance the shrink resistance of wool textiles by facilitating the spreading and adhesion of chitosan onto wool fibers. Wool fibers can be pretreated with hydrogen peroxide in an alkaline solution to eliminate fatty acids from the epicuticle and boost their hydrophilicity. Wool fabrics' shrink resistance can be improved by using low Mw chitosan, and in particular nano-chitosan, which increases the sorption and affinity between chitosan and wool fibers. These additives include anionic surfactant and crosslinker [55]. The scales on wool fibers may also be largely destroyed by pretreatment with protease enzymes, and chitosan application thereafter significantly enhances wool shrink-proofing [56].

11.4.6 Carriers for Textile Finishing

Chitosan can be used as a carrier for other chemicals, like proteins, drugs, antimicrobial agents, hormones, dyes, phase change materials, flame retardants, and fragrances, that do not adhere to textiles easily or are bioactive because of its special polymeric cationic character and gel and film shaping characteristics. Chitosan serves as a barrier wall for solids or liquids in a process known as microencapsulation or nanoencapsulation. These encapsulated liquid or solid substances can be infused into fabrics as minute solid particles [57, 58]. For medications, scents, and other functional finishing agents, encapsulation enables a controlled and sustained release strategy. The encapsulation technique allows textiles to be polished without some finishing chemicals' cytotoxicity. El-Feky used silver sulfadiazine-loaded chitosan nanocapsules in wound dressings to prevent infection, which might last for more than 24 hours [59].

Chitosan has the potential to be employed in the field of wastewater treatment as a means of transporting metal ions, auxiliaries, and dyes, while also imparting functional properties to textiles [60]. Chitosan, an inherent compound, possesses notable merits in comparison to contemporary technologies when employed in filtration systems due to its eco-friendly nature and its ability to eliminate or counteract microorganisms and hazardous metal ions found in wastewater, such as Ag (I), Cu (II), Cd (II), Pb (II), Zn (II), and Mn (II), via chelation [2].

Chitosan nonwoven mats have favorable characteristics for filtration purposes due to their elevated surface area to mass ratio. Chitosan has a strong attraction toward several types of dyes, such as dispersion, direct, acid, reactive, vat, sulfur, and naphthol dyes, owing to the presence of hydroxyl groups and cationic amino, as well as its structural resemblance to cellulose [2]. In acidic conditions, the cationic amino groups of chitosan have the ability to adsorb anionic dyes, auxiliaries, and other chemicals.

Simultaneously, the amino and hydroxyl groups of chitosan function as the principal sites for binding dyes by chelation [2, 60].

11.5 Processing of Textiles with Chitosan

11.5.1 Chitosan Is a Sizing and Desiring Agent Used in the Processing of Textiles

The traditional method of pretreating fibers for resistance to warp breakage brought on by highly stressed weaving processes is called sizing. Sizing agents often establish a coating on yarn by adhering to its surface [61]. It offers the necessary durability and hydrophilicity for an effective weaving procedure. To ensure a smooth finishing process and to prevent issues with the future processes of treating the fibers or fabric, polyvinyl alcohols, and acrylics, along with any synthetic sizing agents, must be thoroughly rinsed away at the end. In terms of biodegradability and effectiveness, chitosan is a better choice than synthetic substances. Chitosan exhibits solubility in pH values below 6.0 due to the deprotonation of more than 50% of its molecules, which occurs at a pKa of 6.5, leading to insolubility [62].

Concentrated hydrochloric acid (0.1, 0.2, and 0.5–1 N) is used to hydrolyze chitosan at high temperatures (30–90 °C), resulting in soluble chitosan with a decreased apparent viscosity. The hydrolyzing activity of HCl was catalyzed by the activation energy, which is correlated with temperature. Additionally, because the hydrolysis of chitosan takes around 60 minutes to complete, the nitrogen level drops from 7.5% to 6.5%. It is believed that the frequent assault of concentrated HCl on chitosan, which results in reduced molecular weight, is caused by chain scission at 1–4 glycosidic links [18]. Furthermore, after 60 minutes, the dissolution of short-chain chitosan exhibits a lower and higher apparent viscosity of 110 MPa, which is most likely caused by the persistence of crystallinity following degradation. A transformation from rod to coil transition is thought to be the cause of the conformational shift [63]. The percentages of size removal and addition are typically used to represent the stability and desirability processes, respectively.

The add-on % of size is primarily influenced by two important factors: the liquid's apparent viscosity and the concentration of solid solids in that range in size. Upon subjecting the size coating to heat treatments, it is seen that the tensile strength is increased. However, the elongation at break is significantly diminished. The observed effects, namely increased brittleness and stiffness, can be attributed to the diffusion and deposition of particulate matter within the fiber filament. In contrast, it has been observed that just water heated to a temperature of 90 °C exhibits a removal efficiency of 100% for hydrolyzed chitosan. A modified version of chitosan called carboxymethylated chitosan is easily soluble in water at a neutral pH and may be employed as a sizing agent. On the reverse side, a type of size agent is produced by an extrusion process when combined with wax or starch [64]. The lack of efficacy of chitosan as a sizing agent can be attributed to its elevated viscosity at temperatures beyond room temperature, particularly when the concentration of chitosan surpasses 2% in an acidic environment. This phenomenon leads to a further rise in the chitosan concentration, and so does the chitosan solution's viscosity. By hydrolyzing chitosan with 0.5 N HCl at a minimum of 90 °C for two hours, carboxymethylated chitosan can be produced. After neutralization with $NaCO_3$, the sodium chloride salt must be removed by filtration and washing. As a sizing agent, the final form is employed [42].

11.5.2 Chitosan: A Binder in the Printing Process

To print on fabric, you need a pigment or dye, a binder, and a thickening. Binders are mostly used to keep liquids' viscosities steady, improve droplet formation, give sufficient adherence to textile surfaces by producing films, and bind the molecules of color or pigment. The majority of binders used in industrial applications are made from vinyl acetate-acrylate copolymers, styrene-butadiene, or styrene-acrylate [65]. Due to its inferior performance compared to commercial binders, only a small number of researchers have experimented with chitosan as a binder and thickening. The preferred quantity of water and acetic acid are used to create a 3% (w/w) chitosan solution in the first phase of production. A 3% homogenous pigment paste for printing is created by adding the pigment to the solution and stirring. Increased acetic acid content is used when pigment dispersion is poor. A very high concentration of chitosan (3%) in acetic acid also produces an extremely viscous solution [65]. When compared to commercial Alcoprint pastes (BASF), the process's results show that chitosan is less efficient and effective as a binder. After treatment with more chitosan and higher curing temperatures, there is a greater tendency for yellowing. Additionally, various batches with identical process conditions have variable yellowing propensities.

When the viscosity is tested, the chitosan solution also has a share-thinning characteristic. Additionally, when the sharing rate is less than 50 per, it has a lower viscosity than commercial Alcoprint pastes (BASF). Conversely, at greater sharing rates, a nearly identical viscosity profile is discovered, but the yield point is lower than that of the commercial paste. It suggests that very little power or tension is needed for the chitosan to begin flowing. Additionally, chitosan solution has a wet pick-up ratio of 75% compared to 85% for commercial paste. After the K/S values are determined, employing a chitosan solution yields a 50% reduced color intensity. It is most likely due to the colorant aggregation in the difficult-to-maintain chitosan solution. In addition, chitosan-based printed fabric has longer bent lengths and more stiffness than conventional Alcoprint pastes. In the end, there is no discernible difference between chitosan and commercially tested binders in terms of wash fastness and spreading behavior [66].

11.5.3 As a Finishing Agent for Sturdy Textile Presses

Due to the abundance of hydroxyl groups, cotton fabric is hydrophilic. As a result, when cotton cellulose is swollen, it loses crystallinity and becomes flexible. Cellulose undergoes plastic deformation due to molecular movement during the washing process. To put it simply, it makes fabric wrinkle and shrink. Because of its ability to increase elasticity and stop the displacement of cellulose, cross-linking agents are used to solve the problem by undoing the plastic deformation [67]. Chitosan, which is used to finish cotton, has a molecular structure that is comparable to that of cellulose. Chitosan is broken down during this procedure by an oxidizing agent like hydrogen peroxide (H_2O_2). Chitosan and hydrochloric acid are used to perform the fragmentation process. Hydrogen peroxide is added after 30 minutes of stirring and is then heated to 60 °C for 2 hours. Two phases are separated using a Hoover filter. To neutralize the residue, it is rinsed with water. Subsequently, the filtrate is mixed with ethanol to induce the precipitation of chitosan, which is allowed to occur for a duration of 24 hours. After the formation of each precipitate, it is subjected to a drying process and afterward transferred into a container that ensures airtight storage conditions. The primary use of the broken-down chitosan is to provide a strong press finishing method for cloth. In this procedure, distilled water is added to the broken-up chitosan, and it is agitated until it completely dissolves.

The chain of chitosan is broken up, which makes it easier for it to diffuse into the fabric and create a network of fibers that increases the recovery angle. The ultimate wrinkling quality of the fabric is significantly influenced by the curing time and temperature.

11.5.4 Chitosan: A Textile Finishing Agent with Antistatic Properties

Due to the relatively low moisture content, polymers form a static surface charge. Additionally, a process brought on by an electric field causes the electrical conductivity to play a significant role in the development of static charge along the synthetic filament [68]. In order to prevent the generation of electrostatic charges, increase electrical conductivity, or both, antistatic chemical compounds have an effect on the tendency to accumulate static charges. Since chitosan is able to hold a high moisture content despite changes in humidity, it may be used as an antistatic finishing agent on textiles. Generally speaking, the electrical resistance of polymers is on the order of 1016 Ω cm, which is quite near to the value it appears as a good insulator [68]. While mineral water's electrical conductivity is 103 times greater than that of pure water, the magnitude of deionized water is 108 times less than that of polymers. As a result, after being exposed to a humid environment, textiles treated with chitosan are able to absorb a sizable quantity of water from the air, increasing electrical conductivity as a result of a decreased tendency to generate static charges. Chitosan is a great antistatic agent because of its good moisture content characteristics [69].

11.5.5 Chitosan: A Textile Fiber That Is Flame- and Fire-Resistant

It is strongly advised to use flame-retardant fabrics, especially in workplaces where there is a significant risk of fire or flame. When the flame is prevented from spreading and put out at the same time as removing the fire source, the flame-retardant property is taken into account. Due to their low toxicity and lack of halogen components, phosphorus-based materials are intriguing and well-known for being flame retardants. Through dehydration and char formation, these substances are highly effective flame retardants for cellulose and its derivatives. Urea and diammonium phosphate are used to treat cellulose during the phosphorylation process [70].

A phosphorous derivative and chitosan might be combined to create an intumescent flame-retardant substance. Chitosan is a polysaccharide that contains nitrogen and is biocompatible, biodegradable, and benign to the environment [71]. The amino and hydroxyl groups on its molecular skeleton give it the ability to create char when utilized as intumescent additives.

11.5.6 Chitosan Is a Hydrophobic Substance Used to Make Textiles That Repel Water

The legs of water striders, butterfly wings, duck feathers, and other organisms have been found to exhibit superhydrophobicity. Water droplets' interactions with a surface can be quantified by measuring the surface's specific energy. It has a lot to do with the roughness and contact angle aspects. Cotton and polyester fibers can be coated with nanoscale chitosan to provide a rough surface. To lower the surface energy, these rough surfaces are further treated with fluoride and silicon [72]. Before being squeezed, the textiles are immersed for one minute in a chitosan solution containing acetic acid (1%). The wet cloth is dried at 80 °C for 5 minutes after being neutralized by ammonia gas for 1 minute [73]. Chitosan dipping and padding permit a thin layer to be formed on the fabric's surface. Later, the pH of the wet cloth is altered by ammonia treatment, and chitosan, which is insoluble in nanoscales, accumulates on the fabric's surface. The SEM investigation demonstrates the deposition of nanofiber-like structures on the fiber surface, producing a rough surface on the microscale cotton fiber structure. As a result of the same procedure with polyester materials, nanoflower-like precipitation has also been seen. Chitosan nanoparticles are deposited to give cotton and polyester fibers the appropriate degree of roughness. Additionally, a polysiloxane emulsion may be utilized to lower the surface energy of the cloth treated with chitosan.

11.5.7 Chitosan: A UV-Blocking Textile Coating

A chitosan-graphene coating is created for the UV-blocking procedure of cotton fabrics. After vigorously swirling for 5 hours, a graphene nanosheet (1–3 nm) is added to the chitosan solution. A cotton cloth is immersed in a chitosan-graphene dispersion for two hours. The cloth goes through two dips and two nips throughout the padding process to achieve an average wet pick-up of 80%. The drying and washing processes are kept separate. The drying process is maintained at 70 °C for 10 min, followed using 10 minutes at 110 °C for the curing phase. The treated cloth with 1% graphene has a 60 times higher UV protection factor (UPF) than untreated cotton. The consideration is given to the uniform dispersion of graphene particles inside the chitosan matrix, without any occurrence of aggregation, in order to get a high UPF value [74].

11.6 Conclusion

Chitosan is used in all phases of textile production, from spinning and weaving to dying, printing, and finishing. Spinning solutions that create polymers can be utilized with chitosan. Wound dressings made from chitosan-based fibers are commonly utilized in the medical field, due to their antibacterial and hemostatic properties. Due to its antibacterial qualities, the addition of chitosan to the sizing solution stabilizes the size solution for weaving. When textiles are pretreated with chitosan before dyeing or printing, the dyes are absorbed more readily, and the color strength and colorfastness are improved. Chitosan, used as a thickener in printing, improves pattern sharpness. The application of chitosan coatings and encapsulating technologies for functional finishing gives fabrics a variety of unique functional qualities, such as antibacterial characteristics, shrink resistance, flame resistance, and scent.

References

1. Amor IB, Hemmami H, Laouini SE, Temam HB, Zaoui H, Barhoum A (2023) Biosynthesis MgO and ZnO nanoparticles using chitosan extracted from Pimelia Payraudi Latreille for antibacterial applications. *World Journal of Microbiology and Biotechnology* 39: 19.
2. Wani MY, Hasan N, Malik MA (2010) Chitosan and Aloe vera: Two gifts of nature. *Journal of Dispersion Science and Technology* 31: 799–811.
3. Amor IB, Hemmami H, Laouini SE, Abdelaziz AG, Barhoum A (2023) Influence of chitosan source and degree of deacetylation on antibacterial activity and adsorption of AZO dye from water. *Biomass Conversion and Biorefinery*: 1–11.
4. Lam NYK, Zhang M, Yang C, Ho CP, Li L (2018) A pilot intervention with chitosan/cotton knitted jersey fabric to provide comfort for epidermolysis bullosa patients. *Textile Research Journal* 88: 704–716.
5. Liu S, Hua T, Luo X, Yi Lam N, Tao X-m, Li L (2015) A novel approach to improving the quality of chitosan blended yarns using static theory. *Textile Research Journal* 85: 1022–1034.
6. Zhou C-E, Kan C-W (2014) Plasma-assisted regenerable chitosan antimicrobial finishing for cotton. *Cellulose* 21: 2951–2962.
7. Sadeghi-Kiakhani M, Safapour S (2015) Eco-friendly dyeing of treated wool fabrics with reactive dyes using chitosanpoly (propylene imine) dendreimer hybrid. *Clean Technologies and Environmental Policy* 17: 1019–1027.
8. Akakuru O, Louis H, Amos P, Akakuru O, Nosike E, Ogulewe E (2018) The chemistry of chitin and chitosan justifying their nanomedical utilities. *Biochem Pharmacol (Los Angel)* 7: 2167–0501.1000241.

9. Ben Amor I, Hemmami H, Laouini SE, Mahboub MS, Barhoum A (2022) Sol-gel synthesis of ZnO nanoparticles using different chitosan sources: Effects on antibacterial activity and photocatalytic degradation of AZO Dye. *Catalysts* 12: 1611.
10. Yilmaz Atay H (2019) Antibacterial activity of chitosan-based systems. *Functional Chitosan: Drug Delivery and Biomedical Applications*: 457–489.
11. Benettayeb A, Ghosh S, Usman M, Seihoub FZ, Sohoo I, Chia CH, Sillanpää M (2022) Some well-known alginate and chitosan modifications used in adsorption: A review. *Water* 14: 1353.
12. Elamri A, Zdiri K, Hamdaoui M, Harzallah O (2023) Chitosan: A biopolymer for textile processes and products. *Textile Research Journal* 93: 1456–1484.
13. Qian L, Zhang K, Guo X, Yu M (2023) What happens when chitin becomes chitosan? A single-molecule study. *RSC Advances* 13: 2294–2300.
14. Cruz RdCAL, Diniz LGM, Lisboa HM, Fook MVL (2016) Effect of different carboxylic acids as solvent on chitosan fibers production by wet spinning. *Matéria (Rio de Janeiro)* 21: 525–531.
15. Hameed D, Manouel T (2020) fibers and textiles produced from chitin and chitosan: a literature study for different production methods. Thesis, Högskolan i Borås, Akademin för textil, teknik och ekonomi.
16. Mauer LJ, Bradley RL (2017) Moisture and total solids analysis. *Food Analysis*: 257–286.
17. Notin L, Viton C, Lucas J-M, Domard A (2006) Pseudo-dry-spinning of chitosan. *Acta Biomaterialia* 2: 297–311.
18. Roy J, Salaün F, Giraud S, Ferri A, Guan J (2017) Chitosan-based sustainable textile technology: Process, mechanism, innovation, and safety. *Biological Activities and Application of Marine Polysaccharides*: 251–278.
19. Zulkifli MZA, Nordin D, Shaari N, Kamarudin SK (2023) Overview of electrospinning for tissue engineering applications. *Polymers* 15: 2418.
20. Sáenz-Trevizo A, Amézaga-Madrid P, Pizá-Ruiz P, Antúnez-Flores W, Miki-Yoshida M (2016) Optical band gap estimation of ZnO nanorods. *Materials Research* 19: 33–38.
21. Van der Schueren L, Steyaert I, De Schoenmaker B, De Clerck K (2012) Polycaprolactone/chitosan blend nanofibres electrospun from an

acetic acid/formic acid solvent system. *Carbohydrate Polymers* 88: 1221–1226.

22. Ohkawa K, Cha D, Kim H, Nishida A, Yamamoto H (2004) Electrospinning of chitosan. *Macromolecular Rapid Communications* 25: 1600–1605.
23. Notin L, Viton C, David L, Alcouffe P, Rochas C, Domard A (2006) Morphology and mechanical properties of chitosan fibers obtained by gel-spinning: Influence of the dry-jet-stretching step and ageing. *Acta Biomaterialia* 2: 387–402.
24. Singh R, Shitiz K, Singh A (2017) Chitin and chitosan: Biopolymers for wound management. *International Wound Journal* 14: 1276–1289.
25. Saha S, Sarkar P (2016) Arsenic mitigation by chitosan-based porous magnesia-impregnated alumina: Performance evaluation in continuous packed bed column. *International Journal of Environmental Science and Technology* 13: 243–256.
26. Liu Y, Xiang K, Chen H, Li Y, Hu Q (2015) Composite vascular repair grafts via micro-imprinting and electrospinning. *AIP Advances* 5: 041318.
27. Toskas G, Cherif C, Hund R-D, Laourine E, Mahltig B, Fahmi A, Heinemann C, Hanke T (2013) Chitosan (PEO)/silica hybrid nanofibers as a potential biomaterial for bone regeneration. *Carbohydrate Polymers* 94: 713–722.
28. Sebe I, Szabó E, Zelkó R (2023) Advances in drug delivery via electrospun and electrosprayed formulations. *Emerging Drug Delivery and Biomedical Engineering Technologies: Transforming Therapy*: 71.
29. Jung B, Theato P (2013) Chemical strategies for the synthesis of protein–polymer conjugates. *Bio-Synthetic Polymer Conjugates*: 37–70.
30. Pinho ED, Martins A, Araújo JV, Reis RL, Neves NM (2014) Size also matters in biodegradable composite microfiber reinforced by chitosan nanofibers. *MRS Online Proceedings Library (OPL)* 1621: 59–69.
31. Lam NYK, Zhang M, Guo H-F, Ho CP, Li L (2017) Effect of fiber length and blending method on the tensile properties of ring spun chitosan–cotton blend yarns. *Textile Research Journal* 87: 244–257.
32. Zhou BC-E, Kan C-W, Sun C, Du J, Xu C (2019) A review of chitosan textile applications. *AATCC Journal of Research* 6: 8–14.
33. Desai K, Kit K, Li J, Zivanovic S (2008) Morphological and surface properties of electrospun chitosan nanofibers. *Biomacromolecules* 9: 1000–1006.

34. Hild M, Fayez Al Rez M, Aibibu D, Toskas G, Cheng T, Laourine E, Cherif C (2015) Pcl/chitosan blended nanofibrous tubes made by dual syringe electrospinning. *Autex Research Journal* 15: 54–59.

35. Ferrero F, Periolatto M (2017) Chitosan coating on textile fibers for functional properties. *Handbook of Composites from Renewable Materials* 4: 165À198.

36. Samanta A, Chakraborty S, Guha Roy T (2012) Dyeing of jute with reactive dyes: Optimisation of the process variables and assessment of colourfastness characteristics. *Journal of The Institution o Engineers (India): Series E* 93: 15–24.

37. Morin-Crini N, Lichtfouse E, Torri G, Crini G (2019) Fundamentals and applications of chitosan. *Sustainable Agriculture Reviews 35: Chitin and Chitosan: History, Fundamentals and Innovations*: 49–123.

38. Abdou E, El-Hennawi H, Ahmed K (2013) Preparation of novel chitosan-starch blends as thickening agent and their application in textile printing. *Journal of Chemistry* 2013. https://doi.org/10.1155/2013/595810.

39. Yuen C, Ku S, Kan CW, Choi P (2007) Enhancing textile ink-jet printing with chitosan. *Coloration Technology* 123: 267–270.

40. Piekarska K, Sikora M, Owczarek M, Jóźwik-Pruska J, Wiśniewska-Wrona M (2023) Chitin and chitosan as polymers of the future—obtaining, modification, life cycle assessment and main directions of application. *Polymers* 15: 793.

41. Rajendran R, Radhai R, Balakumar C, Ahamed HAM, Vigneswaran C, Vaideki K (2012) Synthesis and characterization of neem chitosan nanocomposites for development of antimicrobial cotton textiles. *Journal of Engineered Fibers and Fabrics* 7: 155892501200700116.

42. Chen JL, Zhao Y (2012) Effect of molecular weight, acid, and plasticizer on the physicochemical and antibacterial properties of β-chitosan based films. *Journal of Food Science* 77: E127–E136.

43. Mellegård H, Strand S, Christensen B, Granum P, Hardy S (2011) Antibacterial activity of chemically defined chitosans: Influence of molecular weight, degree of acetylation and test organism. *International Journal of Food Microbiology* 148: 48–54.

44. Li B, Shan C-L, Zhou Q, Fang Y, Wang Y-L, Xu F, Han L-R, Ibrahim M, Guo L-B, Xie G-L (2013) Synthesis, characterization, and antibacterial activity of cross-linked chitosan-glutaraldehyde. *Marine Drugs* 11: 1534–1552.

45. Ferrero F, Periolatto M (2012) Antimicrobial finish of textiles by chitosan UV-curing. *Journal of Nanoscience and Nanotechnology* 12: 4803–4810.

46. Habeeba AU, Reshmi C, Sujith A (2017) Chitosan immobilized cotton fibres for antibacterial textile materials. *Polymers from Renewable Resources* 8: 61–70.

47. Li X-Q, Tang R-C (2016) Crosslinked and dyed chitosan fiber presenting enhanced acid resistance and bioactivities. *Polymers* 8: 119.

48. Petkova P, Francesko A, Fernandes MM, Mendoza E, Perelshtein I, Gedanken A, Tzanov T (2014) Sonochemical coating of textiles with hybrid ZnO/chitosan antimicrobial nanoparticles. *ACS Applied Materials & Interfaces* 6: 1164–1172.

49. ŞAHAN G, DEMİR A (2016) A green application of nano sized chitosan in textile finishing. *Textile and Apparel* 26: 414–420.

50. Toskas G, Brünler R, Hund H, Hund R-D, Hild M, Aibibu D, Cherif C (2013) Pure chitosan microfibres for biomedical applications. *Autex Research Journal* 13: 134–140.

51. Chellamani K, Balaji RV, Sudharsan J (2013) Chitosan treated textile substrates for wound care applications. *Journal of Academia and Industrial Research* 2: 97–102.

52. Shabunin AS, Yudin VE, Dobrovolskaya IP, Zinovyev EV, Zubov V, Ivan'kova EM, Morganti P (2019) Composite wound dressing based on chitin/chitosan nanofibers: Processing and biomedical applications. *Cosmetics* 6: 16.

53. Jia X, Duić N (2021) Advanced methods and technologies towards environmental sustainability. *Clean Technologies and Environmental Policy* 23: 709–710.

54. Wang Y, Zhang N, Wang Q, Yu Y, Wang P (2021) Chitosan grafting via one-enzyme double catalysis: An effective approach for improving performance of wool. *Carbohydrate Polymers* 252: 117157.

55. Yang H-C, Wang W-H, Huang K-S, Hon M-H (2010) Preparation and application of nanochitosan to finishing treatment with anti-microbial and anti-shrinking properties. *Carbohydrate Polymers* 79: 176–179.

56. Vilchez S, Manich A, Jovancic P, Erra P (2008) Chitosan contribution on wool treatments with enzyme. *Carbohydrate Polymers* 71: 515–523.

57. Javid A, Raza ZA, Hussain T, Rehman A (2014) Chitosan micro-encapsulation of various essential oils to enhance the functional properties of cotton fabric. *Journal of Microencapsulation* 31: 461–468.
58. Greiner A, Wendorff JH (2007) Electrospinning: A fascinating method for the preparation of ultrathin fibers. *Angewandte Chemie International Edition* 46: 5670–5703.
59. El-Feky GS, Sharaf SS, El Shafei A, Hegazy AA (2017) Using chitosan nanoparticles as drug carriers for the development of a silver sulfadiazine wound dressing. *Carbohydrate Polymers* 158: 11–19.
60. Ngah WW, Teong L, Hanafiah MM (2011) Adsorption of dyes and heavy metal ions by chitosan composites: A review. *Carbohydrate Polymers* 83: 1446–1456.
61. Hahn T, Bossog L, Hager T, Wunderlich W, Breier R, Stegmaier T, Zibek S (2019) Chitosan application in textile processing and fabric coating. *Chitin and Chitosan: Properties and Applications*: 395–428.
62. Islam N, Dmour I, Taha MO (2019) Degradability of chitosan micro/nanoparticles for pulmonary drug delivery. *Heliyon* 5: e01684.
63. Czechowska-Biskup R, Wojtasz-Pająk A, Sikorski J, Henke A, Ulański P, Rosiak JM (2007) Aqueous solutions of hydrochloric acid as simple solvents of chitosan for viscosity-and light-scattering-based molecular weight determination. *Polish Chitin Society*: 87–94.
64. Chevalier E, Chulia D, Pouget C, Viana M (2008) Fabrication of porous substrates: A review of processes using pore forming agents in the biomaterial field. *Journal of Pharmaceutical Sciences* 97: 1135–1154.
65. Banerjee P, Jang E, Huang J, Holley R, Vadnala S, Sheikh A, Trivedi A, Jackson K, Homman VK, Madan D (2021) Thermoelectric performance enhancement of n-type chitosan-Bi 2 Te 2.7 Se 0.3 composite films using heterogeneous grains and mechanical pressure. *Journal of Electronic Materials* 50: 2840–2851.
66. CHATHA SS, Hussain A, Ali S, Saif M, Mallhi A, Sagir M, Naz M (2016) Significance of chitosan to improve the substantivity of reactive dyes. *Journal of the Chilean Chemical Society* 61: 2895–2897.
67. Shirvan AR, Shakeri M, Bashari A (2019) Recent advances in application of chitosan and its derivatives in functional finishing of textiles. *The Impact and Prospects of Green Chemistry for Textile Technology*: 107–133.

68. Huang L, Xiao L, Yang G (2018) Chitosan application in textile processing. *Current Trends in Fashion Technology & Textile Engineering* 4: 32–34.

69. Abdel-Halim E, Abdel-Mohdy F, Al-Deyab SS, El-Newehy MH (2010) Chitosan and monochlorotriazinyl-β-cyclodextrin finishes improve antistatic properties of cotton/polyester blend and polyester fabrics. *Carbohydrate Polymers* 82: 202–208.

70. Annalisa C, Francesca B, Giulio M, Chiara M, Monica P (2016) DNA-chitosan cross-linking and photografting to cotton fabrics to improve washing fastness of the fire-resistant finishing. *Cellulose* 23: 3963–3984.

71. Jiménez-Gómez CP, Cecilia JA (2020) Chitosan: A natural biopolymer with a wide and varied range of applications. *Molecules* 25: 3981.

72. Ivanova NA, Rutberg GI, Philipchenko AB (2013) Enhancing the superhydrophobic state stability of chitosan-based coatings for textiles. *Macromolecular Chemistry and Physics* 214: 1515–1521.

73. Liu Y, Chen X, Xin JH (2008) Hydrophobic duck feathers and their simulation on textile substrates for water repellent treatment. *Bioinspiration & Biomimetics* 3: 046007.

74. Tian M, Tang X, Qu L, Zhu S, Guo X, Han G (2015) Robust ultra-violet blocking cotton fabric modified with chitosan/graphene nanocomposites. *Materials Letters* 145: 340–343.

Chapter 12

Chitin and Chitosan in Wastewater Treatment

Atif Khurshid Wani,[a] Hadia Hemmami,[b,c,d] Atulika Sharma,[e,f] and Sourbh Suren Garg[g]

[a]*School of Bioengineering and Biosciences, Lovely Professional University, Punjab, India*
[b]*Department of Process Engineering and Petrochemical, Faculty of Technology, University of El Oued, El Oued, Algeria*
[c]*Renewable Energy Development unit in Arid Zones (UDERZA), University of El Oued, El Oued, Algeria*
[d]*Laboratory of Applied Chemistry and Environment, Faculty of Exact Sciences, University of El Oued, El Oued, Algeria*
[e]*Department of Chemistry, School of Chemical Engineering and Physical Sciences, Lovely Professional University, Punjab, India*
[f]*Department of Chemistry, Apeejay Institute of Management & Engineering Technical Campus, AIMETC, Jalandhar, Punjab, India*
[g]*Department of Biochemistry, School of Bioengineering and Biosciences, Lovely Professional University, Punjab, India*

atifkhurshid61200216@gmail.com

As hazardous organic and inorganic pollutants are released into the aquatic environment at an increasing rate, water quality is continually declining. The elimination of such contaminants

Chitin and Chitosan: Physical and Chemical Properties
Edited by Shakeel Ahmed, Soumeia Zeghoud, Hadia Hemmami, and Ilham Ben Amor

ISBN 978-981-5129-39-7 (Hardcover), 978-1-003-58977-8 (eBook)
www.jennystanford.com

from the water system is a developing field of study. Recently, biodegradable and biocompatible natural additives have gained attraction, due to their immense potential to lessen contaminants in wastewater. Chitosan and its composites have been recognized as valuable absorbents, owing to their cost-effectiveness, widespread availability, presence of hydroxyl and amino groups, and capacity to effectively eliminate various pollutants from wastewater. The practical use of chitosan-based nano-biocomposites is hindered by many challenges, such as limited selectivity, inadequate mechanical strength, and susceptibility to solubility issues in acidic conditions. Consequently, a variety of modification approaches were investigated in order to enhance the physicochemical characteristics of chitosan for the aim of treatment of wastewater. The efficacy of chitosan-based nanocomposites in the removal of microplastics, pesticides, and metals from wastewater has been well-documented. In recent times, there has been a notable surge in interest surrounding the use of chitosan-doped nanoparticles as nanocomposites for the purpose of water purification. Hence, the utilization of chitosan with appropriate modifications in the creation of such absorbents represents a pioneering approach to address the elimination of detrimental contaminants within aquatic environments, ultimately aiming to ensure the availability of safe drinking water on a global scale. The advancements in the synthesis of sorbents containing sugars, in particular chitosan biopolymers, and their uses in wastewater treatment are covered in this chapter.

12.1 Introduction

From the past to the present day, water has remained one of the essential natural resources for human life that is used for recreational, household, industrial, and agricultural purposes [1]. Because of residential and agricultural activities, rising industrialization, and greater energy requirements, water shortages and water quality degradation are currently major worldwide challenges. There have been reports of many organic and inorganic pollutants that are important water contaminants, and among them, drugs, phenols, oils, pesticides, detergents, greases, fertilizers, microorganisms, microplastics, and heavy metals are the major causes of water contamination. Drinking water of high standard is becoming increasingly difficult to obtain around the world, especially in developing countries. As a result, the scientific community must consider large-scale water filtration on a worldwide basis [2].

Zubair et al. [3] have developed several ways that have been developed to enhance the quality of wastewater. These techniques include electrolysis, nanofiltration, ion exchange, reverse osmosis, and the use of chemical disinfectants.

The rising cost and environmental hazards associated with the use of chemicals in these processes are becoming a major concern around the world [3]. The scientific community is becoming increasingly interested in adsorption-based methods for wastewater treatment, due to their increased efficacy, affordable cost, and the range of accessible adsorbents, including organic adsorbents [5], biosorbents [6], and inorganic adsorbents [7]. Recent years have seen the development and use of biosorption techniques employing biomaterials, mostly natural biopolymers that are sustainable and renewable. It is previously known that various adsorbents, both natural and synthetic, target various substances in water filtration systems. Various adsorbents, including graphene oxide, xylan, fly ash, and lignin, are currently used to treat wastewater (Table 12.1). Additionally, synthetic adsorbents can be employed to remove contaminants from wastewater. Each nanocomposite demonstrated a distinct mode of operation while removing contaminants from wastewater.

Among many biosorbents, chitosan is an abundant biological macromolecule formed from the deacetylation of chitin and is known to play a significant role in sewage treatment [4]. The polymer consists of two distinct monomeric units, namely 2-amino-2-deoxy-D-glucopyranose and 2-acetamido-2-deoxy-D-glucopyranose, as seen in Fig. 12.1. Furthermore, the substance exhibits a notable abundance of hydroxyl (–OH) and primary amine (–NH_2) functional groups, which endow it with active sites that facilitate effective adsorption. Zubair et al. [5] have said that the discrete nature of chitosan facilitates the effective elimination of anionic or cationic impurities, including reactive or direct dyes, acids, and metals.

Table 12.1 Some examples of various natural sorbents in wastewater treatment

Adsorbent	Purpose	Ref.
Sugarcane bagasse	Congo red	[7]
Banana fiber	Methylene blue	[8]
Gram seed husk	Congo red	[9]
Canola stalks	Acid orange 7	[10]
Sawdust	Crystal violet	[11]
Rice husk	Brilliant green	[12]
Sawdust	Congo red	[13]
Sawdust	Dyes and heavy metals	[14]
Groundnut shell	Tartrazine yellow dye	[15]
Fly ash	Anionic dimers	[16]
Graphene oxide	Removal of phenols, antimony	[17]
Banana peel	Oil	[18]
Xylan and lignin	Methylene blue (MB)	[19]
Wood biochar	Organic compounds	[20]
Date seeds	Dissolved organic carbon	[21]
Sulfur-functionalized rice husk	Mercury	[22]
Wheat straw	Nickel, dissolved organic matter	[23]
Olive branches	Lead, zinc, copper, cadmium	[24]
Sawdust	Copper, cadmium	[25]
Oil palm branches	Methyl-mercury, mercury	[26]
Loberiafruit	Lead	[27]
Camel bones nanocomposite	Cadmium, cobalt	[28]

As aforementioned, chitosan has attracted much attention due to its low cost compared to activated carbon and the presence of a high concentration of hydroxyl and amino functional groups, making it a powerful bioadsorbent for adsorbing a variety of aqueous pollutants [6]. Due to its physicochemical characteristics, chemical stability, high reactivity, outstanding chelation behavior with heavy metals, and great selectivity toward contaminants, this natural biomass promises an intriguing replacement for conventional biomaterials. This chapter goes into great length about the properties and applications of chitosan for the treatment of water, with an emphasis on current developments.

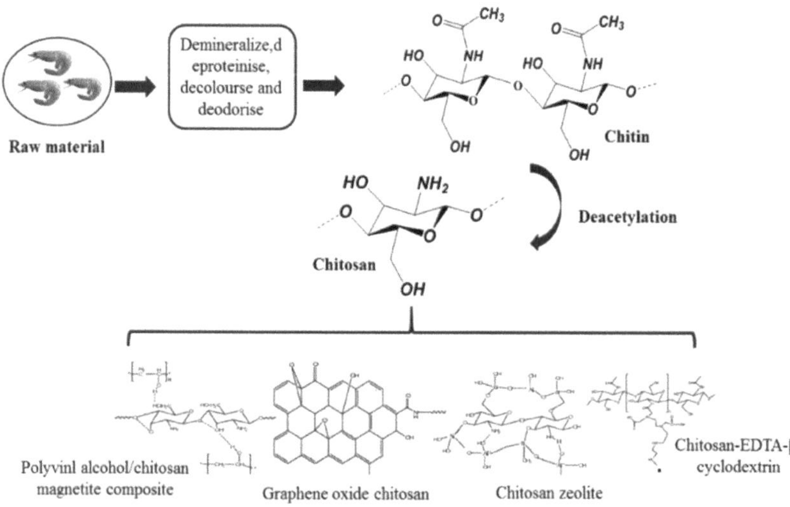

Figure 12.1 Deacetylation-based synthesis of chitosan and its derivatives.

12.2 Textile Industries—Synthetic Dyes and Effluent Wastewaters—Environmental Issues

The usage and discharge of synthetic dyes is a significant issue for the environment and human health related to water pollution. Because of their extensive commercial use, synthetic dyes are one of the significant families of refractory organic chemicals that are extensively distributed in nature [29]. Industries that use more water include those that produce chemicals, pulp and paper, paint, textiles, leather goods, packaging, clothes, food, and pharmacies. The majority (54%) of the world's naturally occurring dye water comes from the textile industry. The most frequent pollutants that may frequently be linked to industrial effluents are dyes because of their high solubility [30]. The rhodamine band Congo Red is a widely used dye; however, its removal from wastewater is extremely important because it has mutagenic and carcinogenic properties. Similar to Cr (VI), which comes from many areas, including leather tanning, textiles, storage batteries, and electroplating, but also poses a serious risk to human health because of its high toxicity and lack of biodegradability. When combined with other industrial pollutants, textile dyes can create highly hazardous and even lethal environmental contaminants that, if released into aquatic systems without adequate treatment, can be fatal to animals [31]. Around 80,000 tons of synthetic dyes are produced by the textile industry each year in India and approximately seven lakh tons worldwide [30]. Color has a significant impact on how the public perceives the quality of the water, leading to more complaints and worries [30]. It is an easy, practical, and efficient procedure to remove these color impurities by adding solid substances known as adsorbents or biosorbents. The utilized adsorbent can be derived from either a mineral, organic, or biological state in nature [30]. Carbon is considered one of the ideal absorbents for industrial use, but its high cost limits its applications. Since 30 years ago, various technologies have been designed to provide effective, economical, and efficient biosorbents for color removal from contaminated water bodies [30]. Figure 12.2 shows the effects of dye effluents on the environment, people, flora, and aquatic life.

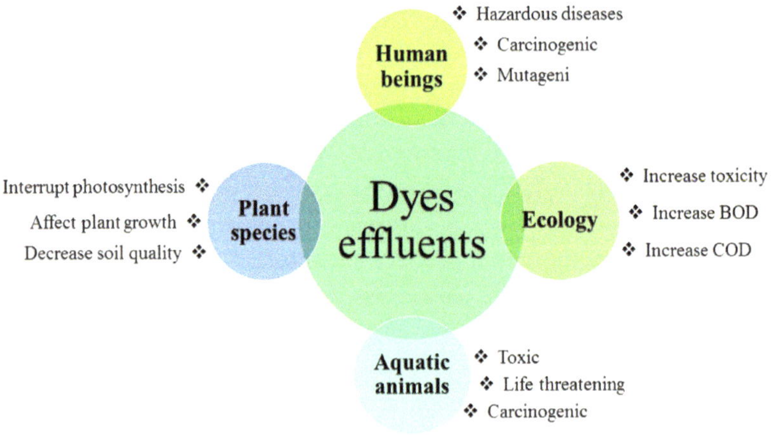

Figure 12.2 The environmental and biological impacts resulting from the discharge of dye effluents [31].

12.3 Function of Chitosan and Nanocomposite Membranes Based on Chitosan in the Treatment of Wastewater

The conventional approach to wastewater treatment encompasses a series of intermittent biological, chemical, physical, and mechanical procedures aimed at eliminating insoluble particles and soluble pollutants from wastewater while also ensuring their legal deactivation. It is one of several machining techniques suggested for dye removal and is often used in the industrial sector for economic reasons. One of several approaches is liquid grading and adsorption processing which are selectable processes, the second (final) step suggested for dye removal is the adsorption process. The liquid grade and the adsorption process can be used to remove different kinds of colored materials.

Activated carbon is frequently employed not just in commercial settings but also for the purpose of eliminating colors and impurities from drinking water sources, including lakes, rivers, reservoirs, and wastewater. The extensive use of activated carbon is limited due to its high cost in terms of both material and processing. To combat these shortcomings, several strategies have been investigated to generate low-cost and successful innovative chitosan-based materials [29]. Over the course of the last two decades, these substances have been employed as biosorbents, specifically for the purpose of eliminating dyes and engaging in metal chelation during the treatment of water and wastewater. Chitosan exhibits a notable propensity for binding with several types of dyes, encompassing reactive dyes, direct acids, metalloids, as well as metals and metalloids. Through a comprehensive investigation, it has been determined that chitosan serves as a highly efficient substitute for traditional sorbents currently employed in processes such as bleaching, metal chelation or extraction, and organic waste elimination. The expansion of chitosan is expected to find prospective markets in the nutraceuticals and cosmetics industry, as well as the water and wastewater treatment sector [29]. Chitosan is a resource that is inexpensive, renewable, non-toxic, and biodegradable and has special qualities, including hydrophilicity. The substance exhibits several notable characteristics that contribute to its effectiveness and environmental friendliness. These characteristics encompass the presence of polyelectrolytes under acidic pH conditions, high reactivity, the ability to coagulate, aggregate, and adsorb, as well as the presence of reactive hydroxyl and amine groups within its macromolecular chain. There are two distinct, significant advantages that support the usage of wastewater treatment. Along with excellent selectivity and binding of contaminants, chitosan has a remarkable propensity to interact with contaminant materials actually present at trace levels. Such materials can be employed in liquid or solid form for the process of adsorption and removal of water-dissolved pollutants from wastewater. One of these processes is bioconjugation to chitosan hydrogel, which is often used for the removal of dyes [29].

12.4 Chitosan as Bioadsorbents

It has been demonstrated that chitosan biosorbents are a reliable, efficient, and economical method of removing contaminants from wastewater. Its cationic properties and coagulation activity make it a suitable biopolymer for wastewater treatment (Fig. 12.3) [32]. Several pieces of evidence have claimed that chitosan-based absorbents have demonstrated strong adsorption capacity for heavy metals, dye elimination, phenolic compounds, and organic debris [33].

Chitosan has emerged as a prominent solution for the mitigation of heavy metal contamination. Numerous studies have shown that chitosan can effectively remove Cr (VI), As (III), As (V), and heavy metal oxyanions (Cu (II), Pb (II), Cd (II), Hg (II), Pd (II), and Pt (IV)) [5]. According to Ghinia et al. [37], chitosan has demonstrated significant efficacy as an adsorbent for zinc and many other metal ions present in wastewater, with the added advantage of being recyclable. For use in composites, the mechanical properties of chitosan are stabilized and improved by grafting, cross-linking, and functionalization. In contrast to poly (4-vinyl pyridine) modified activated carbon (53.7 mg/g) and ethylenediamine-functionalized Fe_3O_4 (61.35 mg/g). The effective elimination of Cr (VI) (67.66 mg/g) was demonstrated in another investigation by Li et al. [34]. A modified form of chitosan, namely magnetic-cyclodextrin chitosan/graphene oxide, is used. Previous research has indicated the efficacy of magnetic chitosan flocculants containing carboxylates and chitosan modified with graphene oxide in the removal of nickel (II) and arsenic (III and V) from artificial wastewater. The efficacy of chitosan and sodium alginate in eliminating copper (II) ions [35], FeO-nanoparticles-chitosan composite beads in eradicating hexavalent chromium (Cr (VI)), and a combination of chitosan and gold nanoparticles with hydrogel beads in removing insecticides have been demonstrated [36]. High fluoride levels from brackish water may be easily removed using chitosan and electrodialysis, according to research. In addition to its diverse uses in the cultivation of *Nannochloropsis sp.* and its efficacy in the sequestration of several heavy metals, including Pb (II), Cu (II), Hg (II), and Cd (II), nano chitosan has gained significant recognition for its extensive utilization in the field of treatment of wastewater. Some examples of other biopolymers used with chitosan to effectively remove pollutants from contaminated wastewater include activated lignin for the removal of methylene blue, alginate for the removal of p-chlorophenol, and gelatin for the removal of Cd (II).

In recent times, there has been significant interest in chitosan and its derivatives due to their notable antibacterial properties. The removal of heavy metals, organic pollutants (including detergents, hydrocarbons, organic dyes, pesticides, phenols, plasticizers, and greases), as well as bacteria, can be achieved by employing chitosan-coated magnetite nanoparticles as chelating agents, absorbents, or bactericidal agents in wastewater treatment processes [37]. According to research, adding ZnO nanoparticles to chitosan enhances the antibacterial activity of water disinfection [38]. In addition, the use of chitosan-zeolite composites has proven to be effective in the elimination of organic humic acids, ammonium, nitrates, dyes, and heavy metal cations from wastewater [33]. Effective oil removal has also been achieved using chitosan biosorbents. Vakili and co-workers conducted a study whereby they observed that both modified and unmodified chitosan exhibited a notable propensity for the removal of diverse colors from wastewater [39]. Methylene blue, acid blue, acid black, acid orange, Congo red, acid red, mordant blue, and eosin are among the extensively investigated organic dyes that have demonstrated effective removal from industrial effluent by the use of chitosan or chitosan derivatives. A brand-new magnetic MB adsorbent made of graphene oxide and chitosan was disclosed by Fan et al. [40]. The feasibility of employing modified chitosan for water desalination, such as chitosan/GO chitosan-derived matrix membranes, has been shown in several studies [41].

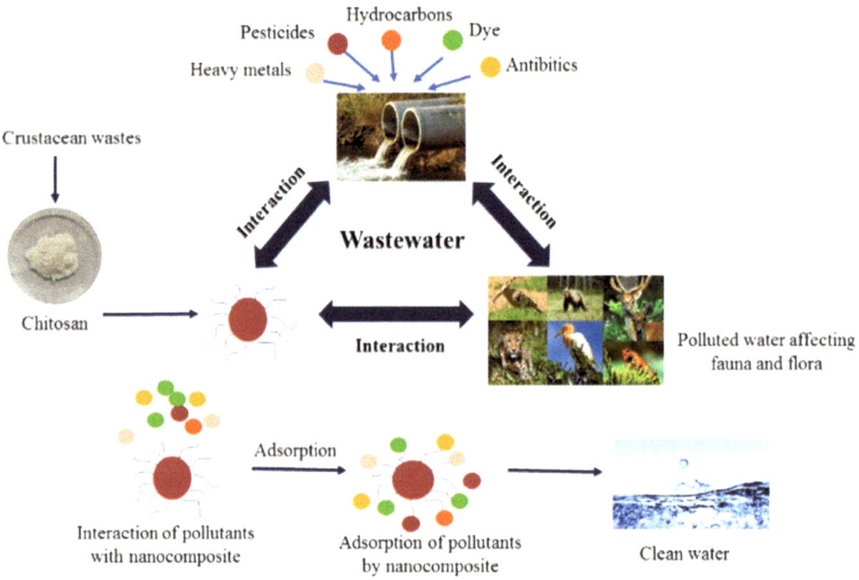

Figure 12.3 Employing chitosan nanomaterial to adsorb contaminants from wastewater.

Based on several studies, it has been shown that developing pollutants, such as perfluorobutane sulfonate and carbamazepine, have the potential for adsorption and subsequent removal from wastewater through the utilization of magnetic chitosan and cross-linked chitosan beads with molecular imprints. Chitosan-based magnetic composite particles have been shown to be effective as adsorbents for the removal of medications such as tetracycline hydrochloride, diclofenac sodium, and hydroquinone by Zhang et al. [42]. The elimination of humic and fulvic acid from drinking water may be achieved in a single stage of co-precipitation by using magnetic chitosan nanoparticles [43]. The utilization of chitosan-based adsorbents, which have undergone modification through the introduction of a crosslinker such as ethylene glycol diglycidyl ether and iminobis-(propylene glycol), has demonstrated effective removal of various inorganic nutrients from water that has been contaminated. Specifically, the chitosan/alumina composite has proven successful in removing nitrate, discarded adsorbents saturated with Cu (II)-chitosan have been effective in phosphate removal, and chitosan beads have shown promise in the removal of boron. In the upcoming years, there will be a huge market for chitosan-based products that perform better and are less expensive.

12.5 Nano-Biocomposites Based on Chitosan

Chitosan, the second most prevalent and consequential natural polymer after cellulose, is produced by the deacetylation process of chitin. Because of its biocompatibility, biodegradability, and non-toxic nature, chitosan is highly sought after for a wide range of general uses. According to Muzzarelli et al. [44] and Zhang et al. [45], chitosan is a commonly utilized vital component for biomedicines that removes hazardous substances from the environment and even the packaging sector. Chitosan offers a lot of benefits, but it also has certain drawbacks, including instability, poor thermo-mechanical properties, and porosity [46]. To enhance the quality of chitosan, several experts have previously been researching various modification techniques. A step toward such a modification system is the creation of chitosan nanocomposites, a complex material with a nanometer-sized size. It has been shown that chitosan-like biopolymers can increase the effectiveness and flexibility of a variety of biomaterials [47]. Nano-biocomposites, often known as green materials, are well recognized for their biodegradability and ecologically favorable characteristics. According to Darder et al. [48], these nano-biocomposites are crucial for biomineralization and bioremediation processes. The creation of cellulose, pectin, starch, chitin, alginate, gum, and chitosan-based nanomaterials are new trends and difficulties for natural biopolymers, according to

Nasrollahzadeh et al. [49]. These nanocomposites are developed by blending inorganic materials with at least one nanometer-scale dimension with natural polymers like chitosan.

These issues are successfully solved by fillers such as clay, carbon nanotubes, and hydroxyapatite nanocomposites [50]. The homogeneous dispersion of nanofillers inside the chitosan matrix can lead to the development of nano-biocomposites with remarkable characteristics, including improved mechanical barrier qualities and greater transparency. The enhanced aspect ratio and surface area of the nano-biocomposites in this scenario can be attributed to the dispersion at the nanoscale level. In contrast to conventional nanocomposites, the nanoscale biocomposites, including chitosan, demonstrate enhanced durability and stability in the form of nano-biocomposite materials but with a little reduction in reinforcing efficiency. As a result of these factors, nano-biocomposites based on chitosan exhibit properties that resemble biocompatible and bioactive matrices and demonstrate enhanced mechanical characteristics [51]. Nanocomposites, characterized by the exact mixing of inorganic fillers with a biopolymeric matrix, have been identified as the biggest hybrid products.

Carbon nanotubes have been shown to be extremely useful materials for improving the properties of polymers. Another study by Yang et al. [56] produced GO/chitosan nanocomposites with greatly increased elongation at break and increased tensile strength and Young's modulus by 64 and 122%, respectively. Nevertheless, the use of chitosan is hindered due to its high moisture permeability and fragility. They were made more durable mechanically, chemically, and thermally by combining chitosan and bentonite. Chitosan, a biopolymer, and bentonite were combined to generate nano-biocomposites by Abdelkrim et al. [52], which were the first to exhibit antibacterial properties. AgNps was combined with chitosan and bentonite to boost antibacterial activity against a variety of microorganisms. By mixing synthetic mica and chitosan, Alba et al. [53] explored chitosan mica nano-biocomposites. The nano-biocomposites were synthesized by an ion exchange process involving a charged mica and a solution of chitosan. Chitosan-based nano-biocomposites play a crucial role in water treatment uses owing to their remarkable attributes such as biodegradability, lack of toxicity, biocompatibility, hydrophilicity, reactivity, and cost-effectiveness [54]. Additionally, chitosan nano-biocomposites show strong antibacterial activity, high stability across a broad pH range, and a longer shelf life.

Creating magnetic chitosan-based nanocomposites for metal ion extraction has been the subject of multiple investigations in recent years [55]. The ability of magnetic chitosan nanoparticle-based nano absorbents to effectively remove metal ions from wastewater has been studied. In contrast to other polymers, the chitosan polymer chain has a higher concentration of hydroxyl and amino groups. This unique composition imparts enhanced selectivity to chitosan, allowing it to effectively melt metals and facilitate the rapid regeneration of sorbents. Therefore, chitosan nanoparticles may eventually serve as washable absorbents for sewage treatment [56].

In this study, N,O-carboxymethyl chitosan-labeled magnetic nanoparticles (NOCC-MNPs) were employed for the synthesis of chitosan-citrate gel beads (CCGBs). These CCGBs exhibited a noteworthy capacity for the adsorption of Cu (II) ions, with an adsorption capacity of 294.11 mg/g. The chelating activities of NOCC were attributed to the attendance of carboxyl, hydroxyl, and amino groups in carboxymethyl chitosan, thereby facilitating the aforementioned finding [57].

The incorporation of diverse reinforcing fillers into the polymer matrix is a feasible and effective approach for augmenting the mechanical characteristics of composite materials [58]. The SiO_2/chitosan composite has exceptional efficacy in the adsorption of heavy metal ions present in aqueous solutions, particularly As(V) and Hg(II), with adsorption capacities of 198.6 mg/g and 204.1 mg/g, respectively. Due to its inherent stability and exceptional mechanical strength, SiO_2 exhibits the capability to effectively impede the agglomeration of the adsorbent. Additionally, the hydroxyl and amino groups that are abundant in chitosan NPs coated on SiO_2 offer functional groups for heavy metal ions [59]. By altering the molecular structure of the polymer and expanding the surface area open to metal adsorption, nanofillers like TiO_2 can improve chitosan adsorption potential [60]. In their study, Razzaz et al. [60] employed a combination of trapping and coating techniques to fabricate chitosan/TiO_2 nanofibers. These nanofibers exhibited notable adsorption capabilities for Pb(II) and Cu(II), ranging from 475.50 to 715.70 mg/g.

12.6 Removal of Various Contaminants Using Chitosan-Based Materials

Chitosan and modified chitosan are extensively utilized in the field of environmental remediation for the purpose of removing and recovering dyes and heavy metals, which are recognized as very prevalent contaminants (see Fig 12.4). They have also been utilized as an antibacterial agent and for desalination processes.

12.6.1 Metal Ions Removal

The occurrence of water pollution due to metal ions is widespread and commonly attributed to industrial activities. Toxicity is a characteristic shown by certain substances, which can have detrimental impacts on organisms. Water and wastewater can undergo the process of biosorption utilizing chitosan in order to facilitate their removal. Heavy metal (including Pb, Cu, Ag, Ni, Cr, As, and others) adsorption has been demonstrated for chitosan and its derivatives, precious metals (such as Pd, Au, and Pt), as well as radionuclides.

In their study, Li et al. [34] synthesized a magnetic-cyclodextrin-chitosan (CC) material that exhibits enhanced separation properties and increased adsorption capacity. Furthermore, the incorporation of graphene oxide (GO) into the carbon composite (CC) was utilized to enhance the adsorption competence for metal elimination by chemical bonding. The findings of the study revealed that the adsorption capacity of magnetic-cyclodextrin-chitosan/graphene oxide (CCGO) for Cr (VI) exhibited a higher removal efficiency (67.66 mg/g) compared to other sorbents like poly(4-vinyl pyridine) modified activated carbon (53.7 mg/g) and Ethylenediamine functionalized Fe_3O_4 (61.35 mg/g). The observed phenomenon can be attributed to the augmentation in the surface area resulting from the greater abundance of hydroxyl and amino groups, in addition to the magnetic characteristics shown by Fe_3O_4. The findings of the study indicate that the efficacy of removal was greater under conditions of low pH. Additionally, it was observed that the adsorption equilibrium of CCGO followed the Langmuir isotherm model.

The authors confirmed that the adsorption performance of CCGO was significantly influenced by the charge concentration and specific surface area, as reported in reference [34]. In this study, the functionalization of chitosan was explored as a means to remove arsenic III and V from aqueous solutions, employing graphene oxide (GO) as a facilitating agent. The kinetics of arsenic adsorption were investigated using a pseudo-second-order model, whereas the adsorption behavior was described using the Langmuir and Freundlich models. The researchers determined that the most favorable pH range for the elimination of As (V) was between 4.3 and 6.5. Additionally, they observed that the adsorption capacity aligned with the Langmuir isotherm model, namely 71.9 mg/g for As (V) and 64.2 mg/g for As (III) [61].

In their study, Li et al. [62] employed carboxylate-rich magnetic chitosan flocculants, which involved the graft copolymerization of chitosan onto magnetite particles, for the purpose of eliminating Ni (II) from synthetic wastewater. Furthermore, the researchers conducted an investigation into the impact of several ionic groups, including hydroxyl, carboxyl, and amino, on the process of metal removal. Fourier transform infrared spectroscopy (FITR), thermogravimetric analysis (TGA), transmission electron microscopy (TEM), X-ray diffraction (XRD), scanning electron microscopy (SEM), and X-ray photoelectron spectroscopy (XPS) methods were employed to examine the efficacy of grafting, magnetic properties, and core-shell architecture. The flocculation of Ni (II) ions by magnetic chitosan with surface grafting was evaluated by the analysis of kinetic data. The equilibrium state, characterized by a 98.3% absorption rate, was attained within a time period of 60 minutes, at pH conditions ranging from 4.0 to 8.0, for the purpose of Ni (II) removal.

The use of itaconic acid as a source of carboxyl groups resulted in enhanced flocculation performance of the carboxylate-enriched magnetic chitosan flocculant compared to the chitosan magnetic flocculant. The authors place significant emphasis on the simplicity of recycling the flocculant, as well as the cost-effectiveness, efficiency, and high adsorption capacity of the separation process [63].

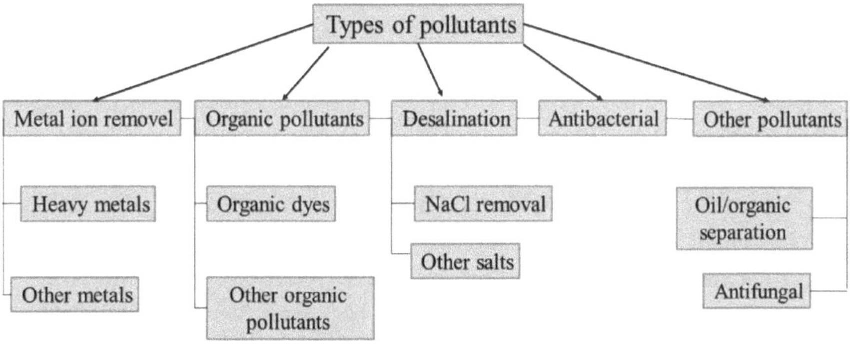

Figure 12.4 The several categories of contaminants that may be effectively eliminated with the use of chitosan and its derivatives.

12.6.2 Organic Pollutants

Every year, a substantial quantity of organic pollutants, amounting to thousands of tons, is introduced into the environment by various businesses such as the chemical, textile, fertilizer, and pharmaceutical sectors. Examples include organic colors, phenols, hydrocarbons, insecticides, plasticizers, detergents, and grease. Chitosan and its derivatives are frequently used to remove colors from water and wastewater when dealing with organic contaminants.

Methylene blue (MB) has been extensively investigated as an organic dye in the context of employing chitosan or chitosan derivatives for the treatment of industrial waste. In their study, Fan et al. [64] introduced a novel adsorbent known as magnetic chitosan/graphene oxide (MCGO) with the capability of extracting methylene blue. The adsorbent (MCGO) was characterized using SEM, FTIR, and XRD techniques to investigate the structural, morphological, and magnetic changes. The examination of the MB adsorption process involved the manipulation of many factors, including pH, adsorption duration, temperature, and initial concentration of MB. Based on the research findings, it has been determined that the optimal pH value for both adsorption kinetics and isotherm is 5.3. The Langmuir isotherm and pseudo-second-order models exhibited the highest degree of conformity with the experimental data. The adsorption capacity of MCGO (95.16 mg/g) was found to be higher compared to magnetic chitosan (60.4 mg/g) for methylene blue. Based on thermodynamic parameters, it was determined that the reaction exhibited characteristics of being both exothermic and spontaneous. Moreover, The process of adsorption of methylene blue took place on the surface of the adsorbent through an ion exchange mechanism [40].

In independent research, the use of alizarin yellow R (AY), a dye with cationic properties, was examined alongside a mixture of anionic and cationic dyes, like methyl violet, for the purpose of adsorption from industrial wastewater. The composite's composition and surface characteristics were assessed by the use of several analytical techniques, including SEM, FTIR, EDX, XRD, and TGA. The elimination efficacy of the magnetic chitosan/GO composite was assessed under various conditions, including pH ranging from 4 to 12, durations ranging from 0 to 80 minutes, doses ranging from 2 to 10 mg, starting dye concentrations ranging from 2 to 30 g/L, and temperatures of 313 K, 303 K, and 323 K. The adsorption kinetics of both dyes similarly conformed to the pseudo-second-order equation. The determination of the equilibrium state of the interaction between the dye and the absorber was conducted utilizing the Freundlich and Langmuir isotherms [65].

12.6.3 Desalination

Materials produced from chitosan have also been applied to the desalination of water. In a similar work, Qian et al., A desalination membrane composed of a chitosan/GO chitosan-derived matrix was fabricated. By altering the GO content, they looked at how the membrane's morphology, wettability, and desalination capability changed. It also looked at how salt and water permeated through the membrane in terms of kinetics and thermodynamics. The permeation flux (30.0 kg/m^2 h) and salt rejection (99.99%) were improved for NaCl (5 wt% aqueous) solution with 1 wt% GO concentration at 81 °C. The outcomes showed that high compatibility between the GO and chitosan polymer increased the durability of the chitosan material. The investigation of the solubility of NaCl in H_2O and the observation of diffusion through the membrane revealed a discernible trade-off between salt/water selectivity and permeability as the concentration of graphene oxide (GO) rose [83]. Deng et al. also modified chitosan for application in water desalination using GO. However, they also reduced GO using titania oxide to improve salt rejection. According to the study, hybrid membrane rejection rates were around 30% higher than those of membranes that exclusively contained GO [66].

Additionally, chitosan has been combined with other biopoly-meric components derived from biomass, including hemicellulose, to create chitosan-based products for the elimination of salt from solutions with specified concentrations. To do this, hemicellulose was first grafted with nitric acid (DTPA, diethylenetriaminepentaacetic acid) and was then cross-linked with chitosan. The primary aim of this investigation was to assess the impact of temperature, reaction duration, and hemicellulose-DTPA concentration on the adsorption of NaCl using the studied adsorbent. The findings indicated that the greatest salt consumption observed was 0.30 g/g. The prediction of the equilibrium of salt adsorption was also conducted utilizing a second-order kinetic model [67].

For the desalination of brackish and seawater, a nanofiltration membrane made of chitosan was created. To boost the membrane's permeate flow, piperazine was added. Different compositional and surface characterization techniques were used to analyze the membrane. The investigation showed that piperazine at a concentration of 25% (w/w) enhanced the permeate flow (60.6 L/m^2/h) nearly twofold without significantly reducing the rejection of Na_2SO_4, which was measured at 89.1% [68].

Numerous studies have shown that chitosan membranes with the right support may effectively separate or remove pollutants. Recently, a loose nanofiltration membrane made of chitosan and catechin was created for the desalination of dyes [69]. The chitosan combine was formed on the surface of a polyacrylonitrile (HPAN) ultrafiltration (UF) membrane to research its effects on dye desorption. The findings revealed that during the dye desalination process, a greater NaCl elimination rate (82%) combined with a 10%-degree loss in dye elimination was detected. The membrane also demonstrated good dye antifouling performance with a flux recovery ratio of 87.8%. In a different work, a chitosan nanofiltration membrane supported by polypropylene fiber was created for desalination purposes [70]. The findings indicated that the membrane exhibited a greater rejection rate at lower pH levels, namely acidic conditions, whereas the rejection rate dropped when exposed to higher pH levels, particularly basic conditions. In their study, Kumar et al. [89] employed a diffusion-induced phase-separation approach to fabricate a polysulfone and chitosan composite ultrafiltration (UF) membrane. The researchers reached the conclusion that the increased efflux seen at low pH levels was a result of the protonation of the chitosan amine (–NH_2) groups.

12.6.4 Antibacterial Effect

The exceptional antibacterial characteristics of chitosan and its derivatives have garnered significant interest during the last decade [71]. One of the potential methods for industrial desalination is membrane distillation. Nevertheless, the membranes exhibit pore wetting or a significant susceptibility to fouling, owing to the existence of diverse hydrophobic or amphiphilic compounds in wastewater. A study done by Wang et al. [72] examined the impact of fouling behaviors and surface-wetting properties on the process of membrane distillation. The study focused on investigating the resistance of composite, omniphobic, and hydrophobic membranes against fouling and wetting. These membranes were composed of a super hydrophilic top surface and a hydrophobic substrate.

Membranes with significantly advanced crude oil content were tested with hypersaline feed solution as a surfactant, with and without Triton X-100. The findings showed that while the pores were wetted when Triton X-100 was present, membranes with super hydrophilic groups on the top surface were more resistant to oil contamination than in the absence of Triton X-100. Additionally, the oil-in-water emulsion was able to easily foul omniphobic membranes, although surfactant-assisted membrane distillation maintained steady performance.

One of the practical possibilities and potential uses in the desalination of saltwater is forward osmosis (FO). Due to its independence from external hydraulic pressure, FO is believed to possess a lower propensity for fouling. In the study conducted by Salehi et al., a layer-by-layer application technique was employed to deposit Chitosan/GO onto a porous support. The support material used was sulfonated polyethersulfone embedded inside a polyethersulfone matrix [73] and generated a very effective FO membrane. The results showed that, in terms of pollution resistance and water permeation, the chitosan/GO-coated membrane performed 24 times better than the thin-film composite membrane. Another research developed a chitosan/GO nanocomposite for antibacterial action using magnetic chitosan and GO. The damaged cell membrane was observed to improve the overall antibacterial effectiveness against *E. coli* [74]. The researchers presented a study on the surface modification of commercially available saltwater thin-film composite (polyamide) membranes by means of geo-functionalized chitosan [75]. The enhanced composite membranes demonstrated enhanced performance in terms of salt rejection, water flow rate, and resistance to fouling in comparison to the unmodified polyamide membrane.

12.6.5 Other Pollutants

Other pollutants, such as CO_2 fixation, algae removal, oil or organic separation, have also been removed using chitosan-based polymers. A bioinspired chitosan-derived mesh remained fabricated to advance the oil/water fouling resistance performance at different pH and hypersaline conditions. According to the study, oil/water combinations may be separated with an efficiency of.99% by using just gravity. According to the authors, treating industrial oil wastes and cleaning up oil spills both benefits from the separation procedure [76]. In a different study [77], chitosan-oxidized cellulose aerogels were created, and they may be employed in the field of water/oil separation or the cleaning of spills involving organic solvents and oils. The pseudo-second-order model is the most appropriate for this type of aerogel, according to the process' kinetic analysis. For a range of oils and organic solvents, the aerogel demonstrated outstanding absorption capability and a superior absorption rate.

A work on mesoporous zeolite/chitosan composite carbon dioxide adsorption or chemical fixation is also documented in the literature. When compared to pure zeolite or chitosan, the composite presented enhanced CO_2 adsorption. The composite also showed impressive catalytic activity for the chemical obsession of CO_2 into cyclic carbonates [78].

Chlorella vulgaris, a freshwater microalga, has been quickly harvested using chitosan as a flocculant. When C. vulgaris was recovered using the different chitosan concentrations, it was shown that chitosan had the maximum effectiveness, with a 92% to 0.4% recovery rate in under three minutes. The highest observed result for harvesting efficiency at pH 6.0 was 99%, multiplied by 0.5%. The results suggest that chitosan has promising potential as a flocculant for the purpose of microalgae harvesting. This is attributed to its notable effectiveness, the minimal dose required, and expedited settling duration [79].

12.7 Mechanism of Adsorption

The comprehension of the adsorption mechanism has significance due to its potential to enhance the efficiency of adsorbents in the elimination of environmental contaminants. Understanding the adsorption mechanism of adapted chitosan as an adsorbent, particularly in the context of heavy metal removal, poses considerable challenges [42]. Only a few investigations have been done on the mechanism of chitosan's interaction with metal ions. The C-position hydroxyl group is thought to facilitate adsorption, whereas the amine group is often considered to be the greatest reactive site for metal ions. The evolution of diverse interactions involving heavy metals is significantly influenced by these two functional groups. Furthermore, the determination of the metal ions, pH of the system, and solution matrix are also influenced by the aforementioned sort of interaction. At neutral pH or in the moderately acidic area, nitrogen possessing two unshared electron pairs functions as a binding site for metal cations [80]. While the amine groups become protonated in highly acidic environments, this causes the chitosan polymers to behave positively. As a result, the chitosan-based substance may also interact with metal anions [81]. The adsorption of heavy metals onto the altered chitosan surface can be attributed to many mechanisms, including non-polar interactions (van der Waals forces), electrostatic, ion exchange, or chemical interactions (chelation or complexation). Numerous surface analyses, including FTIR, XPS, XRD, SEM, and EDX spectroscopy, can be employed to identify multiple functional groups. These analyses play a crucial role in understanding the interaction between raw or modified chitosan and metals. The present study employed FTIR and XPS techniques to examine the underlying mechanism of Cr (VI) adsorption by Fe_2O_3-chitosan beads.

The findings suggested that the modified chitosan's hydroxyl and amino groups could contribute to the metal removal. During the process of elimination, three adsorption events were noticed: chelation, electrostatic interactions, and complexation (as seen in Fig. 12.5). Electrostatic attraction was used to initially sorb Cr (VI) onto the bead, which the chitosan 2 OH group then used to convert Cr (VI) to Cr (III). Two factors—the precipitation of Cr (III) in the hydroxide state and the synthesis of chelates with free amino groups—play a role in Cr (III) immobilization on the surface of the beads. These events also characterize the adsorbent's greater ability to absorb Cr (VI) [82].

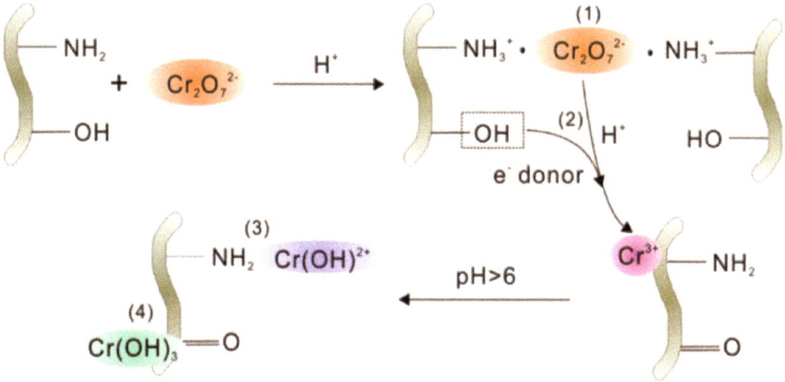

Figure 12.5 Mechanism of interaction between Fe_2O_3-containing chitosan beads and Cr (VI). Adapted and reprinted with permission from [12].

12.7.1 Mechanism of Interaction and Methylene Blue Dye Removal

Chitosan, which has a high concentration of three main functional groups, has exceptional adsorption properties for a variety of different water contaminants, including metal ions and organic dyes [6]. First, the main functional component of chitosan is an amino group ($-NH_2$). Chitosan's $-NH_2$ groups have a single pair of electrons that can be employed to coordinate with positively charged ions. Second, the presence of a significant quantity of hydroxyl groups (-OH) linked to the sugar units facilitates the ability of chitosan to establish hydrogen bonds with other advantageous chemicals. Chitosan is more hydrophilicity, owing to its inclusion of -OH and $-NH_2$ functional groups, making it easier to soluble in water and form hydrogen bonds with beneficial molecules. The acetyl groups ($-COCH_3$) that are joined to the amino groups are the last component of chitosan. They strengthen chitosan's overall structure and improve its capacity to bind to methylene blue dye [6].

Methylene blue is characterized by the presence of a thiazine ring and a heterocyclic ring, including sulfur (S) and nitrogen (N) atoms. The phenothiazine moiety, which is characterized by a planar aromatic ring structure, is covalently bonded to the thiazine ring in this compound. The presence of alternating single and double bonds inside the phenothiazine ring facilitates the delocalization of π-electrons across the conjugate system. The phenylamino group and the thiazine ring in methylene blue play crucial roles as chromophore groups. The distinctive blue hue shown by methylene blue dye has been ascribed to the chromophore groups' capacity to absorb light within the red-orange segment of the electromagnetic spectrum. The color of methylene blue is not directly influenced by the connected dimethylamino group's ($-N(CH_3)_2$) existence. However, it plays a substantial role in enhancing the reactive and electron-donating capabilities of MB. The ability of methylene blue to form hydrogen bonds with other molecules is improved by the presence of the $-N(CH_3)_2$ group. The heightened interaction seen can be attributed to the pronounced electron-donating properties exhibited by the $-N(CH_3)_2$ group, which arise from the presence of two methyl ($-CH_3$) groups attached to the amino nitrogen (–NH–) atom [83].

Building upon the previous discussion, Fig. 12.6 illustrates three primary modalities of interaction that may be employed to comprehend the adsorption mechanism of methylene blue dye by chitosan:

(1) Hydrophobic and "–" stacking interactions: The dye molecules' hydrophobic areas, such as their $-CH_3$ groups, can connect with the chitosan chains' hydrophobic regions, such as their $-CH_3$ groups, through hydrophobic contacts. The methylene blue molecules may bind or adsorb to the surface of the chitosan or to its internal structure as a result of this interaction. Additionally, the aromatic rings in methylene blue and the aromatic rings in chitosan can interact through stacking.

(2) Electrostatic interactions: Chitosan ($-COCH_3$) has negatively charged groups that can interact with the positively charged group in methylene blue (S^+), creating electrostatic attraction among both molecules [43].

(3) Hydrogen bonding: In hydrogen bonding, the nitrogen atoms in the structure of methylene blue dye can act as acceptors. Similarly, chitosan is composed of –OH and $-NH_2$

groups which possess the ability to serve as donors and acceptors of hydrogen bonds, respectively. These three forms of interactions, such as electrostatic attraction, hydrogen bonds, and hydrophobic interactions, are cooperatively taken for the binding and absorption of methylene blue dye by chitosan. The adsorption process can be affected by several elements, including the concentration of the dye, temperature, pH, and the particular properties of the chitosan utilized [83].

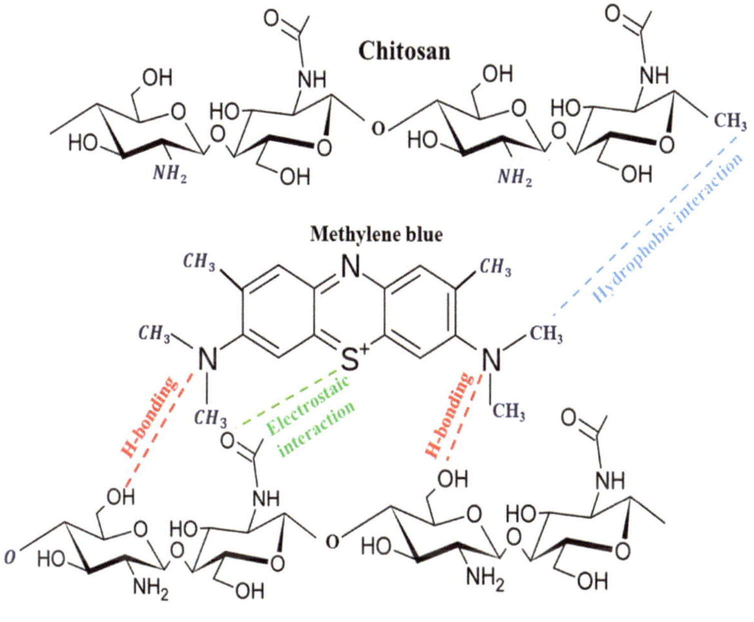

Figure 12.6 Diagram showing hydrophobic, hydrogen bonds electrostatic, and interactions between chitosan and the dye methylene blue. Copyright © 2023, Ilham et al. [83].

12.7.2 Factors Influencing Adsorption Effectiveness

Several important factors affect the rate and efficiency of chitosan as an adsorbent [81]. The adsorption process is primarily influenced by elements such as the material's physical makeup, including its source and composition, as well as its physical features and the manner in which the procedure is applied.

The most significant impact on chitosan's adsorption capability comes from its physical makeup. For the purpose of treating water, chitosan has undergone physical modification to create a variety of materials, including beads and hydrogels [84], magnetic beads [85], nanoparticles swollen beads [86], hollow fibers [87], powder [88], solid fibers [89], flakes [90], membrane [91], sponge-like, and/or honeycomb structures [92].

The origin and chemical makeup of the chitosan used to filter water are the second most crucial factors. Various sources of chitosan, such as shells derived from lobster, crab, and shrimp, have been employed for water filtration purposes [93]. Several different natural materials have been used to extract chitosan. DD, the molar density of amino groups on the surface, and the capacity of metal ions to bind to their amino groups both play a significant role in the capacity of chitosan to adsorb various substances. Chitosanmetal ion adsorption with various levels of deacetylation was examined by Wu et al. [94]. The findings showed that chitosan flakes had corresponding reactive red dye sorption capabilities of 293 (76.3%), 398 (77.9%), and 494 (80.9%) mg/g to crab, lobster, and shrimp sources. The opposite was true in the case of beads (greater DD: poor capacity for sorption). The capacity of chitosan-derived materials to adsorb chemicals is influenced by several physical parameters, like pore size distribution, surface area, elemental composition, density, and pore volume. Adsorption depends heavily on the surface area since a larger surface area makes more moieties accessible for adsorption. According to Li et al. [34], utilizing an adsorbent made of chitosan to remove Cr (VI) has an advantage due to its large surface area. Adsorbent, they added, performed better in removing Cr (VI) at little pH. The research found that compared to raw chitosan, the modified chitosan contained higher hydroxyl and amino groups. The enhanced effectiveness of Cr (VI) adsorption by modified chitosan can be attributed to its greater surface charge concentration and specific surface area, as demonstrated in previous research [34].

pH, degree of agitation, temperature of the solution, and presence of ions or cations are key parameters for chitosan application procedures. The rate-controlling phases are most significantly impacted by these variables. For example, the adsorption rate increases, but the adsorption capacity decreases when the temperature of the active adsorption process increases [95]. This is explained by the desorption rate constant being more sensitive to temperature. The pH of the medium plays a crucial role in the adsorption of metal ions. Specifically, when the pH is very low, around 2.5, it facilitates a greater degree of proton-metal ion adsorption. The pH of the system, together with pKa and the zeta potential of the chitosan amine group, plays an important part since an elevated pH level leads to the occurrence of metallic precipitation [81].

12.8 Conclusion and Future Perspective

Desalination, heavy metal removal, and microbial adsorption decontamination of water and wastewater can all be accomplished using chitosan-based polymers, which also have the advantages of affordable cost, safety, and biocompatibility. In comparison to the current adsorbing materials, the features of chitosan-based compounds give them a bright future. However, it is acknowledged that the main problem preventing chitosan's application on a broader scale is its swelling. Only a small number of companies use material generated from chitosan that is of industrial quality to remove colors and toxic metals in water and wastewater. Therefore, for larger-scale applications, the potential of modified chitosan materials has to be further investigated. There are many research concerns that need to be answered before chitosan can be fully utilized to limit water damage, despite the fact that there is a lot of research being done to use chitosan biosorbents for water and wastewater clean-up. Below we have selected a few of them:

1. To obtain the greatest amount of adsorption, choose an appropriate chitosan-modified adsorbent based on interactions between the adsorbate and the adsorbent.
2. Understanding of the mechanism that increases the pollutant's binding sites throughout the adsorption process, maximizing the adsorbent's ability to absorb the pollutant.
3. Extensive research on regeneration in order to recover the metals and adsorbent and reduce the overall cost.

4. Knowledge of how various contaminants interact with one another throughout the adsorption process.
5. Effective removal of several pollutants at once on a wider scale.
6. Scaling up laboratory-based batch studies will enhance the industrial uses of chitosan-based adsorbents.

Despite these limitations, polymers made from chitosan have immense potential for future water treatment. Because of growing concerns about water pollution, both the public and governments are taking great interest in finding solutions to this urgent problem, and the responsibilities of academics have increased dramatically as a result. The need for renewable resources with enhanced efficiency and cost-effective design is expected to grow in the coming years. With the development of new, cheaper materials coupled with the continuous improvement of water treatment capabilities, scientists now have opportunities for new ideas and collaborations to help identify better solutions to the water problem. More widespread use of chitosan-based polymers as sustainable and regenerative materials may aid future water treatment efforts.

References

1. Venohr M, Langhans SD, Peters O, Hölker F, Arlinghaus R, Mitchell L, Wolter C (2018) The underestimated dynamics and impacts of water-based recreational activities on freshwater ecosystems. *Environmental Reviews* 26: 199–213.
2. Urso M, Ussia M, Pumera M (2023) Smart micro-and nanorobots for water purification. *Nature Reviews Bioengineering* 1: 236–251.
3. Mauter MS, Zucker I, Perreault F, Werber JR, Kim J-H, Elimelech M (2018) The role of nanotechnology in tackling global water challenges. *Nature Sustainability* 1: 166–175.
4. Rhazi M, Desbrieres J, Tolaimate A, Rinaudo M, Vottero P, Alagui A, El Meray M (2002) Influence of the nature of the metal ions on the complexation with chitosan: Application to the treatment of liquid waste. *European Polymer Journal* 38: 1523–1530.
5. Zubair M, Arshad M, Ullah A (2020) Chitosan-based materials for water and wastewater treatment. *Handbook of Chitin and Chitosan*. Elsevier, pp 773–809.

6. Amor IB, Hemmami H, Laouini SE, Abdelaziz AG, Barhoum A (2023) Influence of chitosan source and degree of deacetylation on antibacterial activity and adsorption of AZO dye from water. *Biomass Conversion and Biorefinery*: 1–11.
7. Ibrahim MB, Ibrahim AM, Haruna MA, Yaro MS (2013) Removal of Congo red dye from aqueous solution using sugarcane bagasse and Bambara groundnut shell. *ChemSearch Journal* 4: 22–28.
8. Karthik R, Muthezhilan R, Jaffar Hussain A, Ramalingam K, Rekha V (2016) Effective removal of Methylene Blue dye from water using three different low-cost adsorbents. *Desalination and Water Treatment* 57: 10626–10631.
9. Reddy MS, Nirmala V (2019) Bengal gram seed husk as an adsorbent for the removal of dyes from aqueous solutions–Column studies. *Arabian Journal of Chemistry* 12: 1695–1706.
10. Hamzeh Y, Ashori A, Azadeh E, Abdulkhani A (2012) Removal of Acid Orange 7 and Remazol Black 5 reactive dyes from aqueous solutions using a novel biosorbent. *Materials Science and Engineering: C* 32: 1394–1400.
11. Khattri S, Singh M (2012) Use of Sagaun sawdust as an adsorbent for the removal of crystal violet dye from simulated wastewater. *Environmental Progress & Sustainable Energy* 31: 435–442.
12. Abbas FS (2013) Dyes removal from wastewater using agricultural waste. *Advances in Environmental Biology* 7: 1019–1026.
13. Alam MS, Khanom R, Rahman MA (2015) Removal of congo red dye from industrial wastewater by untreated sawdust. *American Journal of Environmental Protection* 4: 207–213.
14. Akhouairi S, Ouachtak H, Addi AA, Jada A, Douch J (2019) Natural sawdust as adsorbent for the eriochrome black T dye removal from aqueous solution. *Water, Air, & Soil Pollution* 230: 1–15.
15. Asiagwu A, Emoyan O, Ojebah C (2018) Removal of tartrazine yellow dye from aqueous solution using groundnut shell as biomass: Kinetic approach. *International Journal of Engineering Research* 7. DOI : 10.17577/IJERTV7IS050190.
16. Açışlı Ö, Acar İ, Khataee A (2022) Preparation of a surface modified fly ash-based geopolymer for removal of an anionic dye: Parameters and adsorption mechanism. *Chemosphere* 295: 133870.
17. Mahmoud ME, El-Ghanam AM, Mohamed RHA, Saad SR (2020) Enhanced adsorption of Levofloxacin and Ceftriaxone antibiotics from water by assembled composite of nanotitanium oxide/chitosan/nano-bentonite. *Materials Science and Engineering: C* 108: 110199.

18. El-Nafaty U, Muhammad I, Abdulsalam S (2013) Biosorption and kinetic studies on oil removal from produced water using banana peel. *Civil and Environmental Research* 3: 125–136.
19. Fan H, Li F, Huang H, Yang J, Zeng D, Liu J, Mou H (2022) pH graded lignin obtained from the by-product of extraction xylan as an adsorbent. *Industrial Crops and Products* 184: 114967.
20. de Caprariis B, De Filippis P, Hernandez AD, Petrucci E, Petrullo A, Scarsella M, Turchi M (2017) Pyrolysis wastewater treatment by adsorption on biochars produced by poplar biomass. *Journal of Environmental Management* 197: 231–238.
21. Al. Haddabi M, Vuthaluru H, Znad H, Ahmed M (2015) Removal of dissolved organic carbon from oily produced water by adsorption onto date seeds: Equilibrium, kinetic, and thermodynamic studies. *Water, Air, & Soil Pollution* 226: 1–15.
22. Song S-T, Saman N, Johari K, Mat H (2014) Surface chemistry modifications of rice husk toward enhancement of Hg (II) adsorption from aqueous solution. *Clean Technologies and Environmental Policy* 16: 1747–1755.
23. Shen Z, Zhang Y, McMillan O, Jin F, Al-Tabbaa A (2017) Characteristics and mechanisms of nickel adsorption on biochars produced from wheat straw pellets and rice husk. *Environmental Science and Pollution Research* 24: 12809–12819.
24. Alkherraz AM, Ali AK, Elsherif KM (2020) Removal of Pb (II), Zn (II), Cu (II) and Cd (II) from aqueous solutions by adsorption onto olive branches activated carbon: Equilibrium and thermodynamic studies. *Chemistry International* 6: 11–20.
25. Peng H, Gao P, Chu G, Pan B, Peng J, Xing B (2017) Enhanced adsorption of Cu (II) and Cd (II) by phosphoric acid-modified biochars. *Environmental Pollution* 229: 846–853.
26. Kunjirama M, Saman N, Johari K, Song S-T, Kong H, Cheu S-C, Lye JWP, Mat H (2017) Adsorption affinity and selectivity of 3-ureidopropyltriethoxysilane grafted oil palm empty fruit bunches towards mercury ions. *Environmental Science and Pollution Research* 24: 15167–15181.
27. Araújo CS, Almeida IL, Rezende HC, Marcionilio SM, Léon JJ, de Matos TN (2018) Elucidation of mechanism involved in adsorption of Pb (II) onto lobeira fruit (Solanum lycocarpum) using Langmuir, Freundlich and Temkin isotherms. *Microchemical Journal* 137: 348–354.

28. Alqadami AA, Khan MA, Otero M, Siddiqui MR, Jeon B-H, Batoo KM (2018) A magnetic nanocomposite produced from camel bones for an efficient adsorption of toxic metals from water. *Journal of Cleaner Production* 178: 293–304.
29. Crini G, Torri G, Lichtfouse E, Kyzas GZ, Wilson LD, Morin-Crini N (2019) Dye removal by biosorption using cross-linked chitosan-based hydrogels. *Environmental Chemistry Letters* 17: 1645–1666.
30. Kumar S, Ye F, Dobretsov S, Dutta J (2019) Chitosan nanocomposite coatings for food, paints, and water treatment applications. *Applied Sciences* 9: 2409.
31. Qamar SA, Ashiq M, Jahangeer M, Riasat A, Bilal M (2020) Chitosan-based hybrid materials as adsorbents for textile dyes–A review. *Case Studies in Chemical and Environmental Engineering* 2: 100021.
32. Lee CS, Robinson J, Chong MF (2014) A review on application of flocculants in wastewater treatment. *Process Safety and Environmental Protection* 92: 489–508.
33. Nechita P (2017) Applications of chitosan in wastewater treatment. *Biological Activities and Application of Marine Polysaccharides* 1: 209–228.
34. Li L, Fan L, Sun M, Qiu H, Li X, Duan H, Luo C (2013) Adsorbent for chromium removal based on graphene oxide functionalized with magnetic cyclodextrin–chitosan. *Colloids and Surfaces B: Biointerfaces* 107: 76–83.
35. Qin Y, Shi B, Liu J (2006) Application of chitosan and alginate in treating waste water containing heavy metal ions. *Indian Journal of Chemical Technology* 13: 464–469.
36. Dwivedi C, Gupta A, Chaudhary A, Nandi CK (2014) Gold nanoparticle chitosan composite hydrogel beads show efficient removal of methyl parathion from waste water. *RSC Advances* 4: 39830–39838.
37. Abd-Elhakeem MA, Alkhulaqi TA (2014) Simple, rapid and efficient water purification by chitosan coated magnetite nanoparticles. *Journal of Environmental Nanotechnology* 3: 17–20.
38. Motshekga SC, Ray SS, Onyango MS, Momba MN (2015) Preparation and antibacterial activity of chitosan-based nanocomposites containing bentonite-supported silver and zinc oxide nanoparticles for water disinfection. *Applied Clay Science* 114: 330–339.
39. Vakili M, Rafatullah M, Salamatinia B, Abdullah AZ, Ibrahim MH, Tan KB, Gholami Z, Amouzgar P (2014) Application of chitosan and its derivatives as adsorbents for dye removal from water and wastewater: A review. *Carbohydrate Polymers* 113: 115–130.

40. Fan L, Luo C, Li X, Lu F, Qiu H, Sun M (2012) Fabrication of novel magnetic chitosan grafted with graphene oxide to enhance adsorption properties for methyl blue. *Journal of Hazardous Materials* 215: 272–279.

41. Qian X, Li N, Wang Q, Ji S (2018) Chitosan/graphene oxide mixed matrix membrane with enhanced water permeability for high-salinity water desalination by pervaporation. *Desalination* 438: 83–96.

42. Zhang L, Zeng Y, Cheng Z (2016) Removal of heavy metal ions using chitosan and modified chitosan: A review. *Journal of Molecular Liquids* 214: 175–191.

43. Dong C, Chen W, Liu C, Liu Y, Liu H (2014) Synthesis of magnetic chitosan nanoparticle and its adsorption property for humic acid from aqueous solution. *Colloids and Surfaces A: Physicochemical and Engineering Aspects* 446: 179–189.

44. Muzzarelli R, Muzzarelli C (2005) Chitosan chemistry: Relevance to the biomedical sciences. *Polysaccharides I: Structure, Characterization and Use*: pp. 151–209.

45. Zhang H-l, Wu S-H, Tao Y, Zang L-Q, Su Z-Q (2010) Preparation and characterization of water-soluble chitosan nanoparticles as protein delivery system. *Journal of Nanomaterials* 2010: 1–5.

46. Escudero-Oñate C, Martínez-Francés E (2018) A review of chitosan-based materials for the removal of organic pollution from water and bioaugmentation. *Chitin-Chitosan-Myriad Functionalities in Science and Technology* 1. DOI:10.5772/INTECHOPEN.76540.

47. Sharma R, Kumar D (2019) Chitosan-based membranes for wastewater desalination and heavy metal detoxification. *Nanoscale Materials in Water Purification*: 799–814.

48. Darder M, Aranda P, Ruiz-Hitzky E (2007) Bionanocomposites: A new concept of ecological, bioinspired, and functional hybrid materials. *Advanced Materials* 19: 1309–1319.

49. Nasrollahzadeh M, Sajjadi M, Iravani S, Varma RS (2021) Starch, cellulose, pectin, gum, alginate, chitin and chitosan derived (nano) materials for sustainable water treatment: A review. *Carbohydrate Polymers* 251: 116986.

50. Kumari S, Singh RP, Chavan NN, Annamalai PK (2016) Chitosan-based bionanocomposites for biomedical application. *Bioinspired, Biomimetic and Nanobiomaterials* 7: 219–227.

51. Azmana M, Mahmood S, Hilles AR, Rahman A, Arifin MAB, Ahmed S (2021) A review on chitosan and chitosan-based bionanocomposites:

Promising material for combatting global issues and its applications. *International Journal of Biological Macromolecules* 185: 832–848.

52. Abdelkrim S, Mokhtar A, Djelad A, Bennabi F, Souna A, Bengueddach A, Sassi M (2020) Chitosan/Ag-bentonite nanocomposites: Preparation, characterization, swelling and biological properties. *Journal of Inorganic and Organometallic Polymers and Materials* 30: 831–840.
53. Alba MD, Cota A, Osuna FJ, Pavón E, Perdigón AC, Raffin F (2019) Bionanocomposites based on chitosan intercalation in designed swelling high-charged micas. *Scientific Reports* 9: 10265.
54. Ursino C, Castro-Muñoz R, Drioli E, Gzara L, Albeirutty MH, Figoli A (2018) Progress of nanocomposite membranes for water treatment. *Membranes* 8: 18.
55. Jiang Z, Li N, Li P-Y, Liu B, Lai H-J, Jin T (2021) One-step preparation of chitosan-based magnetic adsorbent and its application to the adsorption of inorganic arsenic in water. *Molecules* 26: 1785.
56. Fan H-L, Zhou S-F, Jiao W-Z, Qi G-S, Liu Y-Z (2017) Removal of heavy metal ions by magnetic chitosan nanoparticles prepared continuously via high-gravity reactive precipitation method. *Carbohydrate Polymers* 174: 1192–1200.
57. Mi F-L, Wu S-J, Chen Y-C (2015) Combination of carboxymethyl chitosan-coated magnetic nanoparticles and chitosan-citrate complex gel beads as a novel magnetic adsorbent. *Carbohydrate Polymers* 131: 255–263.
58. Zhang H, Li Y, Shi R, Chen L, Fan M (2018) A robust salt-tolerant superoleophobic chitosan/nanofibrillated cellulose aerogel for highly efficient oil/water separation. *Carbohydrate Polymers* 200: 611–615.
59. Liu J, Chen Y, Han T, Cheng M, Zhang W, Long J, Fu X (2019) A biomimetic SiO_2@ chitosan composite as highly-efficient adsorbent for removing heavy metal ions in drinking water. *Chemosphere* 214: 738–742.
60. Razzaz A, Ghorban S, Hosayni L, Irani M, Aliabadi M (2016) Chitosan nanofibers functionalized by TiO_2 nanoparticles for the removal of heavy metal ions. *Journal of the Taiwan Institute of Chemical Engineers* 58: 333–343.
61. Kumar ASK, Jiang S-J (2016) Chitosan-functionalized graphene oxide: A novel adsorbent an efficient adsorption of arsenic from aqueous solution. *Journal of Environmental Chemical Engineering* 4: 1698–1713.

62. Chen Q, Qi Y, Jiang Y, Quan W, Luo H, Wu K, Li S, Ouyang Q (2022) Progress in research of chitosan chemical modification technologies and their applications. *Marine Drugs* 20: 536.

63. Liu B, Chen X, Zheng H, Wang Y, Sun Y, Zhao C, Zhang S (2018) Rapid and efficient removal of heavy metal and cationic dye by carboxylate-rich magnetic chitosan flocculants: Role of ionic groups. *Carbohydrate Polymers* 181: 327–336.

64. Fan L, Luo C, Sun M, Li X, Lu F, Qiu H (2012) Preparation of novel magnetic chitosan/graphene oxide composite as effective adsorbents toward methylene blue. *Bioresource Technology* 114: 703–706.

65. Gul K, Sohni S, Waqar M, Ahmad F, Norulaini NN, AK MO (2016) Functionalization of magnetic chitosan with graphene oxide for removal of cationic and anionic dyes from aqueous solution. *Carbohydrate Polymers* 152: 520–531.

66. Deng H, Sun P, Zhang Y, Zhu H (2016) Reverse osmosis desalination of chitosan cross-linked graphene oxide/titania hybrid lamellar membranes. *Nanotechnology* 27: 274002.

67. Ayoub A, Venditti RA, Pawlak JJ, Salam A, Hubbe MA (2013) Novel hemicellulose–chitosan biosorbent for water desalination and heavy metal removal. *ACS Sustainable Chemistry & Engineering* 1: 1102–1109.

68. Tang Y-J, Wang L-J, Xu Z-L, Zhang H-Z (2018) Novel chitosan-piperazine composite nanofiltration membranes for the desalination of brackish water and seawater. *Journal of Polymer Research* 25: 1–12.

69. Liu S, Wang Z, Song P (2018) Free radical graft copolymerization strategy to prepare catechin-modified chitosan loose nanofiltration (NF) membrane for dye desalination. *ACS Sustainable Chemistry & Engineering* 6: 4253–4263.

70. Padaki M, Isloor AM, Fernandes J, Prabhu KN (2011) New polypropylene supported chitosan NF-membrane for desalination application. *Desalination* 280: 419–423.

71. Shakeri A, Salehi H, Rastgar M (2017) Chitosan-based thin active layer membrane for forward osmosis desalination. *Carbohydrate Polymers* 174: 658–668.

72. Wang Z, Lin S (2017) Membrane fouling and wetting in membrane distillation and their mitigation by novel membranes with special wettability. *Water Research* 112: 38–47.

73. Salehi H, Rastgar M, Shakeri A (2017) Anti-fouling and high water permeable forward osmosis membrane fabricated via layer by layer assembly of chitosan/graphene oxide. *Applied Surface Science* 413: 99–108.

74. Jiang Y, Gong J-L, Zeng G-M, Ou X-M, Chang Y-N, Deng C-H, Zhang J, Liu H-Y, Huang S-Y (2016) Magnetic chitosan–graphene oxide composite for anti-microbial and dye removal applications. *International Journal of Biological Macromolecules* 82: 702–710.

75. Hegab HM, Wimalasiri Y, Ginic-Markovic M, Zou L (2015) Improving the fouling resistance of brackish water membranes via surface modification with graphene oxide functionalized chitosan. *Desalination* 365: 99–107.

76. Zhang S, Lu F, Tao L, Liu N, Gao C, Feng L, Wei Y (2013) Bio-inspired anti-oil-fouling chitosan-coated mesh for oil/water separation suitable for broad pH range and hyper-saline environments. *ACS Applied Materials & Interfaces* 5: 11971–11976.

77. Li Z, Shao L, Hu W, Zheng T, Lu L, Cao Y, Chen Y (2018) Excellent reusable chitosan/cellulose aerogel as an oil and organic solvent absorbent. *Carbohydrate Polymers* 191: 183–190.

78. Kumar S, Prasad K, Gil JM, Sobral AJ, Koh J (2018) Mesoporous zeolite-chitosan composite for enhanced capture and catalytic activity in chemical fixation of CO_2. *Carbohydrate Polymers* 198: 401–406.

79. Rashid N, Rehman SU, Han J-I (2013) Rapid harvesting of freshwater microalgae using chitosan. *Process Biochemistry* 48: 1107–1110.

80. Guibal E (2004) Interactions of metal ions with chitosan-based sorbents: A review. *Separation and Purification Technology* 38: 43–74.

81. Gerente C, Lee V, Cloirec PL, McKay G (2007) Application of chitosan for the removal of metals from wastewaters by adsorption—mechanisms and models review. *Critical Reviews in Environmental Science and Technology* 37: 41–127.

82. Jiang Y-J, Yu X-Y, Luo T, Jia Y, Liu J-H, Huang X-J (2013) γ-Fe_2O_3 nanoparticles encapsulated millimeter-sized magnetic chitosan beads for removal of Cr (VI) from water: Thermodynamics, kinetics, regeneration, and uptake mechanisms. *Journal of Chemical & Engineering Data* 58: 3142–3149.

83. Ben Amor I, Hemmami H, Laouini SE, Zeghoud S, Benzina M, Achour S, Naseef A, Alsalme A, Barhoum A (2023) Use of insect-derived chitosan for the removal of methylene blue dye from wastewater:

process optimization using a central composite design. *Materials* 16: 5049.

84. Jiao T, Zhao H, Zhou J, Zhang Q, Luo X, Hu J, Peng Q, Yan X (2015) Self-assembly reduced graphene oxide nanosheet hydrogel fabrication by anchorage of chitosan/silver and its potential efficient application toward dye degradation for wastewater treatments. *ACS Sustainable Chemistry & Engineering* 3: 3130–3139.

85. Rezgui S, Amrane A, Fourcade F, Assadi A, Monser L, Adhoum N (2018) Electro-Fenton catalyzed with magnetic chitosan beads for the removal of Chlordimeform insecticide. *Applied Catalysis B: Environmental* 226: 346–359.

86. Omidinasab M, Rahbar N, Ahmadi M, Kakavandi B, Ghanbari F, Kyzas GZ, Martinez SS, Jaafarzadeh N (2018) Removal of vanadium and palladium ions by adsorption onto magnetic chitosan nanoparticles. *Environmental Science and Pollution Research* 25: 34262–34276.

87. Dorraji MS, Amani-Ghadim A, Hanifehpour Y, Joo SW, Figoli A, Carraro M, Tasselli F (2017) Performance of chitosan based nanocomposite hollow fibers in the removal of selenium (IV) from water. *Chemical Engineering Research and Design* 117: 309–317.

88. Andres Y, Giraud L, Gerente C, Le Cloirec P (2007) Antibacterial effects of chitosan powder: Mechanisms of action. *Environmental Technology* 28: 1357–1363.

89. Mathew ML, Gopalakrishnan A, Aravindakumar CT, Aravind UK (2019) Low-cost multilayered green fiber for the treatment of textile industry waste water. *Journal of Hazardous Materials* 365: 297–305.

90. Jagtap S, Yenkie MK, Das S, Rayalu S (2011) Synthesis and characterization of lanthanum impregnated chitosan flakes for fluoride removal in water. *Desalination* 273: 267–275.

91. Tabriz A, Alvi MAUR, Niazi MBK, Batool M, Bhatti MF, Khan AL, Khan AU, Jamil T, Ahmad NM (2019) Quaternized trimethyl functionalized chitosan based antifungal membranes for drinking water treatment. *Carbohydrate Polymers* 207: 17–25.

92. Qi C, Zhao L, Lin Y, Wu D (2018) Graphene oxide/chitosan sponge as a novel filtering material for the removal of dye from water. *Journal of Colloid and Interface Science* 517: 18–27.

93. Younes I, Rinaudo M (2015) Chitin and chitosan preparation from marine sources. Structure, properties and applications. *Marine Drugs* 13: 1133–1174.

94. Min L-L, Yang L-M, Wu R-X, Zhong L-B, Yuan Z-H, Zheng Y-M (2019) Enhanced adsorption of arsenite from aqueous solution by an iron-doped electrospun chitosan nanofiber mat: Preparation, characterization and performance. *Journal of Colloid and Interface Science* 535: 255–264.

95. Chiou M-S, Li H-Y (2002) Equilibrium and kinetic modeling of adsorption of reactive dye on cross-linked chitosan beads. *Journal of Hazardous Materials* 93: 233–248.

Chapter 13

Other Applications of Chitosan

Omaima Aidat[a] and Asma Ben Amor[b,c]

[a]*Food Technology and Nutrition Laboratory, Abdelhamid Ibn Badis University, Mostaganem, Algeria*
[b]*Department of Process Engineering and Petrochemical, Faculty of Technology, University of El Oued, El Oued, Algeria*
[c]*Renewable Energy Development unit in Arid Zones, University of El Oued, El Oued 39000, Algeria*

wissamaidat@gmail.com

It has been more than 40 years since this biopolymer piqued the curiosity of the scientific community worldwide for its range of uses. Chitin and chitosan, when used as functional materials, offer a special combination of properties, including biodegradability to products safe goods, biocompatibility, nontoxicity, antibacterial properties, physiological inertness, chelation of heavy metal ions, gel-forming properties, remarkable protein affinity, and hydrophilicity. These qualities mean that chitin- and chitosan-based materials, which are currently under-utilized, should be widely used in the near future, particularly in applications that are environmentally friendly in systems that operate in biological settings. This biomaterial holds a great deal of potential, and that promise will only grow as new uses and the chemistry to expand its capabilities are researched.

Chitin and Chitosan: Physical and Chemical Properties
Edited by Shakeel Ahmed, Soumeia Zeghoud, Hadia Hemmami, and Ilham Ben Amor

ISBN 978-981-5129-39-7 (Hardcover), 978-1-003-58977-8 (eBook)
www.jennystanford.com

13.1 Introduction

A biopolymer is a substance created by combining several molecules coming from living things. Collagen, alginate, silk, starch, chitin, and elastin are only a few of the many natural polymers that are regarded as biomaterials [1]. The usage of biopolymers mitigates the issues related to synthetic polymers, such as toxicity and non-biodegradability. The capacity of biopolymers to be broken down by immune-stimulating natural enzymes is one of their key characteristics. A significant and promising area of study is the creation of biomaterials utilizing natural polymers.

Chitin is advantageous as a chelating agent, drug carrier, treatment for water, pressure-sensitive tape that is biodegradable, wound healing factor in membranes, and has other benefits for a number of significant applications. Chitin is also environmentally friendly, renewable, biocompatible, biofunctional, and biodegradable [2]. Due to these benefits, this distinctive biomaterial is receiving a lot of attention. Chitin is no longer widely used by the pharmaceutical sector, nevertheless. It has particular uses due to its poor solubility. Due to its greatly extended hydrogen-bonded semicrystalline structure, chitin is extremely hydrophobic and insoluble in both conventional organic solvents and diluted aqueous solutions. Natural chitin's deacetylation and depolymerization, (partial) deacetylation of chitin in the solid form under alkaline circumstances (NaOH concentrate), or enzymatic hydrolysis in the presence of chitin deacetylase are the methods used to create its derivative, chitosan [3].

Chitin and chitosan have a wide range of uses in a variety of industries, including wastewater treatment (dyes, removing metal ions, and as a membrane in purification processes) [4], food (anti-cholesterol and fat binding [5], preservative and food additive, and packaging material), agriculture (seed and fertilizer coating, and controlled agrochemical release), and paper and pulp (gels, surface treatment, nanofibers, membranes, beads, nanoparticles, microparticles, scaffolds, and sponges), which may all be made simply from chitin and chitosan [6]. In this chapter, several applications of chitosan will be touched upon, with a focus on the most important studies in various fields.

13.2 Chitosan in Green Metal Nanoparticle Synthesis

Metal nanoparticle production and functionalization can both be impacted by chitosan (Fig. 13.1). The introduction of the polymer cation chitosan into the reaction solution leads to an electrostatic interaction between the cationic chitosan and the anionic nanoparticles (NPs) (are primarily composed of the negative capping agent) [7], or the surface interaction occurs via absorption of chitosan by metal nanoparticles [8], resulting in a chitosan shell around the nanoparticles. Chitosan can be added before or after the nanoparticles are formed, depending on the production procedure. So, chitosan has an impact on nanoparticle production directly. When synthesizing metal nanoparticles, chitosan can serve as a stabilizing agent, reducing agent, agent that can regulate particle size, and agent that can control particle form.

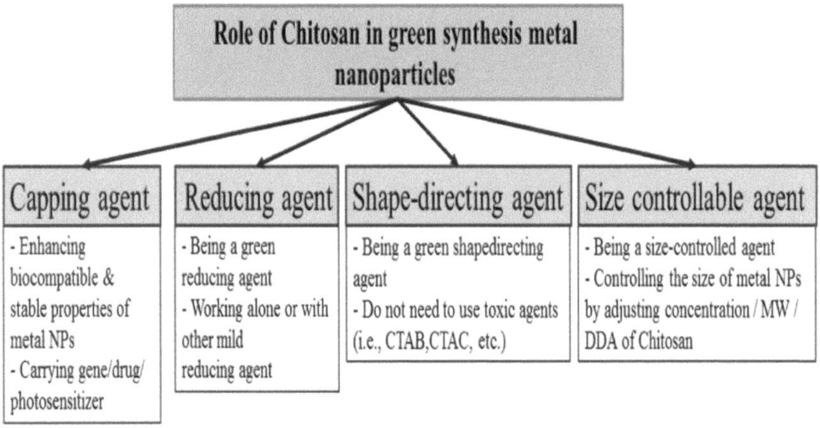

Figure 13.1 Summary of chitosan's function in the green synthesis production of metallic NPs.

13.2.1 As a Stabilizer

The choice of natural polymers as a stabilizer for the manufacture of metal NPs is typical. Chitosan is an excellent stabilizer of metal nanoparticles due to its accessibility, biocompatibility, and strong positive charge [9]. Van der Waals chemical interactions among exposed metal surfaces, as shown in Fig. 13.2a, favor the aggregation of bare metal nanoparticles in solution. Chitosan, on the other hand, is a steric barrier that encloses the metal with a high density of positive charges. Positively charged metal nanoparticles were able to generate homogenous nanoparticle solutions thanks to strong electrostatic interactions, as seen in Fig. 13.2b.

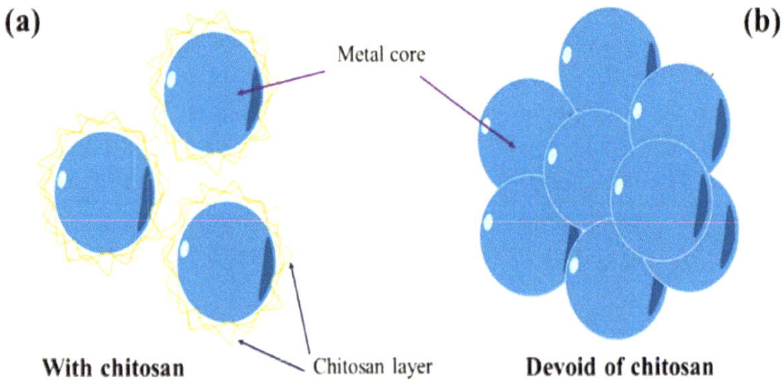

Figure 13.2 The impact of chitosan capping on metal NPs dispersion. (a) Metal NPs collect in solution without the presence of a chitosan capping agent. (b) The effective dispersion of metal NPs in solution using a chitosan capping agent.

Numerous studies have shown chitosan's successful stabilization of metal nanoparticles. As an illustration, chitosan was used as a stabilizer in a green method based on an electrochemical oxidation or complexation process with UV light reduction to create silver nanoparticles (AgNPs) [10]. In a different instance, citric acid and chitosan were used to create gold nanoparticles (AuNPs). The as-prepared AuNPs did not aggregate, and it is stable in the aqueous phase [11]. The water-soluble oligosaccharide (COS) derived from chitosan has found widespread use in the field of metal nanoparticle coatings. These nanoparticles are designed for specific biomedical uses, like targeted drug or gene connection, photographic processing, and artificial tissue creation. One effective method is to use COS as both a stabilizing and reducing agent, enabling the one-step production of gold nanoparticles (AuNPs) used for gene transfer

[12]. The amino groups in COS have positive charges, which increase their affinity for plasmid DNA.

13.2.2 As a Green Reducing Agent

When making NPs chemically, toxic reducing agents are frequently utilized, which releases compounds that are harmful to the environment. Due to the presence of toxic capping that is difficult to extract from the nanoparticles, the chemically produced nanoparticles could not be employed directly for biological purposes. Nanoparticles must be created using environmentally friendly processes and eco-friendly ingredients in order to save the environment and be employed in biomedicine. According to several studies, chitosan can operate as a reducing and stabilizing agent for the environmentally friendly synthesis of AgNPs, CuNPs, and AuNPs [13, 14]. With the help of ultraviolet-visible absorption and X-ray spectrophotometric analyses. In an effort to comprehend the reaction mechanism of Ag ion using chitosan, Carapeto et al. [15]. Experimental data have confirmed that the first steps in the extremely rapid reduction of silver (Ag) in aqueous chitosan solutions occur at ambient temperatures and that this reaction intensifies with increasing temperature. The chemical units present in chitosan, which includes alcohol or glucoside groups, can undergo oxidation, releasing a free electron capable of reducing Ag^+ ions to metallic silver (Ag^0) and consequently inducing the formation of carbonyl groups. The researchers identified the main products formed during this reaction as carbonyl groups, using UV-visible absorption spectra (characterized by an absorption peak at 262 nm, corresponding to an *n transition within a carbonyl group) and X-ray photoelectron spectroscopy (XPS). To manufacture positively charged nanoparticles and ensure that they do not accumulate and are stable for a long time, the silver metal is coated or coated with cationic chitosan [16].

Another hypothesis for the reaction mechanism of chitosan was offered by Wongpreecha et al. [17]. The environmentally friendly method was carried out using an autoclave at a temperature of 120°C and a pressure of 15 psi. Ag^+ ions were bound to electron pairs from nitrogen and/or oxygen present in the chitosan structure in a medium with a pH of 4 (below the pKa of chitosan). Subsequently, under conditions of elevated temperature and pressure, an electron from a lone pair of oxygen.

The performance of metal nanoparticles can be improved by forming a core-shell nanostructure. In order to create core-shell nanostructured nanoparticles for usage in the biomedical industry, chitosan was also employed as a green reducing agent. By using natural chitosan, Wang et al. [18] created a core-shell nanocomposite known as Cu@Pd-chitosan. Cu@Pd-chitosan that was synthesized showed excellent stability, sensitivity, and interference-resistance.

13.2.3 As a Size-Controlling Agent

Size control is another function of chitosan in the green production of metal NPs. According to data from the UV-visible absorption spectra, Kalaivani et al. [19] discovered that the inclusion of chitosan significantly improved the synthesis of AgNPs. Moreover, at greater chitosan concentrations, the size of AgNPs was noticeably reduced (Fig. 13.3). Our most recent investigations once again supported this result. PdNPs [20] and AuNPs [21] are two examples of metal nanoparticles whose sizes shrank when chitosan concentration was raised. To explain how chitosan impacts the size of produced nanoparticles, we put up a theory.

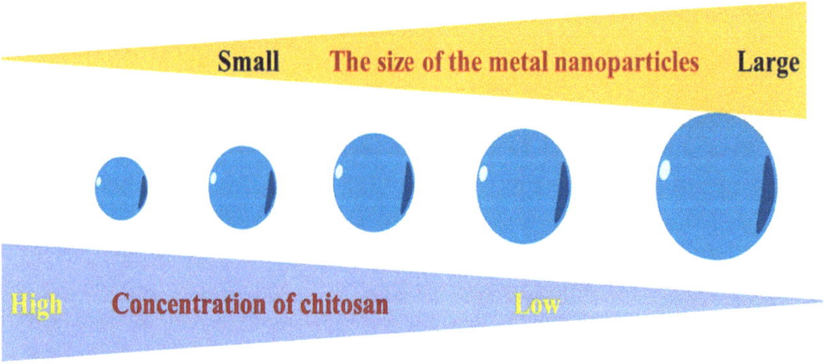

Figure 13.3 The relationship between metal nanoparticle size and chitosan concentration.

When chitosan is present, metal nuclei interact strongly electrostatically with the positively charged chitosan during the production of metal nanoparticles. The greater interaction between chitosan and the metal nuclei results from the aloft concentration of chitosan. The precursors' ability to attach to the metal nuclei is inhibited by this strong contact. Thus, in the presence of a solution containing a lot of chitosan, metal nuclei are unable to develop any further.

13.2.4 As a Shape-Directing Agent

Toxic capping agents, such as trisodium citrate and CTAB, are frequently utilized as the shape-directing agent for metal NPs when using the chemical approach [22]. The best way to improve the biocompatibility of metal NPs for long-lasting biomedical uses is to replace them with natural compounds. AgNPs were electrodeposited on disposable pencil-graphite electrodes using chitosan as a structure-directing agent [23]. AgNPs exhibit distinct morphologies in the presence of chitosan, according to the scientists. In contrast, AgNPs lack chitosan and have an atypical structure. AgNPs have been produced in a variety of morphologies, including hexahedrons, leaves, and dendrites, by manipulating the experimental conditions.

Additionally, altering the chitosan with an anionic ligand, an o-carboxymethyl group, or a cationic N-trimethylamine group can improve its ability to regulate the morphology of nanoparticles. The chitosan's positive and negative charges may be used to influence the formation of various AuNP forms. According to reference [24], thiolate functionalized chitosan may be employed as a soft template to create gold nano chains, nanoflowers, and nanoneedles. The thiol group of chitosan can interact significantly with AuNPs belonging to different assemblies. Additionally, it is demonstrated that, as stated in reference [25], a peptide with an aromatic moiety may be employed to self-build a template for the architecture of gold nanocrystals.

13.2.5 Chitosan as a Multifunctional Agent for Metal Nanoparticle Preparation

Recent studies have shown that the multifunctional agent chitosan may be used to create metal nanoparticles like AuNPs and PdNPs. Chitosan serves as a multifunctional agent in our 2019 development of a unique green approach for the synthesis of flower-shaped porous palladium nanoparticles (PdNPs), including as a stabilizer, size controller, and shape director [20]. The studies were set up with different concentrations of chitosan and vitamin C, and a green reducing agent was employed to convert Pd^{III} to Pd^{0}. Although no proxies appear to form, we consistently obtained PdNPs with a floral shape in all trials. Chitosan was crucial to the production of the flower-shaped PdNPs in that. Following synthesis under the influence of vitamin C, chitosan-coated the surface of the PdNPs by a strong contact with Pd nuclei. Under the influence of chitosan, the palladium nuclei exhibited persistent growth in a non-uniform manner, resulting in the formation of palladium nanoparticles (PdNPs) characterized by a porous floral morphology.

The growth of the chitosan layer is stopped by the reduction process occurring on the surface of the Pd core. Now the PdNPs are fully formed. It is observed that smaller PdNPs were produced when the concentration of chitosan was increased. Smaller PdNPs will be produced, and the chitosan layer will develop more quickly if there is more chitosan in the solution. Based on this idea, the amount of chitosan added may be changed to produce PdNPs of the appropriate size. In previous research [21], chitosan was employed as a multifunctional agent for the green production of gold nanostars (AuNSs). The positively charged nature of chitosan in acidic conditions leads to its binding to the core of AuNSs via robust interactions with the Au nuclei. The anisotropic development of gold nanostructures was facilitated by the tip's strong growth at the chitosan and core's loose contact sites, creating a structure like a star. Chitosan's interaction with Ag nuclei is weaker, and it is unable to inhibit Ag nuclei development in an environment with a higher pH level. As a result, AuNSs failed to develop.

13.3 Chitosan's Function in Cosmetics

Contrary to other biopolymers like collagen or hyaluronic acid, chitosan is not a widely used cosmetic element; yet, interest in this subject is developing. The insufficient water solubility of chitosan is the main reason that affects this. However, because chitosan may serve a variety of beneficial purposes in formulation, there are an increasing number of commodities using it in the market. Although chitosan is not used in many aesthetic applications, its derivatives are used [26]. This biopolymer has two cosmetic uses: film formation and hair fixation. Chitosan can also be used as an emulsion stabilizer, thickener, and rheological modifier, as well as water-binding and skin-wetting (Fig. 13.4) [27]. It prevents water loss by forming a hydrophilic layer on the skin [28].

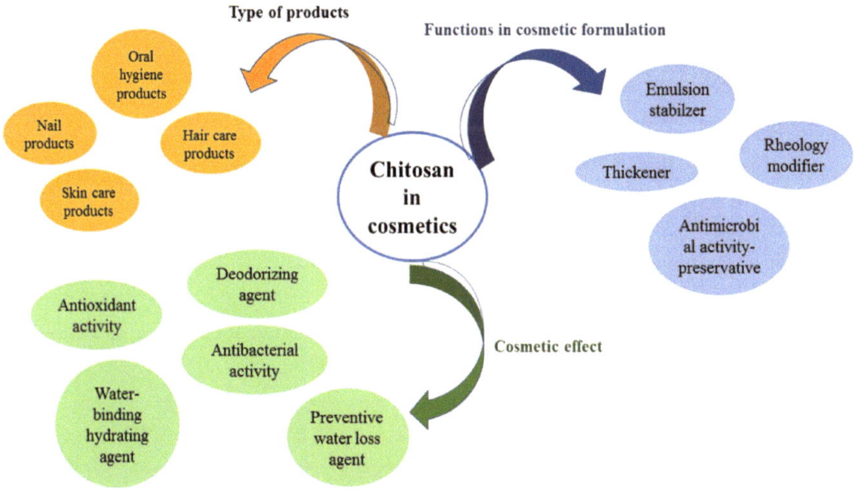

Figure 13.4 Chitosan in cosmetics.

Chitosan's antibacterial properties serve a dual purpose by acting as an active ingredient and reducing the need for preservatives in the formulation. It works well in haircare products since it also has an affinity for keratin [29]. Chitosan may be used to make cosmetic masks that function similarly to wound dressings because of its film-forming capabilities. Every type of cosmetic product, including those for the skin, hair, nails, and oral hygiene, can contain this biopolymer as an ingredient in the form of emulsions, gels, foams, sticks, or aerosols [28, 29].

Achieving an anti-aging effect is not usually the aim of cosmetics. The field of cosmetology aids in the treatment of skin conditions or lessens their impact. Acne, hyperpigmentation, depigmentation, psoriasis, acne, and scarring from surgical procedures are the most prevalent skin conditions that require cosmetology assistance. Self-esteem, social life, quality of life, and overall mental health are all significantly impacted by appearance [30]. The skin may be maintained in a healthy state and look with the right treatment. The previously mentioned beauty masks that resemble wound dressings and the encapsulation of active substances using chitosan are examples of how the cosmetics business is influenced by biomedical sciences [31].

Chitosan can be used in deodorants and antiperspirants because of its antibacterial properties, which serve as a breeding ground for bacteria found in sweat and lower the production of metabolites that cause odor [32]. These qualities make it a popular component in anti-acne cosmetics [33]. In the field of oral healthcare, where bacteria are key players in the formation of dental plaque, antimicrobial activity is a serious issue [29]. Rinses, toothpaste, and chewing gum are all made using chitosan [29, 34] (Table 13.1).

Table 13.1 Chitosan's minimal inhibitory concentration for specific bacterial strains such as those that cause acne [35]

Bacterial strains	MIC of chitosan (μg/mL)
Proteus vulgaris	50
Escherichia coli	80
Salmonella sp.	80
Utibacterium acnes	512
Streptococcus sp.	60
Pseudomonas aeruginosa	32/60
Staphylococcus epidermidis	64
Staphylococcus aureus	16/80
Vibrio cholerae	60

Keratin is the major constituent of the complicated structure of the hair. The outside sheath of the hair is frequently subjected to damaging elements, such as mechanical stress, UV radiation, high modeling temperatures, or chemical elements during dyeing. Chitosan and other biopolymers enhance the consistency of the formulation and the ability of other components to adhere to hair [29]. Furthermore, chitosan's positive charge, which balances the charge of damaged hair, is what gives it its conditioning properties. Cosmetics for conditioning hair that include chitosan can stabilize the structure of the hair by depositing a film there. This results in hair that is thicker and softer in appearance [31]. Time-of-flight secondary ion mass spectrometry (TOF-SIMS) was used by Kojima et al. to examine chitosan's capacity to enter the hair. They contrasted colored hair with natural (undyed) hair. Results revealed that colored hair had more chitosan penetration. This shows and demonstrates the damaging effects of hairstyling procedures and the capacity of chitosan to bind to the structure of hair [36].

Protection from UV rays is another possible use for chitosan in cosmetics. The photoprotective capacity of biopolymer can be inferred from its absorption at wavelengths below 400 nm [29]. The SPF of chitosan gel, however, is quite low [37]. More effective UV filters or ingredients that lessen the impact of UV radiation on the skin may be available in this case, such as chitosan. Pomegranate juice was encapsulated using chitosan nanoparticles, according to Bikiaris et al. The loaded nanoparticles were subsequently added to emulsions. Results indicated that chitosan-containing systems provided better UV protection than the control sample [38].

It is also important to emphasize the utilization of this polysaccharide as a possible component of reusable filtering masks. Numerous face masks are required for personal protection due to the coronavirus epidemic, both among healthcare workers and among civilians. It is advised to wear disposable protective masks on a daily basis. However, this produces a significant quantity of trash. A fresh approach for reusable masks with filters was put out by Choi et al. [36]. They created a membrane filter that was combined with micro- and nanofiber mats made of poly(butylene succinate) (PBS), and they covered it with chitosan nanowhiskers. The tested filter provided a pleasant breathing environment by successfully capturing particulate matter (PM). The suggested filter can be used repeatedly since it is robust. According to the findings of biodegradation experiments, it starts to break down in the soil after a month. It may offer antibacterial and virus-blocking qualities, according to speculation. It offers a viable substitute for disposable filters [39].

13.4 Food

Oil-in-water emulsions were stabilized by chitin nanocrystals for up to one month. Emulsions stabilized by cellulose nanofibril and chitin nanocrystals together lasted six months in storage [40]. The oil droplets were shielded and isolated by the nanomaterials, which also served to stabilize the emulsions by inhibiting the coalescence of the droplets and the formation of a distinct phase. Nanocellulose/nanochitin emulsifiers are useful as sustainably derived stabilizers for food-grade emulsions, including sauces, dressings, and dips, because of the long-term stability they provide [41].

A phlorotannin-nano chitin complex was created when phlorotannin, a kind of polyphenol derived from seaweed, was adsorbed onto the surface of chitin nanocrystals [42]. They prevented biochemical changes and reduced bacterial development in sea bass fillets, extending the shelf life by up to 3 days [42]. In terms of pH changes in fillets, synthesis of secondary metabolites, and microbial proliferation throughout the trial, the complexes functioned better than isolated nanocrystals and phlorotannins [42]. In comparison to an uncoated control, dragon fruit coated with chitosan that was ionically bound to carrageenan lost less weight after harvest over the course of 30 days [43]. Because of their antibacterial qualities, chitosan films also extend the shelf life of meats [44].

13.5 Plastics and Filler Materials

Chitin chains exhibit restricted solubility due to strong hydrogen bonds between them. This has hampered the widespread use of chitin. Research has mainly focused on the chitin derivative chitosan, particularly for applications involving thin films. This is due to chitosan's ability to dissolve in slightly acidic conditions, such as solutions containing 1% acetic acid. Consequently, chitosan has received more attention than chitin in this context [45]. Chitosan thin films are commonly manufactured using a relatively simple solution-casting method. In brief, chitosan-containing solutions are poured onto a surface, and the solvent then evaporates, leaving a thin film as a result. The fragility of these thin chitosan films is attributed to the presence of hydrogen bonds between the chitosan chains [46].

By incorporating different substances, chitosan-based polymers can have their characteristics changed. Chitosan films' hydrogen bonds are broken by adding plasticizers like castor oil or glycerol, making the films more elastic and flexible [46]. Elongation strength and tensile at break are frequently compared to see how additives affect the mechanical characteristics of the film. Despite the lack of plasticizers, it has been demonstrated that starches derived from various plant sources can diminish the tensile strength of chitosan films while simultaneously enhancing elongation at the point of fracture [47]. The mechanical characteristics of polymers made of cellulose, glycerol, and chitosan were similarly impacted by liquid smoke. Liquid smoke, in contrast to starches, inhibited the biodegradation of thin films [48]. Choo et al. [49] found that adding essential oils to chitosan/acetylated starch films decreased their tensile power and increased their elongation at break. Increased antibacterial activities were also associated with higher essential oil concentrations in the films. Mechanical properties are not systematically altered by the incorporation of essential oils. Research by Castro and his team revealed that the introduction of tea tree essential oil (TTEO) into films composed of chitosan and polyvinyl alcohol, up to a concentration of 1.5% by weight, showed no statistically significant influence on tensile strength [50].

On the other hand, it has been shown that adding chitosan to other polymers increases their tensile strength and confers antibacterial qualities where none previously existed [51, 52]. Chitosan films' tensile strength, antibacterial capabilities, and biodegradability were all enhanced by phosphorylated cellulose nano- and microcrystals [53]. The thermal stability of the films was further improved by the inclusion of cellulose nanocrystals [53].

According to Hammi et al. [54], the inclusion of metal oxides during the production of chitosan films might enhance the films' antibacterial capabilities. Films were brittle and cracked when metal oxides were added in excess. Sapei et al.'s [55] subsequent research raised the possibility that the addition of metal oxides can also affect the mechanical characteristics of chitosan plastic. They discovered that adding ZnO NPs to chitosan-banana starch plastics increased their tensile strength while lowering their elongation and water uptake.

The biodegradability of chitosan polymers is one factor contributing to their appeal. In particular, for single-use applications, biodegradability makes chitosan plastics preferable to petroleum-based alternatives that linger in the environment. Researchers have looked into edible films made of plastics containing chitosan for use in food packaging [43]. Films made of chitosan have been shown to extend the shelf life of both meat and vegetables. Chitosan's ability to operate as an antioxidant and vapor barrier, in addition to its antifungal and antibacterial characteristics, has been linked to an increase in shelf life. While the primary emphasis in plastics research is in the realm of thin films, there have been notable instances when 3D items, comparable in size to chess pieces, have been successfully manufactured from chitosan through the employment of injection molding techniques [56]. Biodegradability promotes formulations other than thin films. On that scale, volume fluctuations brought about by solvent evaporation became an issue; Fernandez et al. [56] added wood flour to the chitosan solutions to reduce volume changes during the evaporation process.

The mechanical parameters of several plastic formulations are summarized in Table 13.2 to allow comparison.

Table 13.2 Chitosan or nano chitin-containing polymer films' mechanical characteristics

Ingredients	Tensile power (MPa)	Lengthening at break (%)	Modulus of Young (MPa)	Ref.
Corn starch	1.64 ± 0.1	175 ± 7.07	-	[57]
Corn starch/5% chitin nanowhiskers	2.37 ± 0.04	111 ± 4.24		
Corn starch/1% chitin nanowhiskers	3.69 ± 0.07	179 ± 7.07		
Corn starch/0.5% chitin nanowhiskers	2.79 ± 0.08	176 ± 8.65		
Chitosan film	40	27	1200	[53]
3% microcrystalline cellulose/chitosan	43	20	1350	
Chitosan/3% nanocrystalline cellulose	50	24	1400	
40/60 Chitosan/yellow pumpkin starch, 15% v/v castor oil	6.787 ± 0.274	13.451 ± 3.709	6.093	[47]
Yellow pumpkin starch, 15% v/v castor oil/60/40 chitosan	2.563 ± 1.055	7.285 ± 1.135	5.263	
Starch film	9.54 ± 0.84	51.01 ± 1.32	16.50 ± 1.10	[51]
4% Chitosan NPs/starch	24.91 ± 0.81	34.29 ± 1.69	47.11 ± 2.51	
Starch/1% chitosan NPs	14.74 ± 16.7	46.19 ± 1.71	24.20 ± 1.04	

13.6 Recent Use of Chitosan and Chitin in Plants

13.6.1 Chitin

Chitin may be used directly as a fertilizer to boost crop growth due to its high nitrogen concentration and low C/N ratio. The abundance and architectures of microbial communities are both enhanced by the addition of chitin to the soil. Chitin's potential for use in the biocontrol of plant diseases has also been thoroughly investigated. The pathogen-associated molecular pattern (PAMP) receptor, among other specialized receptors found on the cell plasma membrane, recognizes chitin in plants. Following this detection, chitin can initiate defense reactions against possible bacterial, viral, and fungal pathogens through PAMP-triggered immunity [58]. However, the bulk chitin that is sold commercially is not soluble in the majority of conventional solvents due to its high crystallinity. Despite its useful functions, its inability to dissolve restricts both its practical usage in agriculture and its utility in basic research [58]. Several tactics have been used in an effort to get around this restriction (Table 13.3).

Table 13.3 Recent use of polymers based on chitin

Plant species	Chitin characteristics	Chitin effect	Ref.
Testing the development of pathogens in vitro	Iron/chitin nanoparticles	Growth inhibition of *Salmonella typhi*, *Escherichia coli*, *Pseudomonas aeruginosa*, and *Fusarium oxysporum*	[59]
Nicotiana tabacum L.	Nanokitin bristles	Increased development and germination of seeds, and resistance to *Fusarium spp.*	[60]
Oryza sativa japonica, *Arabidopsis thaliana*	Chitin oligomers	Resistance against *Pseudomonas syringae* and *Alternaria brassicicola*	[61]
Solanum lycopersicum	Protein/$CaCO_3$/ chitin nanofibers or protein/ chitin nanofibers	Plant development and resistance to *Fusarium oxysporum*	[62]
Fragaria x ananassa, Brassica oleracea	Chitin nanofibers	Colletotrichum fructicola and Resistance against *Alternaria brassicicola*	[63]

For instance, the use of alternative-source chitin was tested. Crustacean waste is now the main source of this substance. Testing chitin obtained from the cell wall of Saccharomyces cerevisiae was suggested by Sun and colleagues [58, 64]. There are various benefits to using chitin that is derived from yeast and other fungi, including the fact that (i) unlike chitin from arthropods, it cannot cause allergies in people or animals, and (ii) it possesses physical and chemical characteristics, like nontoxicity, biocompatibility, and biodegradability, that facilitates its use in agriculture [58, 64]. Tomato fruits treated post-harvest with chitin from a yeast cell wall efficiently generated significant resistance against Botrytis cinerea grey mold rot. The authors suggested that the accumulation of reactive oxygen species, callose deposition, increased activity of defense-related enzymes like peroxidase, catalase, and chitinase, superoxide dismutase, as well as increased expression of the corresponding genes, were responsible for the increased disease resistance of tomato fruit treated with chitin [64]. Other approaches involve employing chitin derivatives with reduced molecular weight, making them more soluble. For example, the use of a chitin-derived oligosaccharide derived from prawn shells that have been mechanically ground, autoclaved, and sonicated led to the activation of genes related to nitrogen and carbon metabolism in *Arabidopsis thaliana.* Plants treated with this chitin oligosaccharide showed an increase in fresh weight (10%), root length, and total carbon and nitrogen levels [58] in comparison to untreated plants. Chitin from crustacean shells may also be broken down into nanofibers using a high-pressure water jet system after being pretreated in a grinder. Since chitin nanofibers (CNF) function as a water-soluble material, their usage is more feasible. Egusa et al. [61] were the first to demonstrate that CNF increased resistance to infections in *A. thaliana* and rice when they evaluated them for their elicitor capability. In a recent study, it was observed that strawberry and cabbage plants cultivated in a soil and cellulose nanofiber (CNF) mixture demonstrated a notable decline in the occurrence of spots induced by *Alternaria brassicicola* and a decrease in the severity of lesions caused by Colletotrichum fructicola, in comparison to control plants, upon exposure to fungal infections [63].

In comparison to plants not treated with CNF, the addition of CNF to the soil marginally increased the development of tomato plants and decreased the occurrence of the *Fusarium* wilt disease. Using CNF complexed with both protein and $CaCO_3$ in protein/$CaCO_3$/chitin nanofiber or simply with protein in protein/chitin nanofiber was more successful at inhibiting disease [62].

Similar to this, acidic hydrolysis of prawn chitin can produce nanochitin whiskers (NCW), which are rod-shaped particles with high biological activity and a cationic nature. Two winter wheat types treated with the NCW demonstrated improved production in outdoor pot trials, along with notable improvements in net photosynthesis, intercellular CO_2 concentrations, stomatal conductance, and transpiration rate in flag leaf during the grain filling stage. Wheat that had been subjected to NCW treatment included more grain protein, iron, and zinc [65]. The antifungal effectiveness of NCW was also examined in wheat. According to Liang et al. [66], NCW significantly inhibited the mycelial development and conidial production of *F. graminearum* and *Fusarium pseudograminearum*, the two pathogens that cause crown rot, the disease that causes the most damage to this crop.

NCW promoted tobacco seed germination and seedling growth (including stem length and girth, leaf, area, and number) in indoor and outdoor experiments. Additionally, NCW demonstrated synergistic effects when combined with metalaxyl mancozeb and thiophanate methyl fungicides to suppress tobacco root rot, indicating its potential application in lowering the usage of tobacco plantations with chemical fungicides [60]. Finally, iron was fused with chitin nanoparticles mechanically extracted from *Penaeus semisulcatus* shells. The obtained nanocomposite demonstrated antimicrobial activity against *Salmonella typhi, Escherichia coli, and Pseudomonas aeruginosa,* and employed common antibiotics as positive controls, to a lesser extent against *F. oxysporum* in vitro experiments, indicating its utility against diverse fungal and bacterial diseases [59].

13.6.2 Chitosan

Due to their buildup in the environment and living things, the extensive use of chemicals in agriculture to boost plant yield and resistance against infections might permanently harm the ecosystem. Additionally, it may cause plant diseases to develop resistance, necessitating the introduction of new agrochemicals and/or the need to use more of the current products [67]. Costs and risks inevitably rise accordingly. The use of nanotechnology offers a highly promising solution when exploring new avenues.

Among biomaterials, chitosan stands out as a major player in the field of nanotechnology due to its harmlessness to humans, its biodegradability, and its compatibility with other living beings. Unlike other biopolymers like starch, chitin, gelatin, glucans, and cellulose, chitosan can be easily modified while preserving its natural properties. [68]. Therefore, by modifying its physico-chemical and biophysical characteristics, chitosan has been widely exploited for several uses. Recently, chitosan-based compounds were used in plants to provide a wide range of antibacterial and regulatory functions. Chitosan nanoparticles (chitosan TNP), for instance, were able to act as a germination elicitor of Oryza sativa L. [69] and to positively impact wheat (*Triticum aestivum L.*) seed germination and seedling development [70].

They combine the properties of chitosan with those of nanoparticles, like tiny size, quantum size, and surface and interface impacts. Additionally, these NPs were able to promote the production of pathogenesis-related proteins, boosting tomato resistance to *F. andiyazi* [71], and they were able to suppress the development of *F. oxysporum* in *Zingiber officinale* [72]. The main applications of chitosan TNP complexed with other compounds were plant growth promotion and inducing phytopathogen resistance. Chitosan TNP's ability to entrap nitrogen, phosphorus, and potassium increased the absorption of photosynthesis, nutrients, and the development of treated coffee plants [73] and caused Pisum sativum plants to accumulate starch and key proteins. The ability of chitosan TNP complexed with various metals to promote development and induce resistance was examined (Table 13.4).

While Cu- and Zn-chitosan TNP inhibited *Rhizoctonia solani*-caused cotton seedling illness, zinc-loaded chitosan NP improved durum wheat's gluten content and increased crop production in maize [79]. Cu-chitosan TNP improved the yield, plant development, and nutritional content of onion bulbs as well as the growth and induction of resistance against *Pyricola grisea* blast disease in finger millet plants [80]. Different metal-chitosan nanoparticles (ZnO, CuO, and Ag-chitosan TNP) inhibited the soil- and seed-transmitted Fusarium wilt disease that *F. oxysporum* causes in chickpeas [81]. Application of Ag-chitosan NP prevented Corynespora *cassiicola*'s leaf fall disease on rubber trees and considerably reduced *Phytophthora capsici* development on pepper plants [76], while Lanthanum-chitosan TNP successfully accelerated rice growth and enhanced disease resistance [74].

Table 13.4 Recent use of metal-complexed nanoparticles made of chitosan

Plant species	Metal complexed to chitosan nanoparticles	Effect	Ref.
Oryza sativa L.	La	Promotion of growth and increased disease resistance	[74]
Riticum turgidum var. durum	Zn	Gluten content increases	[75]
Piper nigrum	Ag	Being resistant to *Phytophthora capsici*	[76]
Zea mays	Zn	Promotion of crop yield	[77]
Hevea brasiliensis	Ag	Resistance to *Corynespora cassiicola*	[78]
Gossypium spp.	Cu or Zn	Against *Rhizoctonia solani* resistance	[79]
Eleusine coracana	Cu	Increased plant size and *Pyricola grisea* resistance	[80]

Table 13.5 Recent use of chitosan and NPs complexed with various compounds

Plant types	Molecule complexed to chitosan	Influence	Ref.
Saccarum spp.	S-nitrosoglutathione	Increased resistance to drought	[82]
Solanum lycopersicum	Harpin	Against *Rhizoctonia solani* resistance	[83]
Solanum lycopersicum	Cu	Plant growth stimulation and defense gene expression	[84]
Capsicum annuum	Tripolyphosphate	Promotion of plant growth and accumulation of biomass	[85]
Lactuca sativa L.	Salicylic acid	Plant growth stimulation and increased disease resistance	[86]
Zea mays L.	Salicylic acid	Plant growth stimulation and increased disease resistance	[87]
Hevea brasiliensis	Coumarin	The ability to withstand Alternaria solanii and *Fusarium spp.*	[88]

Additionally, the usage of chitosan TNP complexed with elicitors, phytoregulators, and secondary metabolites was investigated (Table 13.5).

When tomato plants were treated with harpin-chitosan NP, various defensive response genes were expressed, and *R. solani* infection was reduced [83]. Tripolyphosphate-chitosan NP promoted capsicum annuum micropropagation plant development at the right dosage, but hazardous levels abruptly stopped plant growth [85]. Nanoparticles based on salicylic acid and chitosan exert a positive influence on plant development and enhance defense reactions in young lettuce shoots, while nanoparticles of thiamine-chitosan promote growth and induce resistance against *F. oxysporum* in chickpea seedlings [89] and the growth and defense of maize were enhanced by the addition of salicylic acid-chitosan NPs [87]. Chitosan NP, when combined with essential oil from Cymbopogon martini, showed antimycotoxin and antifungal activity against *F. graminearum* in maize grains in a laboratory setting [90]. The growth of *A. solani*, *F. moniliforme*, and *F. oxysporum* was impacted by coumarin-functionalized chitosan derivatives [88]. In wheat, hybrids of vanillin, chitosan, and salicylic acid successfully promoted resistance to *Cochliobolus sativus* and *Puccinia recondita* [91]. A nanoclay made of bentonite and chitosan showed antibacterial activity against the bacteria *F. solani* and *P. syringe* and had an elicitor effect on tomato plantlets [92].

Derivatives of chitosan NPs also influence the responses to abiotic stressors favorably. Cu-chitosan NPs, for instance, boosted sugarcane plants' ability to withstand drought, while S-nitroso glutathione-chitosan NP increased the growth and expression of defense genes in salt-stressed tomato plants [84]. Last but not least, chitosan was also evaluated for use in making a film for covering seeds and fruits (Table 13.6).

Banana fruits were given a longer shelf life and quality preservation by the application of chitosan NPs, whilst fresh blueberries were given quality preservation and microbiological growth inhibition by the coating of chitosan$^+$ procyanidins [97]. The application of a chitosan-based coating mixed with a *Ficus hirta* Vahl fruit extract proved effective in reducing post-harvest losses and improving the preservation of Xinyu mandarins during low-temperature storage [96]. By coating strawberries with a light-activated chlorophyllin-chitosan conjugate, we were able to increase the preservation period of strawberries by limiting the presence of harmful micro-organisms while preserving their appearance and quality [98]. A coating composed of chitosan/nano-silica/sodium alginate was applied to winter jujube fruit, resulting in increased expression of antioxidant-related genes and extended fruit shelf life [99]. Similarly, the use of nano-TiO_2 and chitosan/chitosan/ nano-SiO_2 coated films enhanced the activity of antioxidant enzymes, thus preserving the quality of *Ginkgo biloba* seeds [93].

Table 13.6 Recently using chitosan and nanoparticles to coat seeds and fruits

Plant types	Compound	Effect	Ref.
Ginkgo biloba L.	Chitosan/nano-TiO_2 and chitosan/nano-SiO_2	Maintenance of seed quality	[93]
Zizyphus jujuba Mill. cv Dongzao	Chitosan/sodium alginate/nano-silica	Extending the shelf life of fruit and preventing the formation of germs	[94]
Musa acuminata AAA group	Chitosan	Fruit quality preservation and shelf-life extension using nanoparticles	[95]
Citrus reticulata Blanco	Chitosan + *Ficus hirta* Vahl.	Fruit extract increases fruit storability and reduces post-harvest loss	[96]

13.7 Paper and Pulp Industry

The paper and pulp sector is crucial for the progress of society and the economy. In order to create inexpensive pulp and paper products and additives with higher and enhanced features, such as grease resistance, wet and dry power, etc., there is a huge need for and possibility for substantial study. Chitosan and its byproducts, which are crucial wet-end additives in the production of paper, seem to be effective retention and drainage agents. Chitosan's usage as a papermaking additive to enhance paper's wet and dry power and as a color-fixing agent for colored sheets is the subject of much research [99]. Chitosan's innate ability to form films and its capacity to inhibit the growth of microbes make it ideal for making sheets that are greaseproof and antibacterial [100]. Chitosan/montmorillonite coating preparation was disclosed by Wang et al. [99], in order to create oil-resistant sheets that may be used safely in the food sector.

Papers with improved barrier and antibacterial characteristics were successfully created by sequentially depositing multilayers of carboxymethyl cellulose and chitosan on a paper surface, making them promising materials for food packaging. The multilayers demonstrated antibacterial activity against *S. aureus* and *E. coli*, as well as successfully increased air, water, and oil resistance [99]. In order to remove lignin, unwelcome contaminants, and undesired colors from pulp and paper effluent and to reduce the total organic and chemical carbon-oxygen requirement, chitosan is also utilized as a chelating and complicating factor [101].

13.8 Dentistry

Based on the literature, Fig. 13.5 schematically illustrates how chitosan is used in dentistry.

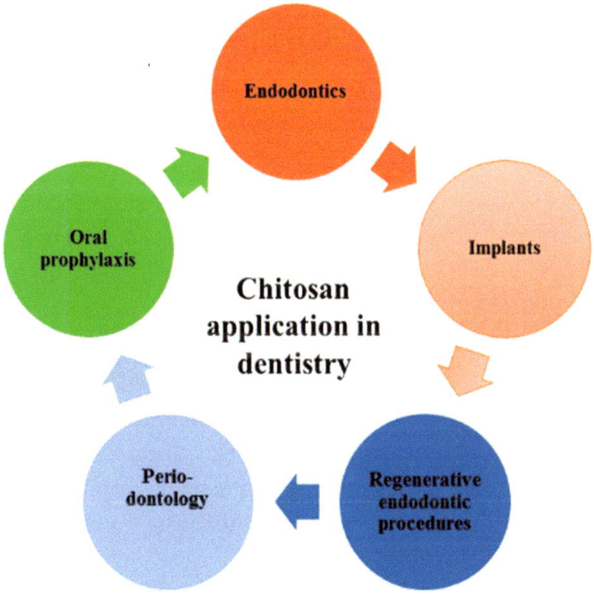

Figure 13.5 Using chitosan in dentistry in accordance with the literature [102–104].

13.8.1 Oral Prophylaxis

Cationic groups in chitosan interact with elements of bacterial cell surfaces that are negatively charged. Gram (-) bacteria are known to be more sensitive to the inhibitory (bactericidal and bacteriostatic) actions of chitosan than gram (+) bacteria. Additionally, the amino groups in chitosan (NH_3^+) may neutralize the acidic environment found in dental plaque, allowing chitosan to keep the pH of plaque above the threshold level needed to prevent enamel demineralization [104].

Entococcus faecalis, *Candida albicans*, *S. mutans*, and *Prevotella intermedia* adhesion and biofilm development while simultaneously stimulating the breakdown of already established bacterial colonies, according to Costa et al. [105]. These findings suggest that the use of a mouthwash containing chitosan could prevent the onset of dental caries, candidiasis, and periodontitis. In addition, chitosan has been shown to interact with hydroxyapatite on tooth surfaces. This particularity of chitosan, especially when combined with nanoparticles, could restrict the formation of biofilms on enamel and dentine and, thus, indirectly reduce the process of dental demineralization. What's more, chitosan has the ability to repair demineralized enamel and halt the progression of demineralization by penetrating tooth enamel [106].

13.8.2 Dental Root Treatment

The root canal treatment procedure is used to decontaminate the root canal system and remove inflamed or necrotic pulp that has been contaminated by bacteria. The antibacterial efficacy of chitosan combined with nanoparticles (CSNP) against pathogens such as *E. faecalis*, *S. mutan*ts, and *P. gingivalis* has been the subject of several studies. Thanks to their potential advantages, CSNPs could be incorporated with calcium hydroxide for endodontic sealants or temporary root canal fillings [106].

Additionally, regenerative endodontic operations have employed chitosan [107].

Porous chitosan scaffolds can be enriched with bioactive substances and growth factors. Under the influence of chemical signals, dental pulp stem cells proliferate and evolve into odontoblasts by increasing the production of specific markers such as dentin matrix acid phosphoprotein, dentin sialophosphoprotein, and alkaline phosphate. Proteins that promote bone growth and transforming growth factor-1 were shown to be bioactive agents capable of enhancing cell adhesion, differentiation, and multiplication of stem cells from dental pulp and odontoblast-like cells within novel chitosan-based scaffolds [102].

13.8.3 Periodontology

Thanks to its appropriate physicochemical characteristics, high biocompatibility, degradability by natural enzymes, and optimal molecular size, chitosan is a suitable carrier for periodontal tissue regeneration (in the treatment of periodontal pockets). Injectable hydrogels combined with chitosan membranes offer promising application possibilities in the fields of tissue engineering and drug delivery systems [108].

In addition, versatile dental mesenchymal stem cells can be cultured on chitosan-based scaffolds, including those derived from fallen human milk teeth and human periodontal ligament cells. By inducing cementoblasts and osteoblasts to generate new tissue structures, this chitosan approach can be successfully employed for tissue regrowth in the periodontium [103].

13.8.4 Implantology

Chitosan-modified implant surfaces are still being studied in terms of osseointegration and healing procedures [109]. To enhance the osteoinductive characteristics of chitosan coatings, the degree of deacetylation (DD%) may be essential. An increase in the DD of chitosan coatings was found by Alnufaiy et al. [110] to drive biomineralization and the development of new osteoblasts.

13.9 Treatment for Burns and Wounds

Jayakumar et al. [111] examined chitin and chitosan-based biomaterials for use in uses for wound dressing. The treatment of burns and wounds is unquestionably one of the most promising uses of chitosan and chitin in medicine. Chitosan possesses adhesive properties that are crucial for the treatment of burns and wounds, in addition to being antifungal, antibacterial, and oxygen-permeable. Many chitin and chitosan derivatives, including membranes, woven fibers, hydrogels, etc., have been patented as a result. Many of these formulations, like the chitin-based Hem Con in the United States and Beschitin in Japan, have already hit the market.

Both polysaccharides aid in the promotion of granulation and cell structure during wound healing. The analgesic properties of chitosan and chitin have both been investigated [112]. Chitin, which is water soluble, and heparin have also been researched as wound healing matrices since it has been shown that they interact with and stabilize growth factors essential for the healing process [112].

13.10 Tissue Engineering

Three-dimensional (3D) supports are frequently used in tissue engineering procedures to facilitate initial cell adhesion and subsequent tissue development. Chitosan resembles glycosaminoglycans (GAGs) are structural components of various human tissues' extracellular matrix. Because of its ability to promote cell adhesion and preserve differentiation activities, chitosan has been extensively applied in the field of engineering with tissues. Substrates such as gelatin-chitosan or collagen-chitosan matrices (some of which are cross-linked with glutaraldehyde) have shown outstanding results in terms of stimulating cell growth and proliferation, particularly in the regeneration and repair of cartilage, trachea, nerves, and bone tissue. In the latter context, a composite including hydroxyapatite has been successfully applied. In addition, membranes consisting of chitosan and chondroitin sulfate have been observed for their ability to encourage chondrogenesis, making them promising materials for cartilage repair [3].

Additionally, chitosan chondroitin sulfate sponges that release platelet-derived growth factors have been studied for bone regeneration [113]. Collagen-GAG-chitosan has been used to create an approximation of human skin [114].

While both compounds were cytocompatible, higher cell adhesion and proliferation were seen when using pure chitosan. Researchers have shown that chitosan-GAG compounds, particularly chitosan-heparin, can control the growth of vascular cells. In a similar vein, it has been discovered that antigenic collagen-hyaluronic acid combinations promote the development of fibroblasts for tissue healing. Porous 3D scaffolds made of chitosan and $Ca_3(PO)_2$ have also been found to be effective for bone regeneration. When loaded with gentamicin, calcium phosphate significantly improved the chitosan matrix and modified the release burst effect. Additionally, excellent cellular biocompatibility was noted. Additionally, the freezing and mixing of chitosan-tricalcium phosphate sponges, sometimes laden with platelet-derived growth factor, has been shown [115]. Encapsulating or isolating pancreatic and hepatic cells is a stimulating technique in tissue engineering.

13.11 Use of Chitosan as Artificial Leather

Individuals who have suffered severe skin loss, usually due to severe burns, experience great distress and are at potential risk of death from severe infections or dehydration. Often, these patients encounter difficulties in recovering from severe contractures as well as extensive and deforming scars. Malette and colleagues conducted a study investigating the healing and fibroblast process of wounds caused by scalpel incisions in the skin and subcutaneous tissue of the abdominal region of dogs [3]. They used chitosan and saline solution as a therapy. A non-antigenic membrane that serves as a biodegradable template for the production of neodermal tissue was the subject of a concept for artificial skin [116, 117]. It would appear that chitosans with structural traits like glucosamine glycans might be taken into consideration for the development of such a substratum for skin restoration [118].

13.12 Opthalmology

Optical clarity, gas permeability, mechanical stability, enough optical correction, notably toward oxygen, immunological compatibility, and wettability are all qualities that chitosan possesses, making it the perfect material for creating the perfect contact lens. The spin casting method is used to create contact lenses using partly depolymerized and purified squid pen chitosan. These lenses are transparent, robust, and have other essential physical characteristics, including tensile strength, modulus, tear strength, elongation, oxygen permeability, and water content. Chitosan is a suitable material for creating ocular bandage lenses due to its antibacterial and wound-healing qualities as well as high film capability [3].

13.13 Conclusion

The potential application of chitosan and chitin in several industrial domains has been recognized, and further research is being considered in that regard. Furthermore, chitin and chitosan have a variety of biological functions that are naturally occurring, and they have a great deal of promise for use in the environmentally friendly production of metal nanoparticles as well as other industrial and medicinal uses.

References

1. Sionkowska A (2021) Collagen blended with natural polymers: Recent advances and trends. *Progress in Polymer Science* 122: 101452.
2. Alhamad AA, Zeghoud S, Amor IB, Hemmami H (2023) Chitosan-based hydrogels for wound healing: Correspondence. *International Journal of Surgery* 109: 1821–1822.
3. Zargar V, Asghari M, Dashti A (2015) A review on chitin and chitosan polymers: Structure, chemistry, solubility, derivatives, and applications. *ChemBioEng Reviews* 2: 204–226.
4. Amor IB, Hemmami H, Laouini SE, Abdelaziz AG, Barhoum A (2023) Influence of chitosan source and degree of deacetylation on

antibacterial activity and adsorption of AZO dye from water. *Biomass Conversion and Biorefinery*: 1–11.

5. Zhang W, Zhang J, Jiang Q, Xia W (2013) The hypolipidemic activity of chitosan nanopowder prepared by ultrafine milling. *Carbohydrate Polymers* 95: 487–491.
6. Pokhrel S, Yadav PN, Adhikari R (2015) Applications of chitin and chitosan in industry and medical science: A review. *Nepal Journal of Science and Technology* 16: 99–104.
7. Collado-González M, Montalbán MG, Peña-García J, Pérez-Sánchez H, Víllora G, Baños FGD (2017) Chitosan as stabilizing agent for negatively charged nanoparticles. *Carbohydrate Polymers* 161: 63–70.
8. Frank L, Onzi G, Morawski A, Pohlmann A, Guterres S, Contri R (2020) Chitosan as a coating material for nanoparticles intended for biomedical applications. *Reactive and Functional Polymers* 147: 104459.
9. Amor IB, Hemmami H, Laouini SE, Temam HB, Zaoui H, Barhoum A (2023) Biosynthesis MgO and ZnO nanoparticles using chitosan extracted from Pimelia Payraudi Latreille for antibacterial applications. *World Journal of Microbiology and Biotechnology* 39: 19.
10. Cinteza LO, Scomoroscenco C, Voicu SN, Nistor CL, Nitu SG, Trica B, Jecu M-L, Petcu C (2018) Chitosan-stabilized Ag nanoparticles with superior biocompatibility and their synergistic antibacterial effect in mixtures with essential oils. *Nanomaterials* 8: 826.
11. Esther J, Sridevi V (2017) Synthesis and characterization of chitosan-stabilized gold nanoparticles through a facile and green approach. *Gold Bulletin* 50: 1–5.
12. Abrica-González P, Zamora-Justo JA, Sotelo-López A, Vázquez-Martínez GR, Balderas-López JA, Muñoz-Diosdado A, Ibáñez-Hernández M (2019) Gold nanoparticles with chitosan, N-acylated chitosan, and chitosan oligosaccharide as DNA carriers. *Nanoscale Research Letters* 14: 258.
13. Venkatesham M, Ayodhya D, Madhusudhan A, Veera Babu N, Veerabhadram G (2014) A novel green one-step synthesis of silver nanoparticles using chitosan: Catalytic activity and antimicrobial studies. *Applied Nanoscience* 4: 113–119.
14. Dananjaya S, Udayangani R, Oh C, Nikapitiya C, Lee J, De Zoysa M (2017) Green synthesis, physio-chemical characterization and anti-candidal function of a biocompatible chitosan gold nanocomposite as a promising antifungal therapeutic agent. *RSC Advances* 7: 9182–9193.

15. Carapeto AP, Ferraria AM, do Rego AMB (2017) Unraveling the reaction mechanism of silver ions reduction by chitosan from so far neglected spectroscopic features. *Carbohydrate Polymers* 174: 601–609.

16. Bhumkar DR, Joshi HM, Sastry M, Pokharkar VB (2007) Chitosan reduced gold nanoparticles as novel carriers for transmucosal delivery of insulin. *Pharmaceutical Research* 24: 1415–1426.

17. Wongpreecha J, Polpanich D, Suteewong T, Kaewsaneha C, Tangboriboonrat P (2018) One-pot, large-scale green synthesis of silver nanoparticles-chitosan with enhanced antibacterial activity and low cytotoxicity. *Carbohydrate Polymers* 199: 641–648.

18. Wang F, Ding X, Niu X, Liu X, Wang W, Zhang J (2020) Green preparation of core-shell Cu@Pd nanoparticles with chitosan for glucose detection. *Carbohydrate Polymers* 247: 116647.

19. Kalaivani R, Maruthupandy M, Muneeswaran T, Beevi AH, Anand M, Ramakritinan C, Kumaraguru A (2018) Synthesis of chitosan mediated silver nanoparticles (Ag NPs) for potential antimicrobial applications. *Frontiers in Laboratory Medicine* 2: 30–35.

20. Phan TTV, Hoang G, Nguyen TP, Kim HH, Mondal S, Manivasagan P, Moorthy MS, Lee KD, Junghwan O (2019) Chitosan as a stabilizer and size-control agent for synthesis of porous flower-shaped palladium nanoparticles and their applications on photo-based therapies. *Carbohydrate Polymers* 205: 340–352.

21. Phan TTV, Ahn S-H, Oh J (2019) Chitosan-mediated facile green synthesis of size-controllable gold nanostars for effective photo-thermal therapy and photoacoustic imaging. *European Polymer Journal* 118: 492–501.

22. Zhao L, Jiang D, Cai Y, Ji X, Xie R, Yang W (2012) Tuning the size of gold nanoparticles in the citrate reduction by chloride ions. *Nanoscale* 4: 5071–5076.

23. Preethi S, Sangaranarayanan M (2020) Shape-controlled electro-deposition of silver using chitosan as structure-directing agent on disposable pencil graphite electrodes: Low-cost electrocatalysts for the detection of hydrogen peroxide and hydrazine hydrate. *Journal of Solid State Electrochemistry* 24: 2773–2788.

24. Ding Y, Gu G, Xia X-H, Huo Q (2009) Cysteine-grafted chitosan-mediated gold nanoparticle assembly: From nanochains to micro-cubes. *Journal of Materials Chemistry* 19: 795–799.

25. Tomizaki K-Y, Kishioka K, Kobayashi H, Kobayashi A, Yamada N, Kataoka S, Imai T, Kasuno M (2015) Roles of aromatic side chains and template effects of the hydrophobic cavity of a self-assembled peptide nanoarchitecture for anisotropic growth of gold nanocrystals. *Bioorganic & Medicinal Chemistry* 23: 7282–7291.

26. Venkataramani D, Tsulaia A, Amin S (2020) Fundamentals and applications of particle stabilized emulsions in cosmetic formulations. *Advances in Colloid and Interface Science* 283: 102234.

27. Aflakseir A, Jamali S, Mollazadeh J (2021) Prevalence of body dysmorphic disorder among a group of college students in Shiraz. *Zahedan Journal of Research in Medical Sciences* 23: e95247.

28. Baek J, Ramasamy M, Willis NC, Kim DS, Anderson WA, Tam KC (2021) Encapsulation and controlled release of vitamin C in modified cellulose nanocrystal/chitosan nanocapsules. *Current Research in Food Science* 4: 215–223.

29. Aranaz I, Acosta N, Civera C, Elorza B, Mingo J, Castro C, Gandía MDLL, Heras Caballero A (2018) Cosmetics and cosmeceutical applications of chitin, chitosan and their derivatives. *Polymers* 10: 213.

30. Raza ZA, Khalil S, Ayub A, Banat IM (2020) Recent developments in chitosan encapsulation of various active ingredients for multi-functional applications. *Carbohydrate Research* 492: 108004.

31. Chaiwong N, Phimolsiripol Y, Leelapornpisid P, Ruksiriwanich W, Jantanasakulwong K, Rachtanapun P, Seesuriyachan P, Sommano SR, Leksawasdi N, Simirgiotis MJ (2022) Synergistics of carboxymethyl chitosan and mangosteen extract as enhancing moisturizing, antioxidant, antibacterial, and deodorizing properties in emulsion cream. *Polymers* 14: 178.

32. Kuo C-W, Chiu Y-F, Wu M-H, Li M-H, Wu C-N, Chen W-S, Huang C-H (2021) Gelatin/chitosan bilayer patches loaded with cortex Phellodendron amurense/Centella asiatica extracts for anti-acne application. *Polymers* 13: 579.

33. Bikiaris ND, Michailidou G, Lazaridou M, Christodoulou E, Gounari E, Ofrydopoulou A, Lambropoulou DA, Vergkizi-Nikolakaki S, Lykidou S, Nikolaidis N (2020) Innovative skin product emulsions with enhanced antioxidant, antimicrobial and UV protection properties containing nanoparticles of pure and modified chitosan with encapsulated fresh pomegranate juice. *Polymers* 12: 1542.

34. Davydova V, Kalitnik A, Markov P, Volod'Ko A, Popov S, Ermak I (2016) Cytokine-inducing and anti-inflammatory activity of chitosan and its

low-molecular derivative. *Applied Biochemistry and Microbiology* 52: 476–482.

35. Kim J-H, Yu D, Eom S-H, Kim S-H, Oh J, Jung W-K, Kim Y-M (2017) Synergistic antibacterial effects of chitosan-caffeic acid conjugate against antibiotic-resistant acne-related bacteria. *Marine Drugs* 15: 167.
36. Choi S, Jeon H, Jang M, Kim H, Shin G, Koo JM, Lee M, Sung HK, Eom Y, Yang HS (2021) Biodegradable, efficient, and breathable multi-use face mask filter. *Advanced Science* 8: 2003155.
37. Mourya V, Inamdar NN (2008) Chitosan-modifications and applications: Opportunities galore. *Reactive and Functional Polymers* 68: 1013–1051.
38. Schmitz C, González Auza L, Koberidze D, Rasche S, Fischer R, Bortesi L (2019) Conversion of chitin to defined chitosan oligomers: Current status and future prospects. *Marine Drugs* 17: 452.
39. Jimtaisong A, Saewan N (2014) Utilization of carboxymethyl chitosan in cosmetics. *International Journal of Cosmetic Science* 36: 12–21.
40. Huan S, Zhu Y, Xu W, McClements DJ, Bai L, Rojas OJ (2021) Pickering emulsions via interfacial nanoparticle complexation of oppositely charged nanopolysaccharides. *ACS Applied Materials & Interfaces* 13: 12581–12593.
41. Lv S, Zhou H, Bai L, Rojas OJ, McClements DJ (2021) Development of food-grade Pickering emulsions stabilized by a mixture of cellulose nanofibrils and nanochitin. *Food Hydrocolloids* 113: 106451.
42. Yu S, Duan M, Sun J, Jiang H, Zhao J, Tong C, Pang J, Wu C (2022) Immobilization of phlorotannins on nanochitin: A novel biopreservative for refrigerated sea bass (Lateolabrax japonicus) fillets. *International Journal of Biological Macromolecules* 200: 626–634.
43. Nguyen T, Boonyaritthongchai P, Buanong M, Supapvanich S, Wongs-Aree C (2021) Edible coating of chitosan ionically combined with κ-carrageenan maintains the bract and postharvest attributes of dragon fruit (Hylocereus undatus). *International Food Research Journal* 28: 682–694.
44. Song D-H, Hoa VB, Kim HW, Khang SM, Cho S-H, Ham J-S, Seol K-H (2021) Edible films on meat and meat products. *Coatings* 11: 1344.
45. Hossin MA, Al Shaqsi NHK, Al Touby SSJ, Al Sibani MA (2021) A review of polymeric chitin extraction, characterization, and applications. *Arabian Journal of Geosciences* 14: 1870.

46. Smith DR, Escobar AP, Andris MN, Boardman BM, Peters GM (2021) Understanding the molecular-level interactions of glucosamine-glycerol assemblies: a model system for chitosan plasticization. *ACS Omega* 6: 25227–25234.
47. Hasan M, Rahmayani R, Munandar (2018) Bioplastic from chitosan and yellow pumpkin starch with castor oil as plasticizer. In: *IOP Conference Series: Materials Science and Engineering*. IOP Publishing, p. 012087.
48. Fathanah U, Lubis M, Nasution F, Masyawi M (2018) Characterization of bioplastic based from cassava crisp home industrial waste incorporated with chitosan and liquid smoke. In: *IOP Conference Series: Materials Science and Engineering*. IOP Publishing, p. 012073.
49. Choo KW, Lin M, Mustapha A (2021) Chitosan/acetylated starch composite films incorporated with essential oils: Physiochemical and antimicrobial properties. *Food Bioscience* 43: 101287.
50. Castro JI, Valencia-Llano CH, Valencia Zapata ME, Restrepo YJ, Mina Hernandez JH, Navia-Porras DP, Valencia Y, Valencia C, Grande-Tovar CD (2021) Chitosan/polyvinyl alcohol/tea tree essential oil composite films for biomedical applications. *Polymers* 13: 3753.
51. Babaee M, Garavand F, Rehman A, Jafarazadeh S, Amini E, Cacciotti I (2022) Biodegradability, physical, mechanical and antimicrobial attributes of starch nanocomposites containing chitosan nanoparticles. *International Journal of Biological Macromolecules* 195: 49–58.
52. Kongkaoroptham P, Piroonpan T, Pasanphan W (2021) Chitosan nanoparticles based on their derivatives as antioxidant and antibacterial additives for active bioplastic packaging. *Carbohydrate Polymers* 257: 117610.
53. Blilid S, Kędzierska M, Miłowska K, Wrońska N, El Achaby M, Katir N, Belamie E, Alonso B, Lisowska K, Lahcini M (2020) Phosphorylated micro-and nanocellulose-filled chitosan nanocomposites as fully sustainable, biologically active bioplastics. *ACS Sustainable Chemistry & Engineering* 8: 18354–18365.
54. Hammi N, Wrońska N, Katir N, Lisowska K, Marcotte N, Cacciaguerra T, Bryszewska M, El Kadib A (2018) Supramolecular chemistry-driven preparation of nanostructured, transformable, and biologically active chitosan-clustered single, binary, and ternary metal oxide bioplastics. *ACS Applied Bio Materials* 2: 61–69.
55. Sapei L, Padmawijaya KS, Sijayanti O, Wardhana P (2017) Study of the influence of ZnO addition on the properties of chitosan-banana

starch bioplastics. In: *IOP Conference Series: Materials Science and Engineering*. IOP Publishing, p. 012044.

56. Fernandez JG, Ingber DE (2014) Manufacturing of large-scale functional objects using biodegradable chitosan bioplastic. *Macromolecular Materials and Engineering* 299: 932–938.
57. Qin Y, Zhang S, Yu J, Yang J, Xiong L, Sun Q (2016) Effects of chitin nano-whiskers on the antibacterial and physicochemical properties of maize starch films. *Carbohydrate Polymers* 147: 372–378.
58. Winkler AJ, Dominguez-Nuñez JA, Aranaz I, Poza-Carrión C, Ramonell K, Somerville S, Berrocal-Lobo M (2017) Short-chain chitin oligomers: Promoters of plant growth. *Marine Drugs* 15: 40.
59. Gomaa EZ (2018) Iron nanoparticles α-chitin nanocomposite for enhanced antimicrobial, dyes degradation and heavy metals removal activities. *Journal of Polymers and the Environment* 26: 3638–3654.
60. Zhou Y, Jiang S, Jiao Y, Wang H (2017) Synergistic effects of nanochitin on inhibition of tobacco root rot disease. *International Journal of Biological Macromolecules* 99: 205–212.
61. Egusa M, Matsui H, Urakami T, Okuda S, Ifuku S, Nakagami H, Kaminaka H (2015) Chitin nanofiber elucidates the elicitor activity of polymeric chitin in plants. *Frontiers in Plant Science* 6: 1098.
62. Egusa M, Parada R, Aklog YF, Ifuku S, Kaminaka H (2019) Nanofibrillation enhances the protective effect of crab shells against Fusarium wilt disease in tomato. *International Journal of Biological Macromolecules* 128: 22–27.
63. Parada RY, Egusa M, Aklog YF, Miura C, Ifuku S, Kaminaka H (2018) Optimization of nanofibrillation degree of chitin for induction of plant disease resistance: Elicitor activity and systemic resistance induced by chitin nanofiber in cabbage and strawberry. *International Journal of Biological Macromolecules* 118: 2185–2192.
64. Sun C, Fu D, Jin L, Chen M, Zheng X, Yu T (2018) Chitin isolated from yeast cell wall induces the resistance of tomato fruit to Botrytis cinerea. *Carbohydrate Polymers* 199: 341–352.
65. Xue W, Han Y, Tan J, Wang Y, Wang G, Wang H (2017) Effects of nanochitin on the enhancement of the grain yield and quality of winter wheat. *Journal of Agricultural and Food Chemistry* 66: 6637–6645.
66. Liang R, Li X, Yuan W, Jin S, Hou S, Wang M, Wang H (2018) Antifungal activity of nanochitin whisker against crown rot diseases of wheat. *Journal of Agricultural and Food Chemistry* 66: 9907–9913.

67. Kumaraswamy R, Kumari S, Choudhary RC, Pal A, Raliya R, Biswas P, Saharan V (2018) Engineered chitosan based nanomaterials: Bioactivities, mechanisms and perspectives in plant protection and growth. *International Journal of Biological Macromolecules* 113: 494–506.

68. Divya K, Jisha M (2018) Chitosan nanoparticles preparation and applications. *Environmental Chemistry Letters* 16: 101–112.

69. Divya K, Vijayan S, Nair SJ, Jisha M (2019) Optimization of chitosan nanoparticle synthesis and its potential application as germination elicitor of Oryza sativa L. *International Journal of Biological Macromolecules* 124: 1053–1059.

70. Li R, He J, Xie H, Wang W, Bose SK, Sun Y, Hu J, Yin H (2019) Effects of chitosan nanoparticles on seed germination and seedling growth of wheat (Triticum aestivum L.). *International Journal of Biological Macromolecules* 126: 91–100.

71. Chun S-C, Chandrasekaran M (2019) Chitosan and chitosan nanoparticles induced expression of pathogenesis-related proteins genes enhances biotic stress tolerance in tomato. *International Journal of Biological Macromolecules* 125: 948–954.

72. Athawale V, Paralikar P, Ingle AP, Rai M (2018) Biogenically engineered nanoparticles inhibit Fusarium oxysporum causing soft-rot of ginger. *IET Nanobiotechnology* 12: 1084–1089.

73. Ha NMC, Nguyen TH, Wang S-L, Nguyen AD (2019) Preparation of NPK nanofertilizer based on chitosan nanoparticles and its effect on biophysical characteristics and growth of coffee in green house. *Research on Chemical Intermediates* 45: 51–63.

74. Liang W, Yu A, Wang G, Zheng F, Hu P, Jia J, Xu H (2018) A novel water-based chitosan-La pesticide nanocarrier enhancing defense responses in rice (Oryza sativa L) growth. *Carbohydrate Polymers* 199: 437–444.

75. Deshpande P, Dapkekar A, Oak M, Paknikar K, Rajwade J (2018) Nanocarrier-mediated foliar zinc fertilization influences expression of metal homeostasis related genes in flag leaves and enhances gluten content in durum wheat. *PLoS One* 13: e0191035.

76. Luan LQ, Xo DH (2018) In vitro and in vivo fungicidal effects of γ-irradiation synthesized silver nanoparticles against Phytophthora capsici causing the foot rot disease on pepper plant. *Journal of Plant Pathology* 100: 241–248.

77. Choudhary RC, Kumaraswamy R, Kumari S, Sharma S, Pal A, Raliya R, Biswas P, Saharan V (2019) Zinc encapsulated chitosan nanoparticle

to promote maize crop yield. *International Journal of Biological Macromolecules* 127: 126–135.

78. Nhien LTA, Luong ND, Tien LTT, Luan LQ (2018) Radiation synthesis of silver nanoparticles/chitosan for controlling leaf fall disease on rubber trees causing by *Corynespora cassiicola*. *Journal of Nanomaterials* 2018. https://doi.org/10.1155/2018/7121549.
79. Abd-Elsalam KA, Vasil'kov AY, Said-Galiev EE, Rubina MS, Khokhlov AR, Naumkin AV, Shtykova EV, Alghuthaymi MA (2018) Bimetallic blends and chitosan nanocomposites: Novel antifungal agents against cotton seedling damping-off. *European Journal of Plant Pathology* 151: 57–72.
80. Sathiyabama M, Manikandan A (2018) Application of copper-chitosan nanoparticles stimulate growth and induce resistance in finger millet (Eleusine coracana Gaertn.) plants against blast disease. *Journal of Agricultural and Food Chemistry* 66: 1784–1790.
81. Kaur P, Duhan JS, Thakur R (2018) Comparative pot studies of chitosan and chitosan-metal nanocomposites as nano-agrochemicals against fusarium wilt of chickpea (Cicer arietinum L.). *Biocatalysis and Agricultural Biotechnology* 14: 466–471.
82. Silveira NM, Seabra AB, Marcos FC, Pelegrino MT, Machado EC, Ribeiro RV (2019) Encapsulation of S-nitrosoglutathione into chitosan nanoparticles improves drought tolerance of sugarcane plants. *Nitric Oxide* 84: 38–44.
83. Nadendla SR, Rani TS, Vaikuntapu PR, Maddu RR, Podile AR (2018) HarpinPss encapsulation in chitosan nanoparticles for improved bioavailability and disease resistance in tomato. *Carbohydrate Polymers* 199: 11–19.
84. Hernández-Hernández H, Juárez-Maldonado A, Benavides-Mendoza A, Ortega-Ortiz H, Cadenas-Pliego G, Sánchez-Aspeytia D, González-Morales S (2018) Chitosan-PVA and copper nanoparticles improve growth and overexpress the SOD and JA genes in tomato plants under salt stress. *Agronomy* 8: 175.
85. Asgari-Targhi G, Iranbakhsh A, Ardebili ZO (2018) Potential benefits and phytotoxicity of bulk and nano-chitosan on the growth, morphogenesis, physiology, and micropropagation of Capsicum annuum. *Plant Physiology and Biochemistry* 127: 393–402.
86. Martin-Saldaña S, Chevalier MT, Iglesias MJ, Colman SL, Casalongué CA, Álvarez VA, Chevalier AA (2018) Salicylic acid loaded chitosan microparticles applied to lettuce seedlings: Recycling shrimp fishing industry waste. *Carbohydrate Polymers* 200: 321–331.

87. Kumaraswamy R, Kumari S, Choudhary RC, Sharma S, Pal A, Raliya R, Biswas P, Saharan V (2019) Salicylic acid functionalized chitosan nanoparticle: A sustainable biostimulant for plant. *International Journal of Biological Macromolecules* 123: 59–69.

88. Yang G, Jin Q, Xu C, Fan S, Wang C, Xie P (2018) Synthesis, characterization and antifungal activity of coumarin-functionalized chitosan derivatives. *International Journal of Biological Macromolecules* 106: 179–184.

89. Muthukrishnan S, Murugan I, Selvaraj M (2019) Chitosan nanoparticles loaded with thiamine stimulate growth and enhances protection against wilt disease in Chickpea. *Carbohydrate Polymers* 212: 169–177.

90. Kalagatur NK, Nirmal Ghosh OS, Sundararaj N, Mudili V (2018) Antifungal activity of chitosan nanoparticles encapsulated with Cymbopogon martinii essential oil on plant pathogenic fungi Fusarium graminearum. *Frontiers in Pharmacology* 9: 610.

91. Popova E, Domnina N, Kovalenko N, Sokornova S, Tyuterev S (2018) Influence of chitosan hybrid derivatives on induced wheat resistance to pathogens with different nutrition strategies. *Applied Biochemistry and Microbiology* 54: 535–539.

92. Merino D, Mansilla AY, Casalongué CA, Alvarez VA (2018) Preparation, characterization, and in vitro testing of nanoclay antimicrobial activities and elicitor capacity. *Journal of Agricultural and Food Chemistry* 66: 3101–3109.

93. Tian F, Chen W, Cai'E W, Kou X, Fan G, Li T, Wu Z (2019) Preservation of Ginkgo biloba seeds by coating with chitosan/nano-TiO_2 and chitosan/nano-SiO_2 films. *International Journal of Biological Macromolecules* 126: 917–925.

94. Kou X, He Y, Li Y, Chen X, Feng Y, Xue Z (2019) Effect of abscisic acid (ABA) and chitosan/nano-silica/sodium alginate composite film on the color development and quality of postharvest Chinese winter jujube (Zizyphus jujuba Mill. cv. Dongzao). *Food Chemistry* 270: 385–394.

95. Lustriane C, Dwivany FM, Suendo V, Reza M (2018) Effect of chitosan and chitosan-nanoparticles on post harvest quality of banana fruits. *Journal of Plant Biotechnology* 45: 36–44.

96. Chen C, Nie Z, Wan C, Chen J (2019) Preservation of Xinyu tangerines with an edible coating using Ficus hirta Vahl. fruits extract-incorporated chitosan. *Biomolecules* 9: 46.

97. Mannozzi C, Tylewicz U, Chinnici F, Siroli L, Rocculi P, Dalla Rosa M, Romani S (2018) Effects of chitosan based coatings enriched with procyanidin by-product on quality of fresh blueberries during storage. *Food Chemistry* 251: 18–24.
98. Luksiene Z, Buchovec I (2019) Impact of chlorophyllin-chitosan coating and visible light on the microbial contamination, shelf life, nutritional and visual quality of strawberries. *Innovative Food Science & Emerging Technologies* 52: 463–472.
99. Piekarska K, Sikora M, Owczarek M, Jóźwik-Pruska J, Wiśniewska-Wrona M (2023) Chitin and chitosan as polymers of the future—obtaining, modification, life cycle assessment and main directions of application. *Polymers* 15: 793.
100. Hassan EA, Hassan ML, Abou-Zeid RE, El-Wakil NA (2016) Novel nanofibrillated cellulose/chitosan nanoparticles nanocomposites films and their use for paper coating. *Industrial Crops and Products* 93: 219–226.
101. Morin-Crini N, Lichtfouse E, Torri G, Crini G (2019) Applications of chitosan in food, pharmaceuticals, medicine, cosmetics, agriculture, textiles, pulp and paper, biotechnology, and environmental chemistry. *Environmental Chemistry Letters* 17: 1667–1692.
102. Farea M, Husein A, Halim AS, Abdullah NA, Mokhtar KI, Lim CK, Berahim Z, Mokhtar K (2014) Synergistic effects of chitosan scaffold and TGFβ1 on the proliferation and osteogenic differentiation of dental pulp stem cells derived from human exfoliated deciduous teeth. *Archives of Oral Biology* 59: 1400–1411.
103. Sukpaita T, Chirachanchai S, Suwattanachai P, Everts V, Pimkhaokham A, Ampornaramveth RS (2019) In vivo bone regeneration induced by a scaffold of chitosan/dicarboxylic acid seeded with human periodontal ligament cells. *International Journal of Molecular Sciences* 20: 4883.
104. Visveswaraiah PM, Prasad D, Johnson S (2014) Chitosan A novel way to intervene in enamel demineralization-An in vitro study. *International Journal of Current Microbiology and Applied Sciences* 3: 617–627.
105. Costa E, Silva S, Madureira A, Cardelle-Cobas A, Tavaria F, Pintado M (2014) A comprehensive study into the impact of a chitosan mouthwash upon oral microorganism's biofilm formation in vitro. *Carbohydrate Polymers* 101: 1081–1086.
106. Fakhri E, Eslami H, Maroufi P, Pakdel F, Taghizadeh S, Ganbarov K, Yousefi M, Tanomand A, Yousefi B, Mahmoudi S (2020) Chitosan

biomaterials application in dentistry. *International Journal of Biological Macromolecules* 162: 956–974.

107. Atila D, Chen C-Y, Lin C-P, Lee Y-L, Hasirci V, Tezcaner A, Lin F-H (2022) In vitro evaluation of injectable Tideglusib-loaded hyaluronic acid hydrogels incorporated with Rg1-loaded chitosan microspheres for vital pulp regeneration. *Carbohydrate Polymers* 278: 118976.
108. Ho M-H, Hsieh C-C, Hsiao S-W, Thien DVH (2010) Fabrication of asymmetric chitosan GTR membranes for the treatment of periodontal disease. *Carbohydrate Polymers* 79: 955–963.
109. Hamedi H, Moradi S, Hudson SM, Tonelli AE (2018) Chitosan based hydrogels and their applications for drug delivery in wound dressings: A review. *Carbohydrate Polymers* 199: 445–460.
110. Alnufaiy BM, Lambarte RNA, Al-Hamdan KS (2020) The osteogenetic potential of chitosan coated implant: An in vitro study. *Journal of Stem Cells & Regenerative Medicine* 16: 44.
111. Radoor S, Jayakumar A, Narayanan A, Karayil J, Parameswaranpillai J, Siengchin S (2023) Biobased materials in wound dressings. In: *Advanced Applications of Biobased Materials*. Elsevier, pp. 379–394.
112. Sasmal PK, Ganguly S (2023) Polymer in hemostasis and follow-up wound healing. *Journal of Applied Polymer Science* 140: e53559.
113. Park YJ, Lee YM, Lee JY, Seol YJ, Chung CP, Lee SJ (2000) Controlled release of platelet-derived growth factor-BB from chondroitin sulfate–chitosan sponge for guided bone regeneration. *Journal of Controlled Release* 67: 385–394.
114. Pleguezuelos-Beltrán P, Gálvez-Martín P, Nieto-García D, Marchal JA, López-Ruiz E (2022) Advances in spray products for skin regeneration. *Bioactive Materials* 16: 187–203.
115. Iannazzo D, Pistone A, Salamò M, Galvagno S (2017) Hybrid ceramic/polymer composites for bone tissue regeneration. In: *Hybrid Polymer Composite Materials*. Elsevier, pp. 125–155.
116. Mohamed A, Xing MM (2012) Nanomaterials and nanotechnology for skin tissue engineering. *International Journal of Burns and Trauma* 2: 29.
117. Amor IB, Emran TB, Hemmami H, Zeghoud S, Laouini SE (2023) Nanomaterials based on chitosan for skin regeneration: An update. *International Journal of Surgery* 109: 594–596.
118. Yuan L, Yao Q, Liang Y, Dan Y, Wang Y, Wen H, Yang Y, Dan W (2021) Chitosan based antibacterial composite materials for leather industry: A review. *Journal of Leather Science and Engineering* 3: 1–18.

chitosan-coated niosomes 295–296
chitosan coatings 199, 306–307, 333, 343, 409, 413
chitosan degradation 181
chitosan derivatives 21, 23, 235, 240, 284, 290, 360, 366, 414
  carboxymethyl 22
  coumarin-functionalized 409
  cross-linked 192
  methyl 22
  thiolated 22
chitosan fibers 186–187, 323–324, 326–330, 333
  cross-linked 333
  semi-crystallized 323
chitosan films 148, 164–165, 185–186, 256, 260, 263–265, 270, 272, 296, 400–403
  acid-enriched active 272
  deacetylated 242
  low-weight 263
chitosan flakes 376
chitosan fungal residues 193
chitosan gel 183, 309, 399
chitosan grafting 22
chitosan hydrogels 184, 197, 298, 358
chitosan hydrolysis 201, 337
chitosan hydroxide 190
chitosan macromolecules 261, 263
chitosan membranes 186, 268, 369, 413
chitosan microspheres 195, 294
chitosan monomers 180
chitosan mucoadhesion 128, 150
chitosan nano-biocomposites 363
chitosan nanocomposites 361
chitosan nanofibers 328, 335
chitosan nanofiltration membrane 369
chitosan nanoparticles 151, 196–197, 301–302, 333, 342, 363, 399, 403, 407–409
  zinc-loaded 407
chitosan nanowhiskers 400
chitosan-natural biopolymers blends 270
chitosan oligomers 150, 154, 194
  sulfated 148
chitosan-oxidized cellulose aerogels 371
chitosan penetration 399
chitosan phosphate 198–199
chitosan phthalate 130
chitosan plastics 402
chitosan polymers 107, 194–195, 298, 368, 372, 402
chitosan-protein interactions 134
chitosan proteins 148
chitosan purification 41–42
chitosan resins 193
chitosan sensitivity 161
chitosan sulfate 189
chitosan-synthetic polymer blends 270
chitosan-synthetic polymers 270
chitosanase enzymes 181
chitosanases 130, 144
chitosomes 306–307
  luteolin-loaded 307
cholesterol 152, 182, 306
choline chloride 71–72
cinnamyl-chitosan 290–291
CNS, *see* central nervous system
coatings 10, 196, 242, 258–259, 266, 306, 337, 410
  chitosan-based 409
  chitosan-graphene 342

collagen-chitosan matrices 414
composites, chitosan-zeolite 359
compounds, chitosan-based 377, 407
copper 191, 336, 354, 358–359, 361, 363–364, 408
cosmetics 22, 33, 35, 42, 82–83, 144, 223, 284, 397–399
cotton 199, 330, 333, 342
cotton fabric 198–199, 339, 342
cotton fibers, chitosan-bound 333
crab shells 17, 65, 71, 163, 241–242
Crohn's disease 22, 292–293
cross-polarization 239–240
curcumin 294

D-glucosamine 64, 201
DBD, *see* dielectric barrier discharge
deacetylation degree 17, 19, 51–53, 64, 70, 100, 111, 128, 161–162, 179, 183, 222, 260, 262, 265, 413
deep eutectic solvents (DESs) 66, 71–73, 75, 83–84
demineralization 14, 16, 45–49, 51, 66–68, 70, 72, 74–75, 77, 96, 242, 412
  chemical 50, 66–67, 77
deproteinization 14, 47, 50, 54, 66, 68–69, 72–77
deprotonation 180, 337
DESs, *see* deep eutectic solvents
dielectric barrier discharge (DBD) 68
diffuse reflectance UV-visible (DRUV) 231
diffuse reflection spectroscopy (DRS) 231
dimethyl ethyl 298
disease resistance 405, 408
DNA 161, 184, 199, 266
DRS, *see* diffuse reflection spectroscopy
drug administration 182, 286, 290, 292, 297
drug delivery 21, 176–177, 194, 196, 283, 285–286, 292, 300, 311
  oral 290
drug delivery systems 82, 194, 284, 292–293, 413
DRUV, *see* diffuse reflectance UV-visible
dyes 11, 113, 133, 336, 338, 343, 355, 357–359, 367, 369, 375, 390
  anionic 191, 330, 336
  synthetic 355–356

electrospinning 186, 288, 326, 328
enzymes 40, 51, 53–54, 73–74, 76, 78, 83, 110, 129–130, 132, 144, 184, 290, 295, 334
  bacterial 159, 292
  proteolytic 50
epicuticle 37, 335
epifluorescence microscopy 305–306
essential oils 129, 271–272, 401, 409
ethanol 14, 48, 70, 73, 79, 306, 324, 340
ethylenediamine 70, 365

FAB, *see* fast atom bombardment
fabrics 35, 329–333, 336–338, 340, 343
chitosan-based printed 339
chitosan/cotton blend 329
fast atom bombardment (FAB) 235
flame-retardant effect 198–199
flame retardants 177, 197, 199–200, 336, 341
flocculants 366, 371
carboxylate-enriched magnetic chitosan 366
chitosan magnetic 366
fluconazole 309–310
FO, *see* forward osmosis
folic acid 23
food packaging 113, 163, 165, 185, 255–274, 402, 411
forward osmosis (FO) 370
fungal biomass 39, 41, 53–54
fungal cell wall 36, 39–40
fungal cells 162
fungal deacetylase 53
fungi
chitosan-resistant 162
chitosan-sensitive 161

galactose 9–10
gastrointestinal tract 290, 292–293
gel permeation chromatography (GPC) 97, 125, 246
gelatin-chitosan 414
gels 17, 20, 113, 120, 124, 175, 183–184, 295–296, 298, 302–303, 305, 308, 336, 390, 398
chitosan-based 298
glucose 2, 6, 8, 75, 97, 147, 270
glycine 190, 193
glycogen 2–3, 6, 9
glycoproteins 10, 39, 111, 129, 146, 150
glycosaminoglycans 111, 129, 146, 414
gold nanoparticles (AuNPs) 359, 392–396
gold nanostars (AuNSs) 396
GPC, *see* gel permeation chromatography
graphene oxide 189, 353–354, 359–360, 364–365, 368

hair 398–399
healing 196, 292, 415
healthcare products 144–145
heavy metal ions 363, 389
heavy metals 35, 113, 183, 353–354, 358–359, 364, 372
hemicellulose 368
hetero-chitosans 155
hydrogels 13, 123, 195, 285–286, 288, 290, 296, 298, 376, 414
cross-linked catechol-chitosan-based 298
hydrophilicity 166, 334–335, 337, 357, 363, 373, 389
hygroscopicity 112
hypocholesterolemic 145

IMAC, *see* immobilized metal affinity chromatography
immobilized metal affinity chromatography (IMAC) 193–194
insects 8, 10, 36, 42–43, 45–47, 54, 76, 230
ionic liquids 64, 66, 69–71, 75, 78–79, 83, 120, 125–126, 325
ionization 127, 179, 236–237

keratinocytes 148, 185

Langmuir isotherm model 365
Langmuir isotherms 367
larvae 42, 45
layer-by-layer (LbL) 198–199
LbL, *see* layer-by-layer
LCST, *see* lower critical solution temperature
lentinan 41
lignin 353–354, 411
lipid-chitosan interactions 135
lipids 14, 37, 43, 82, 132–133, 135, 152–153, 256, 271–272
lipopolysaccharides 159
liver 6, 22–23, 286
lobsters 37, 222, 226, 376
lower critical solution temperature (LCST) 122

magnetic chitosan 360, 363, 365, 367, 370
magnetic chitosan flocculants 359
  carboxylate-rich 365
magnetic chitosan nanoparticles 361
magnetic-cyclodextrinchitosan 364
magnetic nanoparticles 196–197
  chitosan-labeled 363
MALDI, *see* matrix-aided laser desorption ionization
matrix-aided laser desorption ionization (MALDI) 233–234
MB, *see* methylene blue
medication delivery systems 113, 283–284, 288, 290, 292, 296–312
medicine 33, 64, 82, 96, 129, 285, 287–288, 290, 293, 295, 303, 308, 414
metal chelation 192, 357
metalloids 357
methylene blue (MB) 134–135, 354, 359, 366–367, 373–375
moxifloxacin 303
mucin 150–151, 305
mucoadhesion 128, 134, 150–151, 304
mucoadhesive properties 128
mucopolysaccharide 10
mucoproteins 10
mucosubstances 3, 10, 24
mucus 10, 150

N-deacetylation 222–223, 229
NADESs, *see* natural deep eutectic solvents
nano-biocomposites 361–362

chitosan-based 352, 362
chitosan mica 362
nano-chitosan 335, 359
nanochitin 65, 79–81, 83–84
nanochitin whiskers 406
nanochitosan 334
nanocomposites 195, 286, 301, 352–353, 362
nanocrystals 79–81, 84
acid decomposition 81
nanofibers 79–81, 120, 124, 186, 195, 243–244, 288–289, 300, 326, 364, 390, 405
chitosan/TiO_2 364
oxidized chitosan-modified 300
nanoparticles 79, 151, 182, 192, 195, 285, 289–290, 292, 294–295, 301, 305–307, 390–393, 407, 409–410, 412
chitosan-coated magnetite 359
chitosan-doped 352
chitosan-silver 334
folate-chitosan 294
hybrid antibacterial ZnO/chitosan 333
metal-chitosan 407
nasal cavity 303–305, 307
nasal drug delivery system 303, 305
natural deep eutectic solvents (NADESs) 71–73
nitrogen 8, 10, 69, 108, 129, 180, 225, 270, 341, 372–373, 393, 405
NMR, *see* nuclear magnetic resonance
nonwoven textiles 330
nuclear magnetic resonance (NMR) 104, 125, 200, 223–224, 239
nystatin 298

oligosaccharides 96, 237
ophthalmic delivery system 294–295

pesticides 352, 359
petroleum 256–257
piperazine 368
polymers, natural 33, 53, 257, 259, 270–271, 283, 285, 362, 390–391
polysaccharides 1–24, 41, 64, 70, 102, 113, 136, 143, 162, 166, 256, 258, 270, 341, 399
polyvinyl alcohol 193, 199, 259, 270, 337, 401
porous chitosan microspheres 195
potassium hydroxide 46–47, 186
potassium permanganate 47–48
prawn shells 2, 11, 17, 51, 71, 75, 285, 405
proteases 50, 77, 96–97, 146
protein hydrolysates 50

salicylic acid 408–409
shellfish 65, 70
shells 11, 24, 63, 65–66, 68–69, 71–73, 97, 222, 376, 406
mollusks 34
silver nanoparticles (AgNPs) 187, 192, 334, 362, 392–395
sodium hydroxide 46, 52, 68, 199–200, 324
squid pens 34, 38, 46, 124
starch/chitosan blends 185

tacrolimus 301
tea tree essential oil 401
textile industry 321, 323, 355–356
textile processing 328, 337, 339, 341
textiles 64, 82, 144, 244, 321–343, 355–356, 366
  nonwoven chitosan 330
  wool 334–335
therapeutic applications 22, 293
thermal insulation 111
thermal stability 69, 111, 125, 185, 196, 198, 240–241, 259, 274, 402
transdermal drug delivery system 299, 301
transglutaminase 184
trichloroacetic acid 121, 324–325
trifluoroacetic acid 327

ulcerative colitis 292–293
UPF, *see* UV protection factor
urea 71, 120, 122–123, 127, 325
UV absorbance of films based on chitosan 256
UV protection factor (UPF) 342
UV radiation 185, 268–269, 399
UV spectroscopy 201
UV-Vis spectroscopy 104, 223, 231–232

vaginal drug delivery system 307, 309
viscosity 17, 19–20, 48, 51, 83, 96, 98–100, 111, 123, 147, 153, 182, 245, 327, 337–339

wastewater 64, 144, 336, 352–353, 355, 357–360, 363–364, 366, 369, 377
wastewater treatment 11, 14, 33, 35, 64, 82–83, 223, 323, 336, 351–378, 390
water desalination 360, 367–368
water treatment 192, 363, 378
wool 330, 334–335
wound dressings 83, 111, 113, 311, 336, 342, 397–398, 413
wounds 413–415

X-ray diffraction (XRD) 124, 201, 225, 365, 367, 372
XRD, *see* X-ray diffraction
xylan 9, 270, 353–354

yeasts 12, 18, 33–35, 129, 154, 183, 222, 266–268, 285, 322, 404